Northwest Vista
Learning Resour   Y0-DWJ-452
3535 North Ellison Drive
San Antonio, Texas 78251
210.348.2300

Violence on Campus
Defining the Problems, Strategies for Action

Allan M. Hoffman, EdD, CHES
Dean and Professor
College of Health Sciences
University of Osteopathic Medicine and Health Sciences
Des Moines, Iowa

John H. Schuh, PhD
Professor of Educational Leadership and Policy Studies
Iowa State University
Ames, Iowa

Robert H. Fenske, PhD
Professor of Higher Education
Division of Educational Leadership and Policy Studies
Arizona State University
Tempe, Arizona

AN ASPEN PUBLICATION®
Aspen Publishers, Inc.
Gaithersburg, Maryland
1998

Library of Congress Cataloging-in-Publication Data

Violence on campus: defining the problems, strategies for action/ [edited by] Allan M. Hoffman, John H. Schuh, Robert H. Fenske.
p. cm.
Includes bibliographical references and index.
ISBN: 0-8342-1096-7
1. Campus violence—United States. 2. Campus violence—United States—Prevention. 3. Universities and colleges—United States—Safety measures. 4. Universities and colleges—Security measures—United States. 5. n-us. I. Hoffman, Allan M. (Allan Michael) II. Schuh, John H. III. Fenske, Robert H.
LB2345.V56   1998
378.1'9782—dc21
98-27590
CIP

Copyright © 1998 by Aspen Publishers, Inc.
All rights reserved.

Aspen Publishers, Inc., grants permission for photocopying for limited personal or internal use. This consent does not extend to other kinds of copying, such as copying for general distribution, for advertising or promotional purposes, for creating collective works, or for resale. For information, address Aspen Publishers, Inc., Permissions Department, 200 Orchard Ridge Drive, Suite 200, Gaithersburg, Maryland 20878.

Orders: (800) 638-8437
Customer Service: (800) 234-1660

About Aspen Publishers • For more than 35 years, Aspen has been a leading professional publisher in a variety of disciplines. Aspen's vast information resources are available in both print and electronic formats. We are committed to providing the highest quality information available in the most appropriate format for our customers. Visit Aspen's Internet site for more information resources, directories, articles, and a searchable version of Aspen's full catalog, including the most recent publications: **http://www.aspenpublishers.com**
Aspen Publishers, Inc. • The hallmark of quality in publishing
Member of the worldwide Wolters Kluwer group.

Editorial Services: Jane Colilla
Library of Congress Catalog Card Number: 98-27590
ISBN: 0-8342-1096-7

Printed in the United States of America

1 2 3 4 5

Table of Contents

Contributors xi

Foreword xv

Preface xvii

Chapter 1—Violent Crime in American Society 1
Fernando M. Treviño, Sharon L. Walker, and Gilbert Ramírez

    Introduction 1
    Violence Defined 2
    Violence Measured 3
    Consequences of Violence 3
    Epidemiology of Violence 4
    Crime in the United States 4
    Location of Crimes 11
    Time of Crimes 13
    Conclusion 14

Chapter 2—Campus Vulnerability 17
John H. Schuh

    Introduction 17
    Physical Components 18
    Psychological Components 23
    Police and Security Services 26

Chapter 3—Profile of Students Coming to Campus 29
Robert H. Fenske and Stafford L. Hood

Introduction 29
Nontraditional Students 32
A Background of Violence 34
An International Perspective 35
Escalation in School Violence 36
National and State Data Sources 37
Safe Schools and Individual Rights 38
Federal Concerns 40
Conclusion 49

Chapter 4—Weapon Carrying on Campus 53
Randal W. Summers and Allan M. Hoffman

Introduction 53
Weapons Borne by Students 53
Weapons in Society—Armed and Dangerous? 56
Violence in the United States in Perspective 56
The Perception of Crime 57
Handguns and Homicide 59
Why Carry Guns? 60
Underlying Dynamics of Fear 62
Role of the Media 64
Potential Solutions 65

Chapter 5—Matters of Civility on Campus 69
John H. Schuh

Introduction 69
Student Publications 70
Outside Speakers 72
Hate Speech among Members of the Campus Community 73
Electronic Mail and the Internet 76
Rape Victims 77
Hazing 79
Fraternities 80
Other Issues 81

Chapter 6—Violent Crime in the College and University Workplace 87
Allan M. Hoffman, Randal W. Summers, and Ira Schoenwald

Introduction 87

The Campus as an Academic Sanctuary: Myth or Reality? 88
The Campus as a Workplace: Violence in the Workplace 89
The Campus as a Target: An Extension of Societal Violence 90
Definition of Violence: The Violence Continuum 90
National Statistics for Violence on Campus 91
Problems with Reporting and Interpreting Violence Statistics 92
The Cost of Violence 94
Prevention and Intervention Strategies Related to Employees 97

Chapter 7—Violence at Home on Campus 111
Carolyn J. Palmer

Introduction 111
Why Violence Occurs in Campus Residences 112
Special Problems in Family Housing and Greek Housing 116
The Most Serious and Most Common Forms of Violence 117
Addressing the Nonreporting of Violent Crimes 118
Violence Reduction Measures: Signs of Success 120

Chapter 8—Reducing Racial and Ethnic Hate Crimes on Campus: The Need for Community 123
Robert H. Fenske and Leonard Gordon

Introduction 123
The Nature of Hate Crimes 124
Legal Bases and Reporting Requirements for Hate Crimes 127
Incidence of Hate Crimes 129
Conceptual and Theoretical Bases of Ethnoviolence 133
Racial Tensions on Campus 136
Movement Toward Campus Diversity 138
The Campus Environment Team Approach 139
Out of Campus Racial Conflict: Formulating a Viable Racial
 and Ethnic Diversity Policy 140
Developing Diversity Programming To Achieve an Inclusive
 Campus Community 142
Intergroup Dialogues: Voices of Discovery 143

Chapter 9—Women and Violence on Campus 149
Kay Hartwell Hunnicutt

Introduction 149
Dimensions of Rape on Campus 151
Date Rape 152

Battery and Date Aggression 153
Sexual Harassment 154
Peer Sexual Harassment 155
Sexual Harassment by Male Faculty 156
Student-Professor Consensual Relationship 160
The Federal Effort To Combat Violence Against Women 162
Suggested Strategies 164

Chapter 10—Heterosexism and Campus Violence: Assessment and Intervention Strategies 169
Nancy J. Evans and Sue Rankin

Introduction 169
Heterosexism, Campus Climate, and Anti-LGBT Violence 171
Campus Climate Review 171
Impact of Heterosexism and Anti-LGBT Violence 177
Strategies for Change: A Comprehensive Program of
    Intervention 179
Creating Change 182

Chapter 11—Sexual Harassment of Students: The Hidden Campus Violence 187
Michele A. Paludi and Darlene C. DeFour

Introduction 187
Legal Definition 189
Incidence of Faculty-Student Sexual Harassment 191
How Students Define Their Experiences 192
Professor Abuse of and Denial of the Ivory Power 193
Consensual Relationships 195
Other Forms of Sexual Harassment on College Campuses 196
Impact of Sexual Harassment on Individuals 197
Methodological Issues in Research on Sexual Harassment 198
Dealing with Sexual Harassment on College Campuses 200

Chapter 12—Substance Abuse and Violence 207
David S. Anderson and Carol Napierkowski

Introduction 207
The Nature of the Societal Problem 208
Drugs and the Human Body 208
Classification of Drugs 209

The Extent of the Problem 211
The Nature of the Campus Problem 214
Linking Substance Use and Violence: Some Theories 215
College Student Development Context 216
A Different Perspective 217
Campus-Based Efforts 218
A Model for Campus Prevention Efforts: The Multidimensional
  Framework Initiative 222
Campus Implementation Steps 224
Some Final Considerations 225

**Chapter 13—Crisis Management Resulting from Violence on Campus:
Will the Same Common Mistakes Be Made Again? 229**
J. Victor Baldridge and Daniel J. Julius

Introduction 229
Imagine These Scenarios 230
Crisis Events 231
More Crises? 231
Common Mistakes: The Big Nine 234
Building Crisis-Response Capabilities 240
Transforming "Crisis" Activities to "Routine" Management 245

**Chapter 14—Communicating about Violence on Campus: An
Integrated Approach 247**
Larry D. Lauer and Carolyn N. Barnes

What Is Integrated Communication? 247
Understanding the News Media 248
Cultivating Media Relationships 249
Campus Awareness and Campus Security Act 250
The Disciplinary Process and the Buckley Amendment 250
Bringing Together Key Players 251
Preparing for a Crisis 252
Your First Response 252
A Crisis Communications Plan 253
The Value of a Communications Plan 255
How To Work with the Media when a Violent Crime Happens
  on Campus 256
A Case Study in Taking an Integrated Approach to Violent
  Crime 259

Chapter 15—The Legal Response to Violence on Campus 273
Kay Hartwell Hunnicutt and Peter Kushibab

Introduction 273
The Federal Response to Violence on Campus 274
State Efforts To Combat Violence on Campus 284
The College and University Response to Violence on Campus 291
Recommendations for Prevention of Crime on Campus from a
  Legal Point of View 295
Appendix 15–A: Campus Environment Team at Arizona State
  University 300
Appendix 15–B: General Campus Security Checklist 318
Appendix 15–C: Campus Security Act—Checklist 320

Chapter 16—Strategies for Dealing with Violence 327
Eugene Deisinger, Charles Cychosz, and Loras A. Jaeger

Introduction 327
Key Roles for Law Enforcement 328
The Role of Human Resource Services 336
The Role of Student Affairs 336
The Incident Response Team 339
Threat Assessment 341

Chapter 17—Conclusion 347
John H. Schuh

Introduction 347
Societal Conditions 347
Violence in America 347
Campus Vulnerability 348
Students Coming to Campus 348
Weapon Carrying 349
Issues of Civility 349
Workplace Violence 350
Violence in Campus Residences 350
Ethnoviolence 351
Women and Campus Violence 351
Heterosexism and Campus Violence 352
Sexual Harassment of Students 352
Substance Abuse and Campus Violence 352

    Crisis Management 353
    Working with the Media 353
    Legal Issues 354
    Strategies for Dealing with Violence 354
    Dealing with Violence on College Campuses 355
    A Final Word 361

Index 363

About the Authors 377

Contributors

David S. Anderson, MA, PhD
Associate Research Professor
Director
Center for the Advancement of Public
 Health
The Institute of Public Policy
George Mason University
Fairfax, Virginia

J. Victor Baldridge, PhD
Senior Partner
Pacific Management
Manager of Programs
Center for Strategic Leadership
University of San Francisco
San Francisco, California

Carolyn N. Barnes
Director of Marketing
Texas Woman's University
Denton, Texas

Charles Cychosz, PhD
Manager
Safety and Health Development
Deparment of Public Safety
Iowa State University
Ames, Iowa

Darlene C. DeFour, MA, PhD
Associate Professor of Psychology
Hunter College
City University of New York
New York, New York

Eugene Deisinger, PhD
Captain
Behavioral Science Unit
Department of Public Safety
Iowa State University
Ames, Iowa

Nancy J. Evans, MFA, MSEd, PhD
Associate Professor
Department of Educational Leader-
 ship and Policy Studies
Iowa State University
Ames, Iowa

Robert H. Fenske, PhD
Professor of Higher Education
Division of Educational Leadership
 and Policy Studies
Arizona State University
Tempe, Arizona

Leonard Gordon, MA, PhD
Associate Dean for Academic
  Programs
Professor of Sociology
College of Liberal Arts and Sciences
Arizona State University
Tempe, Arizona

Allan M. Hoffman, EdD, CHES
Dean and Professor
College of Health Sciences
University of Osteopathic Medicine
  and Health Sciences
Des Moines, Iowa

Stafford L. Hood, MS, PhD
Associate Professor of Counseling/
  Counseling Psychology
Division of Psychology in Education
Arizona State University
Tempe, Arizona

Kay Hartwell Hunnicutt, PhD, JD
Associate Professor
Division of Educational Leadership
  and Policy Studies
College of Education
Arizona State University
Tempe, Arizona

Loras A. Jaeger, MA
Director of Public Safety
Iowa State University
Ames, Iowa

Daniel J. Julius, EdD
Associate Vice President for Academic Affairs
University of San Francisco
San Francisco, California
Senior Fellow
Higher Education Research Institute
Stanford University
Stanford, California

Peter Kushibab, JD
Assistant General Counsel
Maricopa County Community
  College District
Tempe, Arizona

Larry D. Lauer
Associate Vice Chancellor
Communications and Public Affairs
Texas Christian University
Fort Worth, Texas

Carol Napierkowski, MA, PhD
Associate Professor
Department of Counseling and
  Educational Psychology
West Chester University
West Chester, Pennsylvania

Carolyn J. Palmer, PhD
Associate Professor
Department of Higher Education and
  Student Affairs
Bowling Green State University
Bowling Green, Ohio

Michele A. Paludi, MA, PhD
Principal
Michele Paludi & Associates
Schenectady, New York

Gilbert Ramírez, DrPH
Associate Professor and Vice Chair
Department of Public Health and
  Preventive Medicine
Associate Director
Graduate Program in Public Health
University of North Texas Health
  Science Center
Fort Worth, Texas

Sue Rankin, MS, PhD
Senior Diversity Planning Analyst
Coordinator of Lesbian, Gay,
  Bisexual, and Transgender Equity
The Pennsylvania State University
University Park, Pennsylvania

Ira Schoenwald, PhD
Professor of Public Administration
Associate Vice President for Faculty
  Affairs
California State University,
  Dominguez Hills
Carson, California

John H. Schuh, PhD
Professor of Educational Leadership
  and Policy Studies
Iowa State University
Ames, Iowa

Randal W. Summers, PhD
Adjunct Professor
University of Phoenix
Fountain Valley, California
Adjunct Professor
University of LaVerne
LaVerne, California
General Partner
Summers and Associates
Cambria, California

Fernando M. Treviño, PhD, MPH
Professor and Chair
Department of Public Health and
  Preventive Medicine
Executive Director
Graduate Program in Public Health
University of North Texas Health
  Science Center
Fort Worth, Texas

Sharon L. Walker, RN, MS, MPH
Research Scientist/Project Manager
Survey Research Center
University of North Texas
Denton, Texas

Foreword

Campuses are interesting places. More to the point, campuses are communities. Students live there. Faculty and staff work there. The place is alive—24 hours a day, seven days a week, 365 days a year, more or less. For most residential campuses, it is a quite different environment because of the large concentration of young adults in residence.

Society looks to us to educate these students, both those who fit the traditional residential student mold and the increasingly large number who do not. And by education, society expects more than courses, credits, and information gathering. They really do expect us to care for and to nurture, if you will, the "whole" student and to provide a comprehensive, complex environment for optimum growth and development.

It is in the building of this "community for learning" and providing a healthy environment that the concern for violence on the college campus, in whatever form, becomes important. In fact, it becomes *very* important: to parents, to the public, and to politicians. And well it should be. Campuses need to be safe. Campuses are places to learn and grow. A community of violence is incompatible with these requirements.

Still, academic communities exist within larger communities that have become more dangerous. Campus crime has become the subject of congressional hearings and legislation. Making campuses safe has become a preoccupation among administrators, trustees, faculty, and students. Violence, hate crimes, sexual harassment, concealed weapons, civility, vulnerability, and substance abuse (including alcohol abuse)—all of this and more has become part of the higher education agenda.

There may have been a time when the college campus was perceived as a sanctuary, a place where teaching, learning, and social interchange took place in an environment that posed minimal danger to physical safety and the security of personal property. If ever, those days are no more. As with society in general, crime on campus has escalated from an infrequent occurrence to a subject of genuine concern.

In the eyes of the law, a university is an institution for the advancement of knowledge and learning. Individuals who meet the required qualifications and who abide by the institution's rules are permitted to attend and are presumed to have sufficient maturity to conduct their own affairs. Colleges and universities are no longer held liable under the *in loco parentis* doctrine for injuries sustained by students unless the college has not fulfilled certain duties imposed by law.

Still, institutions have devoted increased resources, vigilance, and creativity to ensuring, to the extent possible, a safe environment. Among the more visible have been the design, maintenance, and operation of campus buildings and grounds, including increasingly sophisticated lock and key systems, increased illumination, and security phone boxes. Campus security forces have grown in number and quality, with ever-greater screening, training, and supervision. Data regarding security incidents are gathered and appropriate campus responses are formulated. Especially important, many campuses educate and remind students and other members of the college community on a recurrent basis about security risks and procedures.

I commend those who have worked to assemble this commentary on violence on campus. It is a difficult but important topic that deserves our close attention. Most of all, campuses provide a healthy environment for learning. That is what we do. One part of that healthy environment is managing and preventing violence and dysfunctional behavior—in whatever form. I hope the chapters that follow will lead us toward that goal.

—Stanley O. Ikenberry
President
American Council on Education
Washington, DC

Preface

Concern about violence on U.S. campuses is increasing. Security measures of all types are proliferating at institutions of higher learning across the nation. Dr. Stanley O. Ikenberry, president of the American Council on Education, noted an example in his recent testimony to the U.S. Congress regarding proposed legislation aimed at reducing campus crime. He stated that in 1988 the University of Illinois, Urbana-Champaign campus employed 37 police officers; 10 years later this number grew to 51, and the cost of this police force nearly doubled from $1.8 million to $3.4 million. In addition, the university has spent $750,000 in the last three years on security equipment such as increased lighting and emergency phones.

Are such increases in concern, security measures, and expenditures in direct relation to actual increases in incidence of campus crime? To what extent do these increases respond to fear and anxiety generated by intense media attention rather than crime itself? Answers to these questions are extraordinarily difficult to obtain because of confusion about crime reporting requirements and even such factors as defining the boundaries of the campus for reporting purposes. Most institutions are under pressure to reach out to new constituencies. Are facilities leased to corporations in university research parks and college classes held in distant high schools to be included in crime reporting jurisdictions? In addition, suspicion is widespread that campus crime is underreported by administrators reluctant to sully the image of the campus as a safe haven for learning and social activities.

This book is aimed at providing answers to questions such as the following: What types of violence, crimes, and other threats to safety are growing most rapidly? Are any types of violence "in remission" or declining? Are the data reliable? Has the increased concern about violence created a climate of fear unwarranted by actual violence rates? Finally, what can or should student affairs and academic administrators do to ensure campus safety?

The chapters in this book cover a wide range of topics and represent diverse perspectives. The editors themselves are from diverse backgrounds. Allan Hoffman, originator of the book plan, is a health sciences administrator. The idea for the book evolved from his earlier work on schools, violence, and society. John Schuh was, at the time of the book's early development, a student affairs administrator. (He is currently a professor of higher education.) In both capacities, he has written extensively on college students. Robert Fenske is a professor of higher education and has also written extensively on college students.

The authors who have contributed to this book hail from many walks of professional life; they include legal experts, public administrators, consultants, and scholars. The contributors were selected for their expertise in the specific topics that comprise the overall book. Beyond that, no constraints were placed on them to conform to any single perspective or conceptual framework. Like any collected work formed under these circumstances, this book represents a wide range of approaches and writing styles. Given this diversity, the editors strove to meet two principal goals for the book. The first was comprehensiveness. A thorough review of the literature on campus violence found no comprehensive book on the subject. The second goal was that each chapter would stand on its own merits in terms of depth and scholarship.

No book on campus violence developed in 1998 can be the last word on the subject. Scholarship in this area will be a "work in progress" for some years to come. For example, as this book goes to press, Congress is hard at work debating legislation aimed at increasing campus security by identifying and widely publicizing campus crime offenders. However, this initiative directly counters student constitutional rights to privacy under the "Buckley Amendment." The U.S. Department of Justice is energetically revamping and expanding campus crime reporting. Campus security offices are developing plans to cope with rising rates of vandalism, hate crimes, rape, and other problems.

Although there is widespread concern about violence on campus, factual information on which to base plans and strategies is much harder to come by. Many of the chapters in this book call for increased efforts to gather information. One of the first victims of rising fears in a near vacuum of factual information is the ability to plan carefully and thoughtfully. This book is intended to help student affairs and academic administrators develop strategies to cope with and, it is hoped, reduce violence on campus.

CHAPTER 1

Violent Crime in American Society

Fernando M. Treviño, Sharon L. Walker, and Gilbert Ramírez

INTRODUCTION

America's history has been rife with turbulence. The violent resistance to the mother country in the American Revolution served as an example for later forceful actions taken by Americans for critical national causes. The most upright and honorable citizens as well as the criminal and disorderly have used the instrument of violence. In two major national crises, the American Revolution and the Civil War, violence was employed to found and to preserve the nation (Graham & Gurr, 1979).

America's infatuation with violence extends to the media, sports, politics, the military, and even church and school (Brendtro & Long, 1995). This phenomenon was aptly illustrated by a recent cursory review of newspaper listings for broadcast and cable movies; more than 50% were explicitly violent by the wording of their titles or indicated violent content through the descriptive comments (Langer, 1997). Sports networks, "live" police shows, dramatic police series, commercials, cartoons, and frequent news broadcasts furnish the viewer a vast array of violent images.

Violent acts may be viewed as the manifestation of learned behaviors. Such learning begins very early (Canada, 1995). Children raised in violent households learn and incorporate into their behavior what they witness within their families (Palermo, 1994). Canada (1995) describes the new generation as the "handgun generation." Many of today's young people grow up under the conditions of war. Observations from his early life in ghetto neighborhoods and later experiences teaching in Boston and New York led Canada to use the term *continuing traumatic stress syndrome* to describe what happens to people who continue to live under war conditions. Brendtro and Long (1995) describe sources of stress for children as a myriad conglomerate of influences from home, society, and schools, with resultant behavior manifested through violent acts because they (the children) cannot disengage from confrontations. Within the span of one decade, the nature of violent acts has changed from the fist, stick, and knife to the gun (Canada, 1995).

The World Health Organization recognized violence as a public health issue in the mid 1980s. Violence, as a global public health problem, has been a leading cause of death and disability. In the United States, the Centers for Disease Control and Prevention acknowledged violence as a priority public health concern. The Public Health Service's *Healthy People 2000: National Health Promotion and Disease Prevention Objectives* identified objectives for improving the nation's health over the next decade. It emphasized the need to prevent the violence that is so drastically affecting the lives of children and adolescents. Objectives related to violence include reductions in the homicide rate, weapon-related deaths, assault injuries, physical fighting among youths, and weapon-carrying by youths. Preventive strategies include expanding conflict resolution education in schools and comprehensive violence prevention programs (U.S. Department of Health and Human Services, 1990).

VIOLENCE DEFINED

The use of force is the key factor in any definition of violence. Rosenberg (1995) identifies violence as the use of force with the intent to harm either oneself or another person or group. Siann (cited in Sigler, 1995) elaborates on the use of great physical force or intensity. He states that while physical force is often impelled by aggressive motivation, it may occasionally be used by individuals engaged in a mutual violent interaction that is regarded by both parties as intrinsically rewarding. He observes that aggression involves the intention to hurt or emerge superior to others, does not necessarily involve physical injury, and may have a variety of motives (Sigler, 1995).

Palermo (1994) distinguished between primary aggression and reactive aggression. *Primary aggression* is a goal-directed, hostile self-assertion that is destructive in character. *Reactive aggression* is more often part of an emotional reaction brought about by frustrating life experiences. Violence in our society is representative of these two types of aggression: programmed, organized aggression and impulsive aggression (Palermo, 1994).

The term *violence* describes a variety of destructive personality traits and antisocial behaviors. The implication of violence is that a force is inflicted on someone for which some violation is the result (Twitchell, 1989). Violence can be viewed as the ultimate tool of control or power.

VIOLENCE MEASURED

With violence defined through terms of aggression and force, the challenge of measuring violence becomes evident. The numbers of arrests for aggravated assaults, murders, rapes, and robberies are available in reported crime statistics. Such figures provide a good assessment of the rate of violence, which is essential in order to create preventive and controlling programs (Palermo, 1994). However, only the outcome—the consequence of escalated aggression that results in a physical confrontation—is readily apparent. Violent acts viewed on television can be counted. Trauma center reports can be analyzed for wound description. Morbidity and mortality data yield death and disability summaries.

Although the volume of quantifiable information is impressive, many crimes, including violent ones, are underreported (Palermo, 1994). Only an estimated 16% of rapes are reported to the police ("Rape statistics," 1997). Furthermore, these sources reveal little about the circumstances of violence—especially the reasons for violence. Nor are the effects of violence as an agent for emotional, psychological, and social change for the individual, the family, and the community known in concrete, measurable detail.

CONSEQUENCES OF VIOLENCE

A distinguished international panel on the understanding and control of violent behavior suggested that the major consequences of violence included physical injury, psychological costs, monetary and other costs, and mortality. They found no national data on psychological consequences and limited information related to economic costs. The direct monetary costs associated with physical injury would include emergency care, continuing medical treatment, physicians' fees, medication costs, and rehabilitation. In 1987, the average medical costs per victim injury were estimated as $5,370 for murder, $616 for rape, $527 for assault, and $344 for robbery. For that same year, the total measurable costs were estimated to be $2 billion for murder, $1.6 billion for rape, $10.1 billion for assault, and $6.3 billion for robbery (Reiss & Roth, 1994).

Indirect costs attributable to violence include expenditures for law enforcement, adjudication, victim services, and correctional facilities (Reiss & Roth, 1993). In fiscal year 1992, federal, state, and local governments spent $94 billion for civil and criminal justice—an increase of 59% over expenditures for 1987 (Bureau of Justice Statistics, 1997).

Additional consequences of violent behavior include the disruption of families; the disintegration of neighborhoods, particularly in the inner city; the need to fortify schools, homes, and businesses; and the deterioration and abandonment of community resources (Reiss & Roth, 1993). Parks and playgrounds have become dangerous zones in many neighborhoods. Personal safety is no longer taken for granted.

EPIDEMIOLOGY OF VIOLENCE

How violent is America? Is violence increasing or decreasing? Are some populations at greater risk for being victims of violence? Who is more likely to initiate an act of violence? Is violence equally distributed across the United States, or are there areas where violence is more likely to occur? Does violence occur more often during certain times of the year? These questions about person, place, and time are public health questions. Violence data can be obtained from numerous sources. In the United States, the annual Federal Bureau of Investigation (FBI) report on crime is an excellent resource. However, this is not the only source of violence data and it has some limitations, particularly with reporting by race/ethnic groups. The FBI currently reports all crime data using the race/ethnic categories of white, black, other race, and unknown. Researchers wanting specific information about other groups, such as Hispanics or Asians, must seek out other sources such as local and regional law enforcement agencies and published reports from other violence-focused groups. Most of the statistics provided in this overview of violence in America come from the 1995 FBI *Uniform Crime Report* (UCR).

CRIME IN THE UNITED STATES

To report on crime in the United States, the FBI uses a crime index that includes crimes of violence and property. *Violent crimes* include murder and nonnegligent manslaughter, forcible rape, robbery, and aggravated assault; all violent crimes involve force or threat of force. *Property crimes* include burglary, larceny-theft, and motor-vehicle theft; the objective of such offenses is to take money or property, but there is no force or threat of force against the victim. Arson statistics are also collected and reported but are not included in the overall crime index; arson involves the destruction of property, and its victims may or may not be subjected to force. The crime index is reported both as a total number of offenses and as a rate (number of offenses per 100,000 population) (FBI, 1995).

Both violent crime and property crime have declined since 1991 (Figure 1–1). This phenomenon has been reported in the media.

Violent crime peaked in 1991 at a rate of 758 offenses per 100,000 people. The rate in 1995 for violent crimes was 685 per 100,000, which represents a 9.7% reduction since 1991; despite this reduction, the 1995 rate still represents a 46%

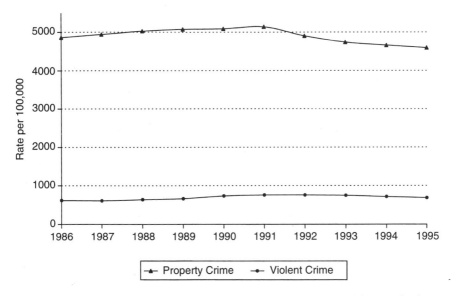

Figure 1-1 Crime in the United States, 1986-1995. *Source:* Reprinted from Federal Bureau of Investigation, 1995, *Uniform Crime Report,* U.S. Department of Justice, Washington, DC.

increase over the 1976 rate of 468 offenses per 100,000 people. Property crime also decreased from a peak of 5,140 offenses per 100,000 in 1991 to 4,593 offenses per 100,000 in 1995 (a 10.6% decrease). The 1995 property crime rate was 4.9% lower than the 1976 rate (4,820 per 100,000). Figure 1-2 depicts violent crimes in the United States from 1986 to 1995.

There is a notable difference between rates of aggravated assault or robbery and rates of forcible rape or murder/nonnegligent manslaughter. In 1995, aggravated assaults accounted for approximately 61% of violent crimes, robberies accounted for 32%, forcible rapes 5%, and murders/nonnegligent manslaughter offenses 1%. The percentage changes from 1991 to 1995 for aggravated assaults, robberies, forcible rapes, and murder/nonnegligent manslaughter offenses were -3.5%, -19.0%, -12.3%, and -16.3%, respectively. The comparable percentage changes from 1976 to 1995 were +79.4%, +10.8%, +39.5% and -7.3%, respectively. Although murder/nonnegligent manslaughter rates were lower in 1995 than 1991 or 1976, the total number of offenses increased during these years. In 1976, there were 18,780 murder/nonnegligent manslaughter offenses, in 1991 there were 24,700, and in 1995 there were 21,600.

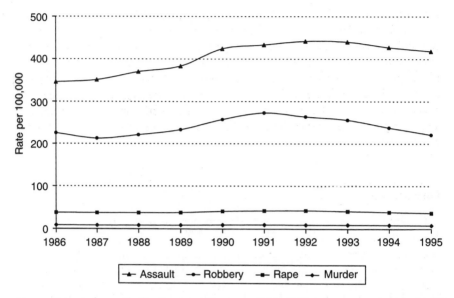

Figure 1–2 Violent Crime in the United States, 1986–1995. *Source:* Reprinted from Federal Bureau of Investigation, 1995, *Uniform Crime Report,* U.S. Department of Justice, Washington, DC.

Aggravated Assault

Aggravated assault is defined by the FBI as "an unlawful attack by one person upon another for the purpose of inflicting severe or aggravated bodily injury" and "is usually accompanied by the use of a weapon or by means likely to produce death or great bodily harm" (FBI, 1995, p. 31). In the reporting of aggravated assault offenses, "attempts" are also included on the basis that "it is not necessary that an injury result when a gun, knife, or other weapon is used which could and probably would result in a serious personal injury if the crime were successfully completed" (FBI, 1995, p. 31). In 1995, 33% of aggravated assaults were committed with blunt objects or other dangerous weapons; personal weapons such as hands, fists, and feet were used in 26% of assaults; 23% of assault offenses involved the use of firearms; and the remaining 18% involved the use of knives or other cutting instruments.

Robbery

Robbery, as defined by the FBI, is "the taking or attempting to take anything of value from the care, custody, or control of a person or persons by force or threat of force or violence and/or by putting the victim in fear" (FBI, 1995, p. 26). Losses from robbery in 1995 were estimated at $507 million; the average value of property stolen was $873 per robbery. Average losses ranged from $400 for robberies of convenience stores to $4,015 for robberies of banks. More than half (54%) of all robberies in 1995 occurred on streets or highways; 21% occurred at commercial and financial establishments; 11% occurred at residences, and the remainder were miscellaneous types. The weapons of choice for robberies in 1995 were firearms and strong-armed tactics; each accounted for 41% of robbery offenses. Knives or cutting instruments were used in 9% of all robberies, and other dangerous weapons used for the remainder.

Forcible Rape

Forcible rape, as defined by the FBI for its calculation of the annual crime index, is "the carnal knowledge of a female forcibly and against her will; assaults or attempts to commit rape by force or threat of force are also included but statutory rape (without force) and other sex offenses are excluded" (FBI, 1995, p. 23). In 1995, 87% of the rapes reported in the crime index were rapes by force; the remaining 13% were assaults or attempts to commit forcible rape. Complaints of rape found to be false or baseless are excluded from the crime counts; the number of "unfounded" complaints, as determined through investigation to be false, is higher for forcible rape than for any other crime included in the crime index.

Murder and Nonnegligent Manslaughter

Murder and nonnegligent manslaughter are defined simply as the willful killing of one human being by another. Not included in this count are deaths caused by negligence, suicide, or accident; justifiable homicides; or attempts to murder. The latter two are counted as aggravated assaults. Firearms were used in 7 of 10 murders in the United States in 1995; this figure is consistent with statistics from previous years. Of murders where type of weapon was reported, 59% involved the use of handguns, 5% involved shotguns, 3% involved rifles, and 5% involved other or unknown firearms. Knives or cutting instruments were used in 13% of the murders, personal weapons (such as hands, fists, and feet) in 6%, blunt objects (such as clubs and hammers) in 5%, and other dangerous weapons (such as poison and explosives) in 4%.

Crimes Against Persons

One of the limitations of the FBI's *Uniform Crime Report* is its relative lack of information on the persons affected by crimes. Of the 21,597 murders and nonnegligent manslaughter offenses reported in 1995, sufficient supplemental data were obtained on 19,655 cases to provide a profile of the race and sex of the victims (Figure 1–3; FBI, 1995).

Combined across races, males represented 77% of all murder and nonnegligent manslaughter victims. Black males were 1.14 times more likely than white males to be murder victims; white females were 1.5 times more likely than black females to be victims.

Figures 1–4 and 1–5 reflect the distribution by age and sex for murder/ nonnegligent manslaughter victims and offenders, respectively, for cases in which age and sex were known (337 victims, or 1.7%, had missing data and are not represented in the graph; 7,797 offenders, or 34.8% had missing data and are not represented). Males aged 20 to 24 were in the group most likely to be murder victims or offenders. Females at greatest risk for being a murder victim were aged 30 to 34 years; females at greatest risk for being an offender were aged 20 to 34.

Figures 1–6 and 1–7 reflect the percentage distribution by age and race for the same offenses (same missing data statistics apply). The age group at greatest risk for being a murder victim or offender, across all races, is 20 to 24. For blacks, people aged 17 to 24 are at significantly greater risk than their counterparts of other races. Blacks aged 20 to 24 are also at greater risk than their counterparts in white and other race groups for being a murder victim.

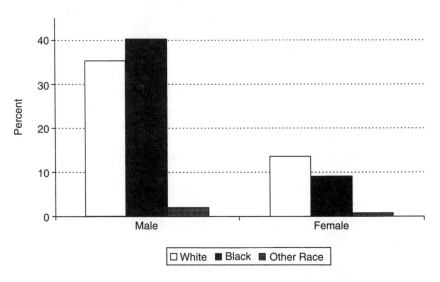

Figure 1–3 Murder and Nonnegligent Manslaughter Victims by Race and Sex, 1995. *Source:* Reprinted from Federal Bureau of Investigation, 1995, Uniform Crime Report, U.S. Department of Justice, Washington, DC.

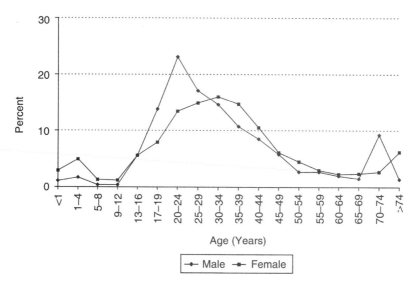

Figure 1-4 Murder and Nonnegligent Manslaughter Victims by Age and Sex, 1995. *Source:* Reprinted from Federal Bureau of Investigation, 1995, *Uniform Crime Report,* U.S. Department of Justice, Washington, DC.

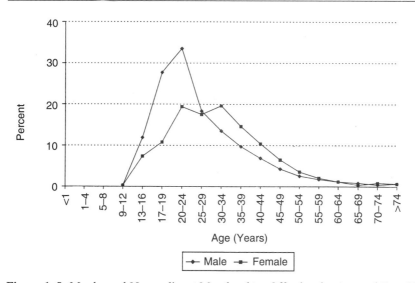

Figure 1-5 Murder and Nonnegligent Manslaughter Offenders by Age and Sex, 1995. *Source:* Reprinted from Federal Bureau of Investigation, 1995, *Uniform Crime Report,* U.S. Department of Justice, Washington, DC.

10   VIOLENCE ON CAMPUS

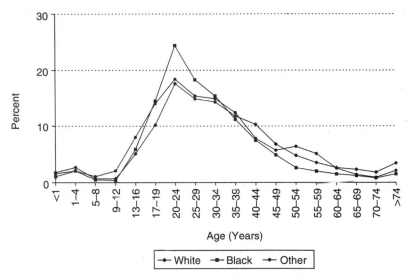

Figure 1-6 Murder Victims by Age and Race, 1995. *Source:* Reprinted from Federal Bureau of Investigation, 1995, *Uniform Crime Report,* U.S. Department of Justice, Washington, DC.

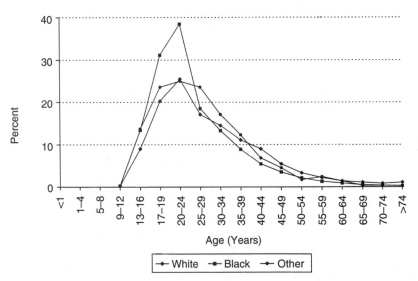

Figure 1-7 Murder Offenders by Age and Race, 1995. *Source:* Reprinted from Federal Bureau of Investigation, 1995, *Uniform Crime Report,* U.S. Department of Justice, Washington, DC.

Who is killing whom? For single victim/single offender murders and where ages are known ($n = 9{,}270$), 3.3% of the murders involved a victim and offender under 18 years of age; 7% involved a victim over 18 and an offender under 18; 9.6% involved a victim under 18 and an offender 18 or over; 80% involved individuals 18 years or older.

Figures 1–8 and 1–9 depict single victim/single offender relationships ($n = 9{,}848$) by race and sex during 1995. White murder victims are most likely to have been murdered by a white offender (84%). The predominance of same-race victim/offender relationships exists for blacks (94%) and for other races (62%). Male victims are more likely to be murdered by a black offender (56%), while female victims are more likely to be murdered by a white offender (53%) (Figure 1–8).

Male offenders are equally likely to murder a white (50%) or black (47%) person. Female offenders, on the other hand, are slightly more likely to murder a black (54%) rather than a white (43%) person. Male and female offenders are both more likely to murder a male than a female (74% and 77%, respectively).

LOCATION OF CRIMES

Violence does not occur equally across different regions of the United States. Figure 1–10 depicts the regional distribution of the nation's violent crimes (murders, rapes, robberies, and assaults). The South clearly appears to be the most

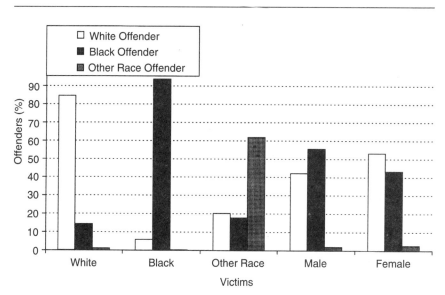

Figure 1–8 Victim-Offender Relationships, 1995: Victims. *Source:* Reprinted from Federal Bureau of Investigation, 1995, *Uniform Crime Report,* U.S. Department of Justice, Washington, DC.

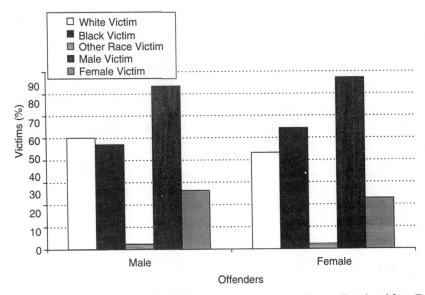

Figure 1–9 Victim-Offender Relationships, 1995: Offenders. *Source:* Reprinted from Federal Bureau of Investigation, 1995, *Uniform Crime Report,* U.S. Department of Justice, Washington, DC.

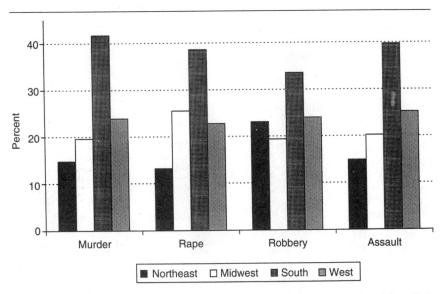

Figure 1–10 Regional Distribution of Violent Crime, 1995. *Source:* Reprinted from Federal Bureau of Investigation, 1995, *Uniform Crime Report,* U.S. Department of Justice, Washington, DC.

violent area; it accounted for 35% to 40% of each of the four types of reported violence. The Midwest and West accounted for 20% to 25% of each category of violent crime. The Northeast region had the lowest percentages of the nation's violent crimes; it accounted for 13% to 15% of the nation's murders, rapes, and assaults and 23% of all robberies.

Figure 1–11 depicts the violent crime rates for Metropolitan Statistical Areas (MSAs), cities outside MSAs, and rural counties. Perhaps surprisingly, murders and rapes occur in equal rates across all three categories. Robberies occur 3 times more often in MSAs than in cities outside MSAs, and 16 times more often in MSAs than in rural counties. The rate of aggravated assault is higher in MSAs than in other areas (1.2 times more likely in MSAs than in cities outside MSAs and 2.5 times more likely than in rural counties), but the difference is not as remarkable as for robberies.

TIME OF CRIMES

Violence does not occur at the same rate throughout the year. Figure 1–12 depicts the distribution of violent crimes by month for 1991 to 1995. For all violent

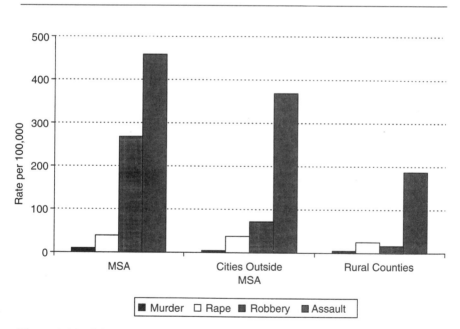

Figure 1–11 Violent Crime in Urban and Rural Areas, 1995. *Source:* Reprinted from Federal Bureau of Investigation, 1995, *Uniform Crime Report,* U.S. Department of Justice, Washington, DC.

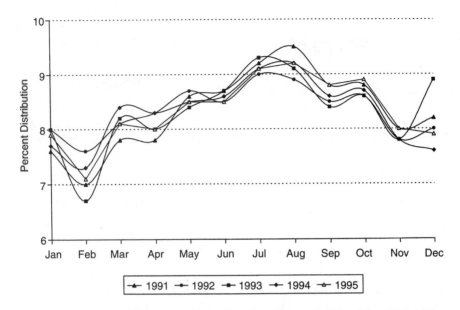

Figure 1–12 Violent Crimes by Month, 1991–1995. *Source:* Reprinted from Federal Bureau of Investigation, 1995, *Uniform Crime Report,* U.S. Department of Justice, Washington, DC.

---

crimes, the peak activity occurs during August (9.1% to 9.5% of the annual violent crimes), and the crime rate is lowest during February (6.7% to 7.6%). Murders occur most often during August and least often during February; forcible rapes occur most often during August and least often during December; robberies occur most frequently during October and least often in February; and aggravated assaults occur most often during July and least often during February.

The FBI's *Uniform Crime Report* also reports data using a different but quite illuminating format: the annual ratio of crime to fixed time intervals. Expressed this way, as a "Crime Clock," the 1995 U.S. crime statistics suggest that one violent crime occurs every 18 seconds! One murder occurs every 24 minutes, one forcible rape occurs every 5 minutes, one robbery occurs every 54 seconds, and one aggravated assault occurs every 29 seconds!

CONCLUSION

History reveals that violence is endemic in the United States and has increased over time. Despite our long association with violence in this country, surprisingly little has been learned about why it occurs and how it can be prevented.

Of all of the disturbing statistics presented in this chapter, perhaps the most alarming statistic is the finding that the group at greatest risk for being a murder victim or offender, across all races, is the group of Americans who are 20 to 24 years of age. This is also the age at which many Americans attend a college or university.

Life in a university is difficult. For many, it represents the first time that individuals are truly alone and without parental or familial supervision. Not only are college students without parental supervision, but also they are without the protection, emotional support, and wisdom that they are accustomed to receiving from their family members and the peers who have served as friends and classmates through high school.

College students experience a strong sense of isolation. Often they find themselves at a great distance from their home in an unfamiliar environment filled with strangers. To make matters worse, the university is a highly stressful environment that constantly puts high demands on students. The university provides constant and sometimes harsh feedback to its students. Surrounded by the best and the brightest, college students often must face their own limitations. Accustomed to excelling in school for 12 years, they now find that they are rather average when compared with more gifted and better educated counterparts.

For many students, college also brings the challenge of dealing with a heterogeneous population. Historically, elementary and secondary education in America has been largely segregated. For many students, the college experience brings the first true interactions with persons of different socioeconomic means; different race and ethnicity; different parts of the state, country, or world; different intellectual abilities; different religions and customs; and sometimes different gender.

The high stress of a college environment is further compounded by peer pressure and ethical and moral decisions about sexual relations and the use of alcohol and other mood-altering substances.

The vast majority of American college students face these stressful and challenging issues at the very time in their lives when they are statistically at greatest risk for being a murder victim or a murderer. It is surprising that so little has been written about violence on the college campus, given this epidemiological recipe for probability of violence.

REFERENCES

Brendtro, L., & Long, N. (1995). Breaking the cycle of conflict. *Educational Leadership* 52 (5), 52–55.

Bureau of Justice Statistics. (1997). *Expenditure and employment statistics* [Internet report]. Washington, DC: U.S. Department of Justice. Available: http://www.ojp.usdoj.gov/bjs/; accessed June 1998.

Canada, G. (1995). *Fist stick knife gun: A personal history of violence in America.* Boston: Beacon Press.

Federal Bureau of Investigation. (1995). *Uniform crime report.* Washington, DC: U.S. Department of Justice.

Graham, H.D., & Gurr, T.R. (Eds.). (1979). *Violence in America: Historical and comparative perspectives.* Beverly Hills, CA: Sage.

Langer, R. (Ed.). (1997, March 9). *TV Magazine.* Dallas: Osborne, Burl.

Palermo, G.B. (1994). *The faces of violence* (Vol. 5). Springfield, IL: Charles C Thomas.

Rape statistics [Internet report]. (1997). Washington, DC: Sexual Assault Information Page; http://www.cs.utk.edu/~bartley/sa/stats.html; accessed June 1998.

Reiss, A.J., Jr., & Roth, J.A. (Eds.). (1993). *Understanding and preventing violence* (Vol. 1). Washington, DC: National Academy Press.

Reiss, A.J., Jr., & Roth, J.A. (Eds.). (1994). *Consequences and control* (Vol. 4). Washington, DC: National Academy Press.

Rosenberg, M.L. (1995). Violence in America: An integrated approach to understanding and prevention. *Journal of Health Care for the Poor and Underserved,* 6 (3), 102–113.

Sigler, R.T. (1995). The cost of tolerance for violence. *Journal of Health Care for the Poor and Underserved,* 66 (2), 124–135.

Twitchell, J.B. (1989). *Preposterous violence: Fables of aggression in modern culture.* New York: Oxford University Press.

U.S. Department of Health and Human Services, Public Health Service. (1990). *Healthy People 2000: National Health Promotion and Disease Prevention Objectives* (DHHS Publication No. PHS 91-50212). Washington, DC: U.S. Government Printing Office.

Chapter 2

Campus Vulnerability

John H. Schuh

INTRODUCTION

Eugene M. Hughes was appointed president of Wichita State University, effective July 1, 1993. At the very first public event of his presidency, a Fourth of July celebration held on campus, two people attending were killed in a shoot-out between gang members. The activity was attended by approximately 30,000 people, and 100 commissioned police officers were on duty to provide security services (Roy & Dorsey, 1993). Among those killed was a woman who was on campus only to attend the evening's festivities and had nothing to do with the gang activity. This tragedy underscores how vulnerable college campuses are, even with thoughtful planning and a large police presence.

As organizations with large numbers of employees, colleges also are vulnerable to increasing workplace violence. The tragedy involving the deaths of four faculty and a student at the University of Iowa painfully illustrates this vulnerability ("U. of Iowa," 1991). According to Baron (1993, p. 3), "Murder in the workplace is now one of the fastest growing types of homicide in the United States." Workplace violence is discussed in Chapter 6 of this book.

In our increasingly violent society, college and university campuses are vulnerable to violent acts. Siegel (1994, p. xi) points out, "I have seen that colleges and universities throughout the nation are struggling as they respond to campus violence." Members of the campus community do not anticipate violent circumstances or acts. As Smith and Fossey (1995) assert, "Campus crime is an anomaly. . . .The heart of the anomaly, of course, lies in our traditional image of campuses as bucolic, tranquil places set aside for our intellectual pursuits" (p. 2). Yet, violence can strike on college campuses, as an incident at the University of Michigan illustrates. In this case, a police officer shot and killed a man who was attacking a woman with a knife. The man and woman shared a university apartment ("Mich. Police Kill," 1997). The woman died from her wounds.

The purpose of this chapter is to discuss selected factors and conditions that contribute to campus vulnerability to violent acts. The chapter is divided into two parts. The first part discusses the physical components and the second part identifies a variety of psychological issues that contribute to campus vulnerability. While the degree of vulnerability is not the same for all campuses, no campus is completely safe. Virtually all institutions have tremendous potential to experience violent acts. It may be nothing more than good fortune that more campuses have not experienced violence.

PHYSICAL COMPONENTS

The Open Campus

One of the primary marketing strategies of institutions of higher education (IHEs) is to invite people to visit campus and use the facilities. Colleges with aggressive recruitment programs try to convince prospective undergraduate students to visit the campus, and activities are planned around individual and group visits (Wright, 1995, p. 12). Among these activities are open houses, lectures, and overnight programs for prospective students and their parents. This is a popular recruiting tool widely recommended by experts in the admission and recruitment process (Abrahamson & Hossler, 1990). Moreover, it is common for signs to be posted along the U.S. Interstate Highway system that point out where colleges and universities are located so that people can find them. In short, IHEs want people to come to their campuses, use the facilities, and enjoy the advantages of the institution. This desire to involve the public in campus activities creates an environment that can lead to violence.

IHEs are eager to invite people to come to campus. Other than regulating parking access, virtually all campuses have no mechanisms for limiting who is allowed on campus. Any person who desires to come to the campus can do so by simply walking or driving to the campus. Few, if any, attempts are made by campuses to regulate who penetrates the perimeter of the campus and, perhaps more importantly, few authorities really know who is on campus at any given time. Presumably, persons who have violence on their minds can get to the campus without being challenged.

Contrast this situation with business and industry. In organizations that are concerned about industrial espionage and other security matters, fences with barbed wire on top secure the perimeter of the property, guards check identification passes before people are allowed to enter, and security officers are on duty. Visitors, and often employees, wear badges so they can be easily identified. These measures, and perhaps other more subtle activities such as closed-circuit television, do not ensure complete security, but it would be difficult to identify entities in modern American society that are more open and easy to access than IHEs. Even government buildings, as a response at least in part to the bombing of the federal building in Oklahoma City, have added security measures. Security measures have been put in place to protect those who work in and use these buildings; IHEs have not responded in the same way.

The dilemma faced by IHEs is similar to that faced by inner-city churches. Efforts to insulate the campus from intrusive crime and violence defeat its purpose as an interactive, societal resource. If a church needs to lock its doors to keep criminals from committing crimes on its premises, that also means that people cannot use the facility for its intended purposes. It changes the purpose of the church from serving the surrounding flock to protecting itself from predators. Similarly, if an IHE has to institute a variety of security measures and limit access to the campus, that works against the purpose of the institution—especially if it is a public institution, which by definition is owned by all citizens.

Access to Buildings

Once visitors arrive on campus, they have a wide variety of buildings available to them including classroom buildings, libraries, and laboratories. Security measures designed to monitor general public access may be found in residence halls and other facilities that require an identification card for access, such as recreation buildings, student health services, or other facilities operated for the benefit of the campus community. But for the most part, it is easy for people to gain access to buildings. Violence can erupt, in part because security measures are so modest, as was the case in a tragedy at the University of Texas at Austin ("Sniper in U. Texas Tower," 1966).

Most importantly, the mindset on most campuses is that they want people to come to the campus and use the facilities. Public institutions do not want to be seen as ivory towers. Rather, they seek to be perceived as assets for all citizens. It is common, for example, for college library services to be available to all citizens. Community colleges and metropolitan universities invite members of the community to use facilities, audit classes, attend concerts, and generally feel welcome on the campus. The University of Louisville, for example, makes it very easy for returning adult students to attend classes on campus. So, after gaining access to campus without challenge, people can use facilities without ever being asked why they are in the building or what their business is.

Length of Operation

Many IHEs strive to provide services and facilities when students need them. This strategy is seen as a means of retaining enrollment and attracting new students. One of the ways that good service can be provided to students and faculty is to make services and facilities accessible during nights and weekends.

One of the best examples of this approach is the campus library. Libraries are open extraordinarily long hours, often providing access 24 hours a day during examination periods. Main libraries, in particular, tend to be among the largest facilities on campus, where users can find many nooks and crannies for isolation and quiet study. This trend has at least two implications for this discussion. First, a person who is interested in sabotaging the library can go virtually unchallenged while carrying out a potential plan of violence against the institution. Someone who is located in a remote study area can go unchallenged for hours. What better opportunity to plan a violent act against the facility? Second, persons going about their academic pursuits in this environment are vulnerable to a violent personal attack. Security, if it exists, is light. There may be no one else on the floor or in the area who can alert security personnel to intervene in the case of a disaster. Remote-control television, a security precaution taken in the private sector to monitor facilities, may not be in place. Consequently, great potential exists for a violent act to occur with nobody available to intervene.

Libraries are not alone in their vulnerability. Classrooms and laboratories also can be the target of a violent act. A dramatic example of the vulnerability of a laboratory building was the explosion of bomb outside a laboratory at the University of Wisconsin, which took the life of a doctoral research assistant (Semas, 1970). No place on campus is immune from violence.

Campus Events

Most IHEs offer a wide variety of events to which members of the general public are invited during a typical academic year. Among these are sporting events, fine arts activities, convocations, and various celebrations. These events can range in size from a few dozen people who are invited to attend a special reception to recognize a retiring faculty member to the 5,233,954 people who attended women's basketball games during the 1995–1996 season (Campbell, 1996). Virtually every day some event is held on campus where members of the campus community and general public are invited.

It is often unknown beforehand who specifically will attend these events. Nearly anyone can purchase a ticket to many events. Very limited security measures are in place for these events where the public is invited and encouraged to participate. In fact, in an era where IHEs are hard-pressed to sustain their levels of governmental support (Schuh, 1993), one of the best ways of increasing public confidence is to hold quality events on campus and demonstrate the richness of the campus environment. Consequently, many events are held on campuses, with obvious security risks.

Access to Administrators and Faculty

Another problem that results from the openness of the campus is that senior leaders, such as presidents and chancellors, and their principal associates are located right on the campus. They attend events and meet with faculty, students, and the general public. Accessibility is seen as a key to success for senior campus leaders, and those who are successful regularly make appearances at all kinds of campus events. According to Kuh and colleagues (1991), "Being accessible to students, affirming the importance of students in speeches, meetings or conversations with community members, knowing students' names, taking time to talk with students about their concerns and interests, and eating with students in their dining halls or interacting with them in their residences and other gathering places are potentially meaningful symbolic actions on the part of institutional leaders" (pp. 360–361). Accordingly, many successful senior leaders make it a point to interact with campus constituencies and be seen on a regular basis.

Contrast the availability of campus leaders with leaders of corporations. Whereas the campus senior leader is well known and available on site, the president of a restaurant chain, automobile manufacturing company, or airline in most cases is located in a different city than most of the consumers of the product (or perhaps a different country in the case of a multinational company), in a corporate headquarters that a typical consumer would be hard-pressed to find. Moreover, it may be difficult for the person bent on violence toward a corporate executive to identify the name of such a person, let alone the person's whereabouts. The person who is unhappy with the college president can find that person easily; it is far more difficult to find the chairperson of the board of a multinational corporation, even if one knew that person's name.

Similarly, the constituency of college leaders is fairly local—the campus and perhaps the community. Graduates are likely to be scattered across a wide geographic area, but students attend classes on the same campus where the senior leaders have offices. Consumers of most products are located in different communities from where corporate headquarters are located.

Faculty members are particularly vulnerable. Consider, for example, the shooting of three faculty members at San Diego State University who reportedly were involved in hearing a graduate student's master's thesis defense. The student allegedly shot them over a failing grade (Haworth, 1996). Wooldredge, Cullen, and Latessa (1995) point out that faculty members often work alone in classroom buildings and are exposed to large numbers of students and staff. They conclude that "college faculty may be more vulnerable than students to campus victimization, given their work routines and the time they spend on campus" (p. 103).

The same case can be made in contrasting campus leaders with governmental leaders. In a large metropolitan area, finding the mayor or city manager may involve a lengthy trip; similarly, finding the state governor most likely will require a trip to another city, and identifying that person's office building may take a bit of work. The point is obvious: senior leaders of IHEs are on site and routinely interact with members of their constituency whereas other leaders in our society are much more removed and interact with constituents in significantly more controlled circumstances.

The consequence is that senior higher education leaders certainly are more vulnerable in a physical sense to violence than their counterparts in government or private enterprise. They are easy to identify; work in prominent offices; and routinely interact with students, faculty, and the general public. If they are the target of a violent act, it is not very difficult for the perpetrator of violence to find them.

Campus Location

The final aspect of the physical dimensions of campus vulnerability has do to with the permanence of the physical structures. Once a campus is built, it usually stays in that place, regardless of how the surrounding area changes over time. A campus founded in an area perceived as safe may discover over time that the neighborhood has changed and that safety and security become more prominent issues as the area deteriorates. Examples of institutions located in or near neighborhoods where safety is a concern include the University of Chicago, Marquette University, and the University of Southern California.

Problems that occur in the city can find their way onto the campus. At San Diego State, for example, a murder suspect hid in a residence hall for 7 hours ("Murder Suspect," 1997). The suspect's car contained rifles that may have been used in a murder. This incident illustrates campuses' vulnerability to criminals. Rapists, murderers, and other criminals can come onto the campus at virtually any time. Their presence places those who are part of the campus community at risk.

Private enterprise, on the other hand, can sell its property and move to another location, perhaps with important tax incentives and other inducements from the community in which it locates. Most of the best jobs in South Central Los Angeles, for example, left the area after the Watts riots of 1965. The best new jobs in Atlanta, Dallas, and Philadelphia are located outside the city proper (Leinberger, 1992). But moving a campus is about as easy as moving a graveyard. A mere proposal, for example, of moving a campus located in a deteriorating neighborhood to the suburbs undoubtedly would be met with howls of protest. Even if the public relations element of the situation could be managed, the cost of building a campus is astronomic. Moreover, the facilities of most colleges and universities are so diverse and specialized, that finding a buyer willing to pay a fair price would be virtually impossible. Who, other than another IHE, could use a facility with classrooms, laboratories, residence halls, libraries, and a football stadium?

Quite obviously, once a campus is built, it stays in its location, and the IHE has to adjust to a changing environment around it. Although this is not a problem for all IHEs, any institution located in a city will have to realize that at some time the nature of the surrounding area could become a problem.

PSYCHOLOGICAL COMPONENTS

The second part of this chapter discusses the psychological factors that contribute to the vulnerability of IHEs. These factors complement the physical circumstances supporting why campuses are vulnerable to violent acts. As was the case with the physical components, these factors and conditions are not entirely discrete; they tend to interact with one another in creating a climate of vulnerability.

Trust and Respect

Campuses have an air of trust about them. All comers are respected for their points of view and, until they demonstrate they do not deserve to be respected, they are welcome on the campus.

In this environment, students are taught to trust and respect those who have points of view different than theirs, regardless of how unorthodox, bizarre, or radical that perspective may be. Debate about issues is encouraged, and points of view that differ from the norm are considered. In fact, there may be no other institution in our society where it is possible to take an unconventional or even offensive point of view. Indeed, one only needs to point to the 1930s when Communist cells were established on college campuses (Brax, 1981), or the 1960s when opposition to U.S. involvement in the Vietnam conflict was particularly strong (*Report of the President's Commission on Campus Unrest,* 1970).

We do not recommend for a moment that free speech codes be abrogated, or that a lack of trust and respect should become the operating norm on college campuses. Far from it. We value the historical place that IHEs have had in American society as places where ideas of all kinds can be examined and debated. However, it is important to point out that this kind of discourse contributes to a climate where violence can occur.

Controversy

Controversy is central to the life of college campuses and has been part of the history of American higher education for decades (Brubacher & Rudy, 1976). For example, speakers who challenge the very fiber of the institution and of society are invited to campus (Shea, 1994). Unless perhaps they have a history of inciting riots, such individuals are invited to come to campus to share their views—at times with student audiences, but at other times with people who are invited to the campus to hear the speech. Ironically, sometimes the cost of bringing the speaker to the campus is borne by the institution directly through campus resources or student fees. Often students and others protest actions of the senior institutional officer when they disagree with his or her actions (Wilson, 1996).

Few organizations in contemporary society are as likely as colleges and universities to invite controversy, often of an abrasively strident nature. In turn, controversial speakers may raise the specter of violence and, unfortunately, there is a history of such people turning controversial topics into calls for violence (Kahn, 1970). Smith and Fossey (1995) assert, "Two of those factors—the civil rights movements and the Vietnam era protests—are especially significant for the phenomenon of campus crime, because their arrival brought, for the first time, widespread international lawbreaking and violence to campuses, albeit lawbreaking and violence that came from ideology born of the highest moral persuasion" (p. 7).

Substance Abuse

It is not hard to establish a link between substance abuse and criminal acts and violence (Kuh, 1994; Smith, 1989). Over the years, college campuses have been the locus for various forms of substance abuse, including alcohol, illegal drugs, and controlled substances. A variety of reports have decried the abuse of substances on college campuses (Kuh, 1994; Wechsler, Kuh, & Davenport, 1996; Wilsnack & Wilsnack, 1982), but little has been accomplished to truly stem the tide of substance abuse on campus. To be sure, not all campuses have experienced or continue to experience substance abuse among their student bodies, but the fact remains that substance abuse on college campuses is a matter of concern.

Whether IHEs are more tolerant of substance abuse (particularly alcohol) than other institutions in society is not certain. It is clear that some elements of IHEs have experienced chronic problems in this area (Fischer, 1987), with no particular progress toward eradicating substance abuse. In fact, there is evidence to suggest that substance abuse is tolerated in some circles (Kuh & Arnold, 1993) and that some elements of campus life do not contribute to the overall development of students (Kuh, Pascarella, & Wechsler, 1996). According to Smith (1989), "Beverage alcohol is widely used and even applauded as a social lubricant and tranquilizer at the same time that alcohol abuse brings many grievous social problems" (p. 121). With a certain level of tolerance toward substance abuse comes a host of problems related to violence that are discussed elsewhere in this volume (see Chapter 12). Among them are such health issues as sexually transmitted diseases, pregnancy, and abortion. What has not been experienced to a great extent, but may be just around the corner, are such issues as open drug dealing, gang violence, robbery, and so on. Substance abuse, particularly involving controlled substances such as narcotics, opens the door to violence in and around the college campus.

At Tennessee State University, for example, a gunfight resulting from a robbery associated with drug dealing resulted in a student's death in a residence hall. According to newspaper reports, "One man died in the dorm room . . . when a deal to buy a $10 bag of marijuana turned into a robbery and then a gunfight, police said" (Ward & Kimbro, 1997). Drugs, quite obviously, can be linked directly to violence.

Stress

The stress college students experience is greater than ever before, given increasing tuition, fewer services, and reduced course offerings (Cage, 1992). The result is that students live and learn in a very stressful environment. Aspects of the college environment that create stress include the following:

- Traditional-age students need to manage their social and sexual development while simultaneously handling academic demands they have never faced before in their academic career.
- In many cases, these same students are living away from home for the first time and are dealing with the stress of handling that adjustment.
- Students of all ages are faced with borrowing tremendous sums of money to pay for their education. The result is that they assume tremendous debt, which they will have to manage upon leaving their college or university. Their only alternative is to work increasingly greater numbers of hours while attending class.

Failure in this environment is obvious. Grades are awarded at the end of each academic term, and students know exactly where they stand. If they do not meet acceptable standards, they may be withdrawn from the institution against their will; in all situations, students are not awarded degrees if they do not complete their work satisfactorily. Even if students can conjure up excuses as to why it is not their fault that they have not completed their academic program successfully, they have to engage in a program of massive self-deception or live with their failures. Regardless of the course of study selected, stress becomes part of students' academic environment. Violence can result when young students encounter such high levels of stress for the first time.

Time at the IHE

With the exception of the so-called perpetual student, most students spend a fairly limited length of time on the campus. After 4 or 5 years, students graduate and move on. Although graduate students may take a bit longer to complete their course of study, they, too, eventually move on to the world of work, another institution, and so on. Contrast this with people who join organizations and serve as employees for 10 or 20 years (Arenofsky, 1994). Because students have a fairly short-term relationship with their colleges and universities, they have less of a psychological investment in the institution. This lack of psychological investment may manifest itself in the form of violence, such as vandalism to campus facilities. Consider the problem of vandalism that residence hall directors must contend with on a continual basis. One presumes that, upon graduation, students will maintain their homes far better than they treat their residences in college.

POLICE AND SECURITY SERVICES

One societal response to violence and crime has been to improve police and security services. Although police services are discussed elsewhere in this book, a comment is appropriate here. One response to dealing with increasing crime and violence has been to provide a greater police presence (Sievert, 1971) or tightened security (Peterson, 1993)—approaches not always met with open arms. Indeed, questions are still raised about whether police officers ought to be highly visible, much less armed, and whether greater security measures are warranted. According to Peak, "The tools of municipal police departments, when observed on post-secondary campuses, may seem anomalous in an environment where discovery and transmission of knowledge are paramount" (1995, p. 238). It is important to remind the reader that the role of campus police agencies still is open to some debate, and the role of the campus police department is different than in a municipality. As Sloan and Fisher (1995) point out, "On a college campus, however, the relationship between the community and the campus police is different. Here, the police are more active, more involved in policy making and share with the community the responsibility for reducing crime and maintaining order" (p. 14). Moreover, the question of whether to provide weapons to police officers is controversial in some quarters. As Powell, Pander, and Nielsen (1994) observe, "a small but vocal group of faculty members, student affairs personnel and administrators is often able to keep guns off campus" (p. 94).

Police and security services are not always welcomed on college campuses. Nonetheless, it is highly likely that many colleges and universities will take measures to strengthen their police services as violence touches more and more campuses.

REFERENCES

Abrahamson, T.D., & Hossler, D. (1990). Applying marketing strategies. In D. Hossler, J.P. Bean & Associates (Eds.), *The strategic management of college enrollments* (pp. 100–118). San Francisco: Jossey-Bass.

Arenofsky, J. (1994). Career decisions. *Career World,* 22, 4–6.

Baron, S.A. (1993). *Violence in the workplace.* Ventura, CA: Pathfinder.

Brax, R.S. (1981). *The first student movement: Student activism in the United States during the 1930s.* Port Washington, NY: Rennikat.

Brubacher, J.S., & Rudy, W. (1976). *Higher education in transition: A history of American colleges and universities,* 1636–1976. New York: Harper.

Cage, M.C. (1992, November 18). More students seek counseling, but less is available. *The Chronicle of Higher Education,* p. A26.

Campbell, R.N. (1996, July 1). Women's basketball attendance tops five-million mark. *The NCAA News,* 33, pp. 1, 12.

Fischer, J.M. (1987). A historical perspective: Alcohol abuse and alcoholism in America and on our campuses. In J.S. Sherwood (Ed.), *Alcohol policies and practices on college and university campuses* (pp. 1–17). Washington, DC: National Association of Student Personnel Administrators.

Haworth, K. (1996, August 16). Gunman kills 3 in engineering classroom at San Diego State U. *The Chronicle of Higher Education News Update.*

Kahn, R. (1970). *The battle for Morningside Heights.* New York: Morrow.

Kuh, G.D. (1994). The influence of college environments on student drinking. In G. Gonzalez & V. Veltri (Eds.), *Research and intervention: Preventing substance abuse in higher education* (pp. 45–71). Washington, DC: U.S. Department of Education, Office of Educational Research and Improvement.

Kuh, G.D., & Arnold, J.C. (1993). Liquid bonding: A cultural analysis of the role of alcohol in fraternity pledgeship. *Journal of College Student Development,* 34, 327–334.

Kuh, G.D., Pascarella, E.T., & Wechsler, H. (1996, April 19). The questionable value of fraternities. *The Chronicle of Higher Education,* p. A68.

Kuh, G.D., Schuh, J.H., Whitt, E.J., & Associates. (1991). *Involving colleges: Successful approaches to fostering student learning and development outside the classroom.* San Francisco: Jossey-Bass.

Leinberger, C.B. (1992). Business flees to the urban fringe. *The Nation,* 225, 10–14.

Mich. police kill man after campus stabbing. (1997, October 15). *The Chronicle of Higher Education,* p. A8.

Murder suspect hides out on campus. (1997, November 7). *The Chronicle of Higher Education,* p. A8.

Peak, K.J. (1995). The professionalization of campus law enforcement. In B.S. Fisher & J.J. Sloan, III (Eds.), *Campus crime: Legal, social and policy perspectives* (pp. 228–245). Springfield, IL: Charles C Thomas.

Peterson, I. (1993, March 6). Authority chose access over tighter security. *The New York Times,* p. 25.

Powell, J.W., Pander, M.S., & Nielsen, R.C. (1994). *Campus security and law enforcement* (2nd ed.). Boston, MA: Butterworth-Heinemann.

Report of the president's commission on campus unrest. (1970). New York: Arno.

Roy, J.C., & Dorsey, R. (1993, July 5). Holiday erupts in violence. *The Wichita Eagle,* p. 1A, 6A.

Schuh, J.H. (1993). Fiscal pressures on higher education and student affairs. In M.J. Barr & Associates (Eds.), *The handbook of student affairs administration* (pp. 49–68). San Francisco: Jossey-Bass.

Semas, P.W. (1970, September 28). Faculty and students fearful and confused after fatal bombing at U. of Wisconsin. *The Chronicle of Higher Education,* p. 8.

Shea, C.J. (1994, March 16). Dealing with virulent speakers. *The Chronicle of Higher Education,* p. A32.

Siegel, D. (1994). *Campuses respond to violent tragedy.* Phoenix, AZ: Oryx Press.

Sievert, W.A. (1971, March 1). With crime increasing, universities are beefing up their police forces. *The Chronicle of Higher Education,* p. 1, 3.

Sloan, J.J., III, & Fisher, B.S. (1995). Campus crime: Legal, social and policy concerns. In J.J. Sloan, III, & B.S. Fischer (Eds.), *Campus crime: Legal, social and policy issues* (pp. 3–19). Springfield, IL: Charles C Thomas.

Smith, M.C. (1989). Students, suds and summonses: Strategies for coping with campus alcohol abuse. *Journal of College Student Development,* 30, 118–122.

Smith, M.C., & Fossey, R. (1995). *Crime on campus.* Phoenix, AZ: Oryx Press.

Sniper in U. Texas tower kills 12, hits 38; Wife, mother also slain; Police kill him. (1966, August 2). *The New York Times,* p. 1, 14.

U. of Iowa mourns shooting victims. (1991, November 13). *The Chronicle of Higher Education,* p. A4.

Ward, G., & Kimbro, P.L. (1997, October 21). Ex-student dead, another injured in TSU drug dispute. *The Nashville Banner* (Internet report, no page numbers).

Wechsler, H., Kuh, G., & Davenport, A.E. (1996). Fraternities, sororities and binge drinking: Results from a national study of American colleges. *NASPA Journal, 33,* 260–279.

Wilsnack, R.W., & Wilsnack, S.C. (1982). Introduction and overview. In J.C. Dean & W.A. Bryan (Eds.), *Alcohol programming for higher education* (pp. 1–4). Carbondale, IL: American College Personnel Association.

Wilson, R. (1996, March 1). Rutgers U. president weathers a siege. *The Chronicle of Higher Education,* p. A23.

Wooldredge, J.D., Cullen, F.T., & Latessa, E.J. (1995). Predicting the likelihood of faculty victimization: Individual demographics and routine activities. In B.S. Fisher and J.J. Sloan, III (Eds.), *Campus crime: Legal, social and policy perspectives* (pp. 103–122). Springfield, IL: Charles C Thomas.

Wright, B.A. (1995). Admissions recruitment: The first step. In R.R. Dixon (Ed.), *Making enrollment management work* (New Directions for Student Services Sourcebook, No. 71). San Francisco: Jossey-Bass.

CHAPTER 3

Profile of Students Coming to Campus

Robert H. Fenske and Stafford L. Hood

INTRODUCTION

Violence on campus is perpetrated by and on several classes of people including students, staff, and faculty as well as other persons on or near the campus who have no formal connection to the institution. This chapter deals with pre-enrollment characteristics and experiences of students, probably the most important of these constituencies in terms of number of violent incidents as either victim or perpetrator. The title of this chapter may be misleading. The following pages delineate a profile of high school and younger students, many but not all of whom go on to college. Those who do come to our campuses do so from an educational environment that has unquestionably become more violent over the past two decades. The excellent book by Hoffman (1996), among others, contains extensive documentation of what has been called "the rising tide of violence" in America's elementary and secondary schools. Extensive databases, described later, quantify the dire effect of violence and crime on the learning environment and on the safety of students and teachers. Since the 1970s, U.S. schools have turned to widespread use of security officers to patrol school grounds and hallways, drug-sniffing dogs, and metal detectors at school entryways. These measures and statistical evidence of increased incidence of violence and crime demonstrate the seriousness of the problem.

Given that the schools that prepare students for our campuses have become increasingly violent, what is the situation in our colleges? The other chapters in this book describe and discuss the current rise of violence in postsecondary educa-

---

The authors would like to thank Jann Contento and Richard R. Sines for their assistance in preparing this chapter. Both are doctoral students in the Division of Educational Leadership and Policy Studies, Arizona State University.

tion. The federal Campus Security Act of 1990 was the first of many federal initiatives aimed at stemming the rise of crime and violence in our colleges. As this book is being written, Congress is debating legislation that would increase campus security by identifying and widely publicizing perpetrators of campus crime and violence but at the same time maintain constitutional rights to privacy of students accused or convicted of crimes.

It might seem that a causal connection could be made between the rise of violence in the pool of potential students in the "feeder" schools and the rise of violence on campus. However, this cause-and-effect relationship has not yet been tested by data. One could argue that, although American society is becoming increasingly violent, students coming to our campuses directly from or soon after high school comprise a societal subset that is much less prone to violence either as perpetrators or victims. After all, these young people are a minority of their age cohort, self-selected to be less inclined to be in contact with violence. They are generally enrolled in the more studious high school academic tracks and have goals and positive plans for a rewarding future. Unlike the stereotype of a dropout from the school system, college-bound high school students could be expected to avoid potentially violent situations that would jeopardize such plans. To some extent, these logical assumptions are borne out by research findings, as will be shown in following sections.

One could further conjecture that the "educational pipeline" contains a filtering system that removes all or most violence-prone youth well before the transition point between high school and postsecondary education, notably at the legal school-leaving age (16 in most states). This conjecture has some empirical support. Dropout studies consistently show linkages among antisocial, disruptive behavior; poor academic performance; and propensity to leave school, voluntarily or involuntarily, before high school graduation. "Zero-tolerance" policies against violence and weapon carrying in schools remove violence-prone youths, at least temporarily. Expulsion from the school, even temporarily, would also discourage persistence in the school system and, presumably, remove troublesome youths from the pool of potential college students. A final argument might be made that time and maturation also reduce violent tendencies among those coming to campus. The high-school-to-college transition is a rite of passage toward responsible adult behavior, toward reasoned discourse as an alternative to "acting out" violent behavior. U.S. Department of Justice Crime Victimization Survey data consistently show that pupils in grades 6 to 10 are significantly more likely to be involved in school crimes and violence such as physical fighting and weapon carrying than pupils in grades 11 and 12 (National Center for Education Statistics [NCES], 1995). Would it not be logical to assume that the same trend extends into the college years?

On the other hand, it could be argued that college-going high school students are inevitably affected by violent incidents in school and other settings, whether as passive observers, victims, or perpetrators. Alan J. Lizotte, professor of criminal justice and executive director of the Consortium for Higher Education Campus Crime Research at the State University of New York at Albany, observed that "national crime data showed that young people were becoming more violent and increasingly likely to be involved in crime," and that "if you have people in high school who are doing drugs and carrying guns around, some of those people are bound to end up attending college" (Lederman, 1995, p. A31).

One aspect of the relationship between violence in high school and on the college campus is discussed in Chapter 12. The section "The Nature of the Societal Problem" describes the relationship between substance abuse and violent behavior. The authors assert that this relationship is present in the high school setting and in overall society, as well as on the college campus. They note that it is inappropriate to place the entire blame for violent behavior linked to substance abuse on the college environment. They cite studies that uncover serious problems of substance and alcohol abuse among high school students.

Although logical assumptions may be made to link behavior before and during college, data are simply not available to draw clear cause-and-effect relationships between violent behavior in kindergarten through 12th-grade (K–12) settings and similar behavior in collegiate settings. We do not know whether elementary and secondary school students involved in violence in their schools are more likely, or less likely, to go on to college. Furthermore, if such students are more likely to come to campus, we do not know the extent to which they are involved in violence on our campuses. Without data that would resolve these questions, we rely on the logical assumption that students coming to campus have inevitably been affected by direct or vicarious exposure to violence and that such exposure can affect their college careers in many ways. It is highly useful for campus staff and administrators to be knowledgeable about such background experiences of incoming students in order to deal effectively with students' expectations, attitudes, values, and behaviors.

This chapter focuses primarily on "traditional-age" students (students 18 to 24 years of age), especially those who come to campus soon after completion of high school. The chapter explores these students' experiences with violence in their former school settings, especially high school and, to a lesser extent, junior high and middle school. It is assumed that contacts with violence in their former school settings are likely to affect their attitudes, expectations toward, and actual experiences with violence on the college campus. Large and informative databases are available on violence in the school settings of traditional students coming to campus. First, however, the chapter considers the "nontraditional" or older students.

NONTRADITIONAL STUDENTS

Nontraditional students are also part of the campus violence picture, but they are a relatively minor part for a number of reasons. Nontraditional students are usually defined as older than the traditional age range of 18 to 24 years, and attend part-time rather than full-time. For these reasons, as well as others, they are usually commuter students. Growth in the number of nontraditional students has been rapid over the past 25 years, but they still represent a minority of all students. Of the 14,279,000 students in 1994, only 6,141,000 (43%) were part-time, and there are relatively few older-than-traditional-age students among the full-timers. In 1994, three fourths of the 8,138,000 full-time students were traditional age. In contrast, 72% of the part-time students were older than traditional age; of these, 61% were women, and the majority were aged 35 or older (NCES, 1996).

Thus, the "typical" nontraditional student is a woman older than 34 who commutes to campus to take one or two courses. For several reasons, such students are unlikely to be as involved in violent crime as traditional-age students, either on an absolute or relative basis. Traditional-age students are much more likely to attend full-time and reside on or near campus. Their social activities (such as evening and weekend parties) often involve drinking and drug use, which, in turn, can result in a high incidence of violent behavior.

In contrast, most older-than-traditional-age students commuting to campus to take one or two courses spend little time on campus outside of that required for classroom attendance. Such abbreviated and focused contact with the campus does not correlate with high involvement with violent behavior. These students are also unlikely to be involved in activist or political behavior, which can lead to being a victim of hate crimes. The profile of a typical nontraditional student evoked by enrollment statistics is that of a "fortyish" woman taking one or two evening courses at a local public community college or urban 4-year college or university. This portrait suggests an individual who is very concerned about personal or family finances, child care, and accessible parking, and who has little propensity for involvement in violence, either as a victim or perpetrator. This is not to say that older students are unaffected by campus violence. For example, women who commute to campus for evening classes may be particularly vulnerable as victims, especially on urban campuses with high levels of violent crime such as robbery and sexual assault.

There is also concern about older students as possible perpetrators. Persons who have been convicted of serious crimes and served long prison sentences are *de facto* likely to be older than 18 to 24. One of the logical ways for such persons to rehabilitate themselves is to seek higher education credentials. In fact, Pell Grants were, until recently, awarded to incarcerated persons to begin degree programs. For example, in August 1993, the public in a large Southwestern city was informed by the city's leading newspaper that a convicted murderer had been admitted to the law school of the public university located in the city. The headline read "Objection: Killer's Admission to Law School Berated" (McCowan, 1993). Sides quickly formed on the issue, and some attacked the law school's admissions policy by, as one politician put it, "pandering to public fears" (Yozwiak, 1993). Investigative reporters responding to such fears soon uncovered other ex-convicts attending the same law school, including another convicted murderer who had been convicted of "helping mastermind the contract killing of a baby and the child's parents" (Leonard, 1994).

The concerns about campus safety were elaborated on in the city's other major newspaper. "There are murderers among us. We just don't know who they are. They are moving along with us in our daily lives, doing the things we do ... except that they are different. Through some deficiency or concealed frailty, they are destined to succumb to a provocation—real or imagined—and slay another human being" ("Life and Death," 1995).

Other commentators pointed out that civil rights restored to convicted persons who served their sentences include education, and that it makes sense to rehabilitate such persons into useful, productive citizens by providing access to educational programs. Nonetheless, the state legislature drastically cut funding for the law school to force it to deny admission of convicted felons (Van Der Werf, 1995). The controversy, which had begun in August 1993 with the original admission of the first murderer, continued until 1998 when the state board of law examiners refused to allow the ex-convict to sit for the state bar exam (McCoy, 1998).

Clearly, as shown in this and many other cases, the possible criminal background of persons enrolling in higher education touches deep fears about the possibility of repeat offenses. As discussed in Chapter 14, such fears raise the issue of the college's or university's "duty to inform" and "duty to protect" students. *Eiseman v. New York* (1987), a landmark case in higher education, involved an ex-convict who was allowed to register in a program specifically aimed at rehabilitating persons released from prison. In September 1975, a man released from prison enrolled in such a program in New York. In June 1976, he murdered two students in their apartment. The family of one of the victims, a female whom he had also raped, sued the state, claiming that the perpetrator should not have been admitted to the college without inquiry, which would have revealed his previous psychiatric problems. The court agreed that the prison physician, as agent of the state, should have informed the college and its students of the felon's medical history, and that the college was also negligent in admitting him. The decision was ultimately reversed by the Federal Court of Appeals, which ruled that the college had no "duty to inform" the ex-convict's fellow students of his medical history.

The general principle upheld in the *Eiseman* case is that the doctrine of *in loco parentis* does not hold in today's colleges. This means that colleges have no legal duty to protect students from potential danger from other students, even from those with previous criminal convictions. Furthermore, the college does not even have the duty to inform students of an admitted student's previous dangerous criminal activity. However, a more recent Kansas State Supreme Court ruling seems to contradict these principles. The notable 1993 case of *Nero v. Kansas State University* resulted in the court's ruling in a split decision that "state colleges and universities have a limited, legal duty to protect their students from crime." The case involved a female student who was sexually assaulted by a male student who had been accused of raping another female student the previous year. The victim claimed that the university should have known that the rapist was likely to strike again, and that it failed to protect her by allowing him to live in her dormitory. A dissenting justice disagreed, stating: "The majority opinion, in essence, requires that the university presume guilt" and that "monitoring students for foreseeable criminal behavior could potentially infringe on the rights of the accused" ("Kansas Court," 1993).

The profile of nontraditional students remains indistinct because of their diversity, including their wide age range, and the limited amount of contact they, as part-time students, typically have with the campus. There is a concern that this group includes individuals who may have been convicted of more serious offenses than offenders among traditional-age students. However, unlike for traditional-age students, comprehensive databases do not exist regarding nontraditional students' exposure to, or participation in, violence or crime prior to enrollment.

A BACKGROUND OF VIOLENCE

The following portrait of traditional students is a statistical profile drawn from the findings of large data sets such as longitudinal surveys and reports of various agencies. But these are not "dry statistics." The data reveal a dynamic picture of students coming to campus from school and societal settings increasingly steeped in a rising tide of crime and violence that pervades schools and neighborhoods across the nation. This societal trend hardly needs explicating; every reader of this book has been affected by the trend in innumerable ways. Home security services and devices, hardly thought of by most citizens 30 years ago, has been one of the fastest-growing industries over the past 15 years. Virtually every political campaign from village council member to the national presidency has had crime and violence as a centerpiece. Ways to stem this trend, other than increased expenditures for law enforcement and incarceration, continue to elude the nation. Most observers are predicting further increases as our youth continue to be saturated from the earliest age of perception with scenes of violence throughout the media. For example,

Television brings a steady volume of vicarious violence into living rooms. Over 97 percent of American households have at least one television set, and it is estimated that young children watch an average of four hours of television a day. Each year, they are likely to watch passively, and typically without adults present, acts of violence at unprecedented levels. The American child will see an estimated 8,000 murders and 100,000 acts of televised violence before leaving elementary school. It's hard to know how viewing violence affects children, but some authorities note that we are raising a generation of children unlike any others (Nelson, Carlson, & Polansky, 1996, pp. 403–404).

Long-time observers of K–12 school environments often paint poignant comparisons between the past and the current scene. Hoffman, in the introduction to his recent volume *Schools, Violence and Society* (1996), draws such a comparison.

On a bright, sunny day with wispy clouds and a gentle breeze, the faint laughter of children can be heard. The school yard, a place where children play with friends, jump rope, throw and kick balls, trade cards, laugh, and just hang out in this peaceful—albeit temporary—sanctuary. Inside the school, a warm welcome smile from the teacher, grateful for the apple from a kind student. The school, a place to read and write and create and learn. In the decades to come the sounds of laughter fade, and the once hospitable sanctuary is transformed into a place of fear. Instead of apples, the children bring guns to school. Instead of games or laughter, there are profanity, assaults, drugs, and alcohol (p. xi).

Incidence of violence in society is related to age, and is especially prevalent among youths at the age of high school completion and college enrollment. The U.S. Department of Justice recently determined that 16-to-19 year olds were 17 times more likely to be victims of a crime involving a handgun than persons 65 and older (U.S. Department of Justice, 1994).

AN INTERNATIONAL PERSPECTIVE

In 1990, the World Health Organization released findings of an international comparison of violent death among youths aged 5 to 24 in the large industrialized countries, during the late 1980s. According to the report, "The United States had the highest overall violent death rate (481 per 100,000 youths), a rate more than twice as high as those of Japan, Italy, and the United Kingdom" (p. 2). The youth homicide rate in the United States was more than 6 times higher than in Canada, the next highest nation, and more than 20 times higher than in Japan, the country with the lowest rate.

Early in 1997, the Centers for Disease Control and Prevention released "an extraordinary score card of youth violence" revealing that "the United States had the highest rates of childhood homicide, suicide and firearms-related deaths of any of the world's 26 richest nations (Havemann, 1997, p. 1). The findings were startling and disconcerting to the nation, especially since the report pointed to high homicide rates among U.S. children. In contrast, many nations reported "that they had no homicides involving children under the age of 15" (p. 1). "The statistics show that the epidemic of violence that has hit younger and younger children in recent years, is confined almost exclusively to the United States" (p. 1). The report identified no causes for the huge gap between the United States and other countries, but many observers point to the ready availability of firearms, especially handguns, and the prevalence of drug and alcohol abuse among younger and younger children.

ESCALATION IN SCHOOL VIOLENCE

One of the inevitable consequences of multiple, uncoordinated data-gathering efforts by numerous public and private entities, is inconsistent and sometimes contradictory findings. The sensational nature of the phenomenon being studied lends itself to selective emphasis of findings by the media and various advocacy groups. Furthermore, the probability of underreporting and even concealment of data most damaging to the reputation and support of public schools renders an accurate, balanced view very difficult, if not impossible, to attain. This section presents findings from what are believed to be the most valid, reliable, and objective sources.

As mentioned earlier, school is no longer a haven from the crime and violence that exists in the neighborhood and larger society. The general level of violence in society has escalated, and such escalation is especially acute among high school–age youth. According to the Bureau of the Census (1996), the rate of arrests of youths aged 17 and younger was 4.1 per 1,000 youths in 1950; this number increased five-fold in the next 5 years to 21.2. That rate more than doubled to 47.0 in 1960 and then climbed to 104.3 by 1970. The nation was appropriately concerned by the fact that more than 1 of every 10 young persons had been arrested by the age of 18, but the school as a setting for behavior leading to arrest was not yet a major concern. However, as the arrest rate climbed beyond 130 per 1,000 by 1981, the schools were inevitably involved. The arrest rate has fluctuated between 115 and 130 since 1981; it stood at 130.3 in 1993. Arrest rate is only one of many indicators, but it correlates strongly with other indicators such as number of crime victims and rate of property damage. The 1980s, which saw the arrest rate escalate to its current high level, also witnessed the advent of school security guards and on-site searches for guns and drugs.

As the level of societal violence rises, it inevitably spills over into the schools, affecting students, teachers, staff, administrators, and parents. Students can be victims or perpetrators, and can switch between these roles in separate incidents.

Most Americans would agree that all schools should provide an environment that is safe for students. Unfortunately, this ideal presents a serious challenge in a society where crime against students and teachers at schools, as well as other threats to security and a sense of well-being, are reported with alarming frequency.

In 1993, the National Center for Education Statistics (NCES) conducted the National Household Education Survey on a national sample of 6th through 12th graders. In this survey, "victimization" is defined as direct personal experience of threats or harm, and includes knowledge or witness of crime or bullying at school (National Data Resource Center, 1995). "The ... results suggest that unsafe conditions at school are a reality for most U.S. students. Half of 6th through 12th grade students personally witnessed some type of crime or victimization at school, and about one out of eight students reported being directly victimized at school" (p. 2). Of the students surveyed, 12% "or about one out of eight, reported having been directly and personally victimized at school during the current school year" (p. 3).

In terms of the number of students affected, a recent survey by the National Association of School Psychologists estimated that about 282,000 students are attacked in secondary schools every month (Onondaga-Courtland-Madison Board of Cooperative Educational Services, 1997). The survey found that teachers are also involved in incidents of violence. "The statistics for teachers are not much brighter. The same source claims that every month 125,000 secondary school teachers (12%) are threatened with physical harm and about 5,200 are physically attacked" (p. 1).

These findings and innumerable others reveal that large and increasing numbers of students and teachers are direct victims of school violence. But from an educational perspective, the most critical victims are learning, trust, and security. Surely, attitudes toward these are strongly shaped in junior and high school and carried forward into college.

NATIONAL AND STATE DATA SOURCES

In 1993, the Metropolitan Life Insurance Company sponsored an extensive survey of teachers. The report from this survey, titled *The American Teacher: Violence in America's Public Schools,* probably produced more concern about school safety for students and teachers than any other single report in the 1990s. The project surveyed public school teachers and students in grades 3 through 12. Louis Harris and Associates, the noted polling organization, gathered data for the study.

The report was widely publicized and triggered survey activity in other private organizations as well as in government, especially at the federal level. The University of Michigan's Survey Research Center gathered data from students in grades 8, 10, and 12 in a national sample of public and private schools. The center has gathered data on this basis since 1975. In 1995, the center published a report entitled *Monitoring the Future, 1994, 1995* that summarized trends over the past 20 years (Survey Research Center, 1995). Like the Metropolitan Life Insurance Company report, *Monitoring the Future* painted a bleak picture of schools in the 1990s. Drug and alcohol usage has been rampant, and weapons carrying continued to increase over the 20-year period. Other national surveys have been conducted by the National School Board Association and the Children's Defense Fund, among many others.

Led by California, a number of states have mandated periodic or occasional surveys of school crime and violence. In addition, some large city systems, such as New York and Los Angeles, have conducted similar surveys. Despite the seeming plethora of data-gathering activity, there is lingering concern about under-reporting and even concealment by school boards and administrators of actual incidents of crime and violence. The obvious reasons for this phenomenon include fear of voter reprisal, loss of confidence, and shift of enrollment from public to private schools or to home schooling. The enrollment shift away from public schools is a rapidly growing movement led, in part, by parental concerns about crime and violence in the public schools.

SAFE SCHOOLS AND INDIVIDUAL RIGHTS

What are the responsibilities of the schools to achieve "safe, disciplined, and alcohol- and drug-free schools" as stated in goal 7 of the National Education Goals Panel (1997a)? Efforts to carry out those responsibilities have stretched school budget resources to the limit. However, some feel that the overt "control" efforts are inadequate and self-defeating, because they create the "school as fortress" image so destructive to an atmosphere that nurtures learning. The continual presence of security guards (sometimes with drug-sniffing dogs) and the requirements of teachers to "wand" students generate questions of fairness and students' constitutional rights to privacy and the assumption of innocence. Ascher (1994) summarizes the problem:

"At the same time, a national uncertainty about how to handle potential conflicts between discipline, safety, and students' rights has made teachers unsure about what parents and the larger society want them to do. Can lockers be searched for weapons without giving students sufficient warning? Is it fair to wand only 'suspicious-looking' students and not others? Do students have a right, as some claim, to carry weapons for their own defense?" (p. 3).

Nonetheless, strong expectations for a safe learning environment are placed on the schools by parents and the courts. The National School Safety Center stated that a review of such expectations leads to the conclusion that the right to safe schools includes the right of students and staff to protection against (1) foreseeable criminal activities, (2) violence or student crime that adequate supervision can prevent, (3) potentially dangerous students who are identifiable, and (4) dangerous persons admitted to school in a negligent manner (National School Safety Center, 1990). From the viewpoint of students who are potential victims, as well as their parents and teachers, these are laudable rights. But enforcement of such victim rights in a system of compulsory public schooling must be balanced against the rights and even legal requirement of the perpetrators to remain in school.

Two recent incidents of school violence and their aftermath reveal how convoluted the path to school safety and a learning-conducive environment can be. Both occurred in Racine, Wisconsin, an industrial city of about 80,000 located south of Milwaukee. In many ways Racine is typical of such American cities, and according to national statistics, incidents such as these are not restricted to large urban areas. According to newspaper reports, on January 12, 1994, a 13-year-old, seventh-grade girl (referred to as R.T.) was mocked by another girl following an after-school talent show tryout. An argument followed outside the school, and R.T. pulled out a pocketknife and cut the other girl across the right ear, upper lip, and the length of her left forearm. The wounds required more than 300 stitches. R.T. had been arrested 4 months earlier for slashing another juvenile girl's face. Under existing school policy, R.T. was permanently expelled. However, the following year a civil rights attorney contacted R.T.'s mother and offered to represent her. The legal process ended in the U.S. District Court, where Judge John Reynolds ruled that the school district had to reinstate R.T. The school district was concerned "about the message that will be sent if the district could not back up its promise of tough consequences for those who use weapons" (Burke, 1996a, p. 1C), and a school board member stated: "It's going to be difficult for any school board to weigh the rights of students who disrupt schools against all those who behave and come to school for an education when courts intervene the way this court has" (p. 1C). R.T.'s attorney then filed suit for monetary damages against the district. R.T.'s family never exercised their right to the district's appeal process. "Instead, they went straight to federal court with the allegation that the girl's civil right to an education had been violated" (p. 11A).

The second incident, also at Racine Unified School District, involved an attack on a teacher. On January 25, 1996, a female high school speech and drama teacher tried to end an altercation between two female students. According to newspaper accounts L.Q., one of the combatants, then turned on the teacher. L.Q. stripped off her shirt down to her bra in order to better attack the teacher. About 200 students pressed around to watch, but no one moved to intervene. When interviewed later, the teacher "broke down in tears several times while talking about the attack" and stated that "the girl deliberately hit me . . . in the back, and when I felt the blow, I covered my head and she kept hitting me about 20 to 25 times. I was in shock and could not move to escape the blows. . . . The emotional damage is ongoing. I was out of school for 11 weeks" (Burke, 1996b, p. 1A). L.Q. was permanently expelled but subsequently joined the lawsuit of R.T. and, represented by the same lawyer, sought reinstatement and monetary damages. After a legal battle that went to the U.S. District Court, both girls were reinstated. The school district also sought and received dismissal of the case by offering monetary settlements, which were accepted by the girls' mothers. L.Q. received $27,720 and R.T. received $45,000 (Burke, 1997).

Clearly, it is difficult to balance the "affirmative duty of protection by public schools" on behalf of victims against the constitutional right to education and the legal requirement of compulsory attendance of the perpetrators. Aside from the machinations of the justice system in resolving these issues, the disruption of the schools by incidents such as these are multiplied by the thousands across the nation.

FEDERAL CONCERNS

One measure of the gravity of the school violence problem is the extent to which the federal government has expressed its concern. This concern has been manifested in legislation, goal setting, and information gathering. In 1986, Congress passed the original Drug-Free Schools and Communities Act for funding in Fiscal Year 1987. This act was superseded by the Safe and Drug-Free Schools and Communities Act in 1994. In the interim between the passage of these acts, Congress also passed the Gun-Free School Zones Act in 1990, the purpose of which was to bar the possession of guns in or near a school. Laws with broader purposes were also passed in 1994, including the Brady Bill and the massive Crime Bill. Debates on both of these general bills included much discussion of the impact they might have on reducing school violence.

The federal government's concern is also manifested in the attention paid to lack of progress toward goal 7 of the National Education Goals Panel (NEGP). Goal 7 is "safe, disciplined, and alcohol-and drug-free schools." The NEGP was created in 1990 to measure and disseminate progress, or lack of it, toward eight national goals derived from the original Goals 2000 project. The NEGP represents all levels of government and was given the status of a federal agency in 1994. Recently, the panel published a report on progress toward the eight goals, most of which center on achievement in academic subjects, since the early 1990s. Of the eight goals, the "safe schools" goal showed the least progress, and even some very significant deterioration in several indicators (National Educational Goals Panel,

1997a). Specifically, "Between 1991 and 1996, the percentage of 10th graders who reported that they had used an illicit drug increased from 24% to 40%" (National Education Goals Panel, 1997b, p. 61). The percentage of 10th graders who were offered or sold drugs at school the previous year nearly doubled (from 18% to 32%) between 1991 and 1996.

Alcohol use remained high, with about two thirds (65%) of 10th graders reporting in 1996 that they had used it the previous year (National Education Goals Panel 1997b). The "percentage of 10th graders who reported that they were threatened or injured at school during the previous year" remained high, dropping only slightly from 40% to 36% between 1991 and 1996 (p. 63). However, the "percentage of public school teachers who reported that they were threatened with physical injury or physically attacked by a student from their school during the previous 12 months" increased by half (from 10% to 15%) between 1991 and 1994 (p. 3). Furthermore, the percentage of secondary school teachers reporting disruptive student behavior interfering with their teaching increased significantly (from 37% to 48%) over the same period (p. 64).

In March 1998, NCES released the report of a national survey of violence in public schools. The survey gathered data on serious violent crime from a nationally representative sample of school principals. The data included only those crimes and violent incidents serious enough to have been reported to local police or other law enforcement agencies. Because of the tendency of school administrators to protect the image of their schools, it may be assumed that underreporting of such serious crimes occurred. The crimes included murder, suicide, rape, assaults with a weapon, and theft, among others (NCES, 1998a). The survey results were not encouraging to those who hope that significant progress is being made in stemming the rise of violence in public schools. The survey found that serious violent crimes tended to be more prevalent in about 10% of the schools.

> During 1996–97, about 4,000 incidents of rape or other types of sexual battery were reported in our nation's public schools. There were about 11,000 incidents of physical attacks or fights in which weapons were used and 7,000 robberies in schools that year. About 190,000 fights or physical attacks not involving weapons also occurred at schools in 1996–97, along with about 115,000 thefts and 98,000 incidents of vandalism (NCES, 1998b).

Clearly, the goal of "safe, disciplined, and alcohol-and-drug-free schools" appears less attainable than ever. The deteriorating situation affects all students, their teachers, and their attitude toward learning. All college-bound high school graduates have been immersed in an educational environment rife with drug and alcohol usage, classroom disruptions, and threats of violence or actual violence toward students and teachers. The effects of such an environment on college-bound students' attitudes and acceptance of violent behavior has not been measured directly, but such experiences are bound to be detrimental.

Measuring progress toward national goals and the need for, and evaluation of, extensive legislation requires extensive and accurate data. Various departments of the federal government have initiated data-gathering efforts focusing on school violence. Illustrative of such efforts are the following:

- *Youth Risk Behavior Survey* (YRBS), Centers for Disease Control and Prevention (1995)
- *Schools and Staffing Survey,* NCES (1994)
- *National Household Education Survey,* NCES (1993)
- *School Crime Supplement of the National Crime Victimization Survey,* NCES (1995)
- *National Crime Victimization Survey,* Bureau of Justice Statistics, U.S. Department of Justice (1997)

In addition to these and other federal data sets, the NCES has developed a series of three longitudinal surveys of high school and junior high school students. The first of these, the National Longitudinal Survey of 1972, gathered data on a national sample of high school seniors. It collected information on plans for post–high school activities, but none of the items related to school environment could be construed as pertaining to violence. In 1972, concern about violence in schools was not yet at the level that warranted attention in a national survey. In 1980, the NCES initiated the High School and Beyond survey of high school sophomores. This survey contained a few items on safety concerns in the school environment, signaling an emerging awareness of school violence. However, in the latest of the three surveys, the National Education Longitudinal Study of 1988 (NELS:88), school safety concerns are prominent and are approached by direct questions. The base-year survey gathered data from a nationally representative sample of nearly 25,000 eighth-grade students in 1988. The project also surveyed administrators and teachers in the respondents' school as well as parents. The students were followed up as 10th graders, as seniors, and in 1994 as college students 2 years after high school graduation. The data set containing all the follow-ups through the third follow-up in 1994 was analyzed to produce the following findings.

We analyzed the data from the students who were high school seniors in 1992. Only seniors in public schools are included, but these represent 87% of the total number (16,883 out of 19,367). The survey included more than 40 variables relating to school safety, violence, drug/alcohol usage, weapons, and other factors related to this chapter. Data for several of the variables most closely related to the topic of this chapter are presented below. Not all seniors were asked all of the questions; therefore the total number included in each table is less than the total number of public school seniors in the overall survey. Tables 3–1, 3–2, and 3–3 present data relating to school safety issues cross-tabulated by urbanicity (urban, suburban, rural) of the schools in which the high school seniors are enrolled. Seniors in urban schools felt less safe in their school than seniors in either suburban or rural schools. Those in rural schools felt slightly safer than seniors in suburban schools (Table 3–1).

The seniors were asked to respond to a question concerning whether they had observed fights between racial/ethnic groups at school. As shown in Table 3–2, about one third (32.7%) of the seniors in urban schools observed such fighting compared to about one fourth (26.4%) of seniors in suburban schools and about one fifth (19.8%) of seniors in rural schools.

Seniors' perception of gangs in their school varied strongly by urbanicity of the school setting. Gangs are perceived as being much more prevalent in urban schools than in the other two settings. Well over one third (36.3%) of the seniors in urban schools agreed or strongly agreed with the statement "There are many gangs in school" compared with less than one fifth (18.0%) and less than one tenth (8.6%) of the seniors in suburban and rural schools, respectively (Table 3–3).

Table 3–1 High School Seniors' Response to Item "Respondent Doesn't Feel Safe at This School"

Urbanicity	Strongly Agree	Agree	Disagree	Strongly Disagree	Total
Urban					
Number	107	318	1,329	789	2,543
Percent	4.2%	12.5%	52.3%	31%	100.0%
Suburban					
Number	119	339	2,214	1,845	4,517
Percent	2.6%	7.5%	49%	40.9%	100.0%
Rural					
Number	76	226	1,646	1,775	3,723
Percent	2.1%	6%	44.2%	47.7%	100.0%
Total					
Number	302	883	5,189	4,409	10,783
Percent	2.8%	8.2%	48.1%	40.9%	100.0%

Source: Reprinted from National Educational Longitudinal Study (NELS 88/94), *National Center for Education Statistics (NCES)*, U.S. Department of Education.

Table 3-2 High School Seniors' Response to Question "Do Fights Occur Between Racial/Ethnic Groups?"

Urbanicity	Strongly Agree	Agree	Disagree	Strongly Disagree	Total
Urban					
Number	194	636	1,227	482	2,539
Percent	7.7%	25.0%	48.3%	19.0%	100.0%
Suburban					
Number	277	928	2,229	1,079	4,513
Percent	5.4%	21.0%	49.4%	24.2%	100.0%
Rural					
Number	166	569	1,759	1,232	3,726
Percent	4.5%	15.3%	47.2%	33.0%	100.0%
Total					
Number	637	2,133	5,215	2,793	10,778
Percent	6.0%	19.7%	48.3%	26.0%	100.0%

Source: Reprinted from *National Educational Longitudinal Study (NELS 88/94),* National Center for Education Statistics (NCES), U.S. Department of Education.

We also analyzed some of the salient variables by gender. Boys were three times more likely than girls to get into a physical fight at school. Only 6% of the girls reported they got into at least one fight compared to 17.6% of the boys (Table 3-4).

High school senior females placed much more importance on school safety than males. In response to the question "How important is a low crime environment" in school, only 18.3% of the females indicated it was not important compared to 31.1% of the males. Conversely, 38.5% of the females indicated a low crime environment was "very important" compared to 25.7% of the males (Table 3-5).

The NELS:88 data available to us for this analysis did not allow for following individual students or groups of students from eighth grade into college. However, it was possible for us to disaggregate students by type of high school program. Specifically, we were able to identify and segregate public high school seniors according to general studies or college preparatory programs. On the natural assumption that college prep students were much more likely to go directly (or within a few years) to college than general studies students, we compared these two groups for several variables.

Table 3–3 High School Seniors' Response to Item "There Are Many Gangs in School"

Urbanicity	Strongly Agree	Agree	Disagree	Strongly Disagree	Total
Urban					
Number	234	683	1,063	548	2,528
Percent	9.3%	27.0%	42.0%	21.7%	100.0%
Suburban					
Number	181	627	2,018	1,680	4,506
Percent	4.0%	14.0%	44.8%	37.2%	100.0%
Rural					
Number	58	265	1,395	2,008	3,726
Percent	1.5%	7.1%	37.4%	54.0%	100.0%
Total					
Number	473	1,575	4,476	4,236	10,760
Percent	4.4%	14.6%	41.5%	39.5%	100.0%

Source: Reprinted from *National Educational Longitudinal Study (NELS 88/94),* National Center for Education Statistics (NCES), U.S. Department of Education.

Table 3–4 High School Seniors' Response to Item "Got into a Physical Fight at School"

Sex	Never	Once or Twice	More Than Twice	Total
Male				
Number	4,341	718	212	5,271
Percent	82.4%	13.6%	4.0%	100.0%
Female				
Number	5,214	292	37	5,543
Percent	94.0%	5.3%	.7%	100.0%
Total				
Number	9,555	1,010	249	10,814
Percent	88.4%	9.3%	2.3%	100.0%

Source: Reprinted from *National Educational Longitudinal Study (NELS 88/94),* National Center for Education Statistics (NCES), U.S. Department of Education.

Table 3–5 High School Seniors' Response to Question "How Important Is a Low-Crime Environment?"

Sex	Not Important	Some Importance	Very Important	Total
Male				
Number	1,452	2,022	1,201	4,675
Percent	31.1%	43.2%	25.7%	100.0%
Female				
Number	962	2,268	2,020	5,250
Percent	18.3%	43.2%	38.5%	100.0%
Total				
Number	2,414	4,290	3,221	9,925
Percent	24.3%	43.2%	32.5%	100.0%

Source: Reprinted from *National Educational Longitudinal Study (NELS 88/94),* National Center for Education Statistics (NCES), U.S. Department of Education.

There were discernible, but not strong, differences between general studies and college prep seniors on their feeling of safety in school. In general, college prep seniors felt somewhat safer at their school; 45% of the college prep students strongly disagreed with the item "Respondent doesn't feel safe at this school" compared with 38.8% of the general studies students (Table 3–6).

There was little difference between general studies and college prep students in the likelihood of having something stolen at school (Table 3–7), but general studies students were more likely to have someone offer to sell them drugs at school than were college prep students (Table 3–8). About one in five general studies students (19.6%) received at least one such offer compared to 13.1% of the college prep students (Table 3–8).

Similarly, college prep students were only slightly less likely to be threatened with physical harm than general studies students. Although the likelihood of such threats was under 20% for both groups, college prep students were less likely (13.6%) to have received one or more threats; 16.1% of the general studies students received such threats (Table 3–9).

The strongest difference between general studies and college prep seniors among the variables analyzed was in the propensity for fighting. Of the general studies students, 13.4% got into at least one "physical fight at school" compared to 6.6% of the college prep seniors (Table 3–10).

Table 3–6 High School Seniors' Response to Item "Respondent Doesn't Feel Safe at This School"

High School Program	Strongly Agree	Agree	Disagree	Strongly Disagree	Total
General					
Number	123	359	2,003	1,569	4,054
Percent	3.0%	8.7%	49.5%	38.8%	100.0%
College Prep					
Number	90	278	1,998	1,932	4,298
Percent	2.0%	6.5%	46.5%	45.0%	100.0%
Total					
Number	213	637	4,001	3,501	8,352
Percent	2.5%	7.5%	48.0%	42.0%	100.0%

Source: Reprinted from *National Educational Longitudinal Study (NELS 88/94),* National Center for Education Statistics (NCES), U.S. Department of Education.

Table 3–7 High School Seniors' Response to Item "Had Something Stolen at School"

High School Program	Never	Once or Twice	More Than Twice	Total
General				
Number	2,786	1,089	187	4,062
Percent	68.6%	26.8%	4.6%	100.0%
College Prep				
Number	3,033	1,177	102	4,312
Percent	70.3%	27.3%	2.4%	100.0%
Total				
Number	5,819	2,266	289	8,374
Percent	69.5%	27.0%	3.5%	100.0%

Source: Reprinted from *National Educational Longitudinal Study (NELS 88/94),* National Center for Education Statistics (NCES), U.S. Department of Education.

Table 3–8 High School Seniors' Response to Item "Someone Offered To Sell Respondent Drugs at School"

High School Program	Never	Once or Twice	More Than Twice	Total
General				
Number	3,264	446	350	4,060
Percent	80.4%	11.0%	8.6%	100.0%
College Prep				
Number	3,744	375	190	4,309
Percent	86.9%	8.7%	4.4%	100.0%
Total				
Number	7,008	821	540	8,369
Percent	83.7%	9.8%	6.5%	100.0%

Source: Reprinted from *National Educational Longitudinal Study (NELS 88/94),* National Center for Education Statistics (NCES), U.S. Department of Education.

Table 3–9 High School Seniors' Response to Item "Someone Threatened To Hurt Respondent at School"

High School Program	Never	Once or Twice	More Than Twice	Total
General				
Number	3,401	511	142	4,054
Percent	83.9%	12.6%	3.5%	100.0%
College Prep				
Number	3,722	502	82	4,306
Percent	86.4%	11.6%	2.0%	100.0%
Total				
Number	7,123	1,013	224	8,360
Percent	85.2%	12.1%	2.7%	100.0%

Source: Reprinted from *National Educational Longitudinal Study (NELS 88/94),* National Center for Education Statistics (NCES), U.S. Department of Education.

Table 3–10 High School Seniors' Response to Item "Got into a Physical Fight at School"

High School Program	Never	Once or Twice	More Than Twice	Total
General				
Number	3,514	430	112	4,056
Percent	86.6%	10.6%	2.8%	100.0%
College Prep				
Number	4,029	244	38	4,311
Percent	93.4%	5.6%	1.0%	100.0%
Total				
Number	7,543	674	150	8,367
Percent	90.1%	8.0%	1.9%	100.0%

Source: Reprinted from *National Educational Longitudinal Study (NELS 88/94)*, National Center for Education Statistics (NCES), U.S. Department of Education.

Overall, seniors more likely to be college-bound (college prep track) were less likely to engage in "risky" behavior and were more concerned with safety than those in the general studies program. However, most of the seniors in the two groups answered similarly, and these two groups were more alike than different. Thus, the overall impression to be gained from these data is that, whatever effects in the school environment would predispose seniors toward risky or violent behavior, those heading toward college are as likely to be affected as those who are not.

CONCLUSION

The two Racine high school female students described earlier may or may not appear on a college campus in the near future, and if they do, may never repeat such violent acts. But the effects of their attacks—not just on the direct victims, but on the school environment—are widespread. For example, consider the 200 students who thronged around the attack on the teacher without intervening. Many of them will soon be on college campuses after having participated in what might be termed "passive violence." Multiply these participants by the many hundreds of such incidents every month in the nation's schools, and a perception of the effects on college campuses begins to form. The picture is obviously not very reassuring for those who would like to believe that the campus can be a sanctuary against the violence that has engulfed society. Schuh, in Chapter 2, offers cogent reasons why the campus must be prepared to deal with increasing incidents of violence. For example, drug use, as it continues at a high level among traditional-age college students, can lead to violence perpetrated by students selling or using drugs. When the campus is in an area affected by organized gangs, their warfare over "turf" for selling drugs on or near campus is likely spill over onto the campus as nonstudents use the campus as a battleground.

This chapter began by cautioning the reader against assuming a cause-and-effect relationship between the increase in violence in K–12 schools and concurrent increases in violence on college campuses. Data do not exist to link violent behavior of specific students in elementary and high school with similar behavior of these same students after they arrive on campus. However, it is not a leap of faith to assume that the environmental experience is linked. The "rising tide of violence" that has engulfed the K–12 schools can be presumed to affect the campus environment. The other chapters in this book support this assumption by describing the rise of campus violence.

Student affairs and other administrators on campus can use the information in this chapter as an "early-warning" system. For example, the National Educational Goals Panel (1997b) found that drug and alcohol use by 10th graders increased from 24% in 1991 to 40% in 1996; offering of drugs for sale to 10th graders at school increased from 18% in 1992 to 24% in 1996; and threats to and physical attacks on teachers increased by half between 1991 and 1994. When youth are immersed in such an environment during their high school years, it's a safe bet that their experiences and behavior in college will be affected to some extent. This "environmental" linkage can perhaps lead to some ways in which campus administrators can deal effectively with violence on campus. Perhaps educational material on the effects of violence could be offered at the initial enrollment of new students. Many, if not most, colleges and universities offer "coping" or "survival" courses to new freshmen. Such courses should include ways for these new students to deal with the violence in their new educational environment.

We analyzed data from NELS: 88 to establish a pattern of similarities or differences in concerns about school safety and violent behavior by school urbanicity; by gender; and, more importantly, between students in college preparatory versus general studies programs. There were more similarities than differences between college preparatory and general studies students, suggesting that overall environmental influences affect both groups about the same. To the extent that college preparatory students are likely to arrive on college campuses shortly after graduation, the assumption can be made that whatever has affected these students in high school is likely to, at least to some extent, influence their expectations and behavior on campus.

Still, it would be helpful to have access to data that would directly track specific students, or at least groups of them, from eighth grade, through high school, and on into college. Such tracking would help to establish or refute the assumed link based on similarities between K–12 and college environments.

Unfortunately, of the numerous variables dealing with school safety, crime, and violence present in the baseline (8th grade) data, as well as the first follow-up (10th grade) and second follow-up (12th grade), data were excluded from the third follow-up data gathered from students who went to college. The potential for directly linking behavior and experiences between K–12 and college thus cannot be explored. Perhaps the NCES or other organizations gathering longitudinal data can fill this gap in the near future.

REFERENCES

Ascher, C. (1994). *Gaining control of violence in the schools: A view from the field.* New York: ERIC Clearinghouse on Urban Education.

Bureau of the Census. (1996). *Statistical abstract of the United States.* Washington DC: U.S. Department of Commerce.

Bureau of Justice Statistics. (1997). *National crime victimization survey.* Washington, DC: U.S. Department of Justice.

Burke, M. (1996a, August 9). Reinstatement: Ruling jars unified officials. *The Racine Journal-Times,* pp. 1A, 11A.

Burke, M. (1996b, October 24). 2nd expelled girl joins lawsuit. *The Racine Journal-Times,* pp. 1A, 11A.

Burke, M. (1997, July 29). Unified settles lawsuit: Expelled students accept payments. *The Racine Journal-Times,* pp. 1A, 7A.

Centers for Disease Control and Prevention. (1995). *Youth risk behavior survey.* Atlanta, GA: Author.

Eiseman v. State of New York, 70 N.Y.2d 175, 511 N.E.2d, 518 N.Y.S.2d 608 (1987).

Havemann, J. (1997, February 7). For children, an epidemic of homicide. *The Washington Post,* pp. 1, 6.

Hoffman, A.M. (1996). *Schools, violence and society.* Westport, CT: Praeger.

Kansas court says colleges have duty to protect students. (1993, October 6). *The Chronicle of Higher Education,* p. A30.

Lederman, D. (1995, February 3). Colleges report rise in violent crime. *The Chronicle of Higher Education,* p. A31.

Leonard, S. (1994, August 3). Texas killer of 3 has been ASU law student. *The Arizona Republic,* pp. A1, A2.

Life and death on America's campuses. (1995, June 1). *The Phoenix Gazette,* p. B8.

McCowan, K. (1993, August 14). Objection: Killer's admission to law school berated image without ASU's help, attorney says. *The Arizona Republic,* pp. A1, A10.

McCoy, M. (1998, February 3). Killer's bid to become a lawyer is set back. *The Arizona Republic,* pp. B1, B3.

Metropolitan Life Insurance Company. (1993). *The American teacher: Violence in America's public schools.* New York: Author.

National Center for Education Statistics. (1993). *National household education survey.* Washington, DC: U.S. Government Printing Office.

National Center for Education Statistics. (1994). *Schools and staffing survey, 1994.* Washington, DC: U.S. Government Printing Office.

National Center for Education Statistics. (1995). *School crime supplement of the national crime victimization survey.* Washington, DC: U.S. Government Printing Office.

National Center for Education Statistics. (1996). *Digest of education statistics.* Washington, DC: U.S. Government Printing Office.

National Center for Education Statistics. (1998a). *Violence and discipline problems in U.S. public schools: 1996–97.* Washington, DC: U.S. Government Printing Office.

National Center for Education Statistics. (1998b). *Incidents of crime and violence in public schools.* Available at: http://nces.ed.gov/pubs98/violence/98030003.html; accessed May 1998.

National Data Resource Center. (1995). *Statistics in brief: Student victimization at school.* Washington, DC: National Center for Education Statistics.

National Education Goals Panel. (1997a). *The National Education Goals report summary: Mathematics and science achievement for the 21st century.* Washington, DC: U.S. Government Printing Office.

National Education Goals Panel. (1997b). *The National Education Goals report: Building a nation of learners, 1997.* Washington, DC: U.S. Government Printing Office.

National School Safety Center. (1990). *School crisis prevention and response.* Washington, DC: Author.

Nelson, J.L., Carlson, K., & Polansky, S.B. (1996). *Critical issues in education: A dialectical approach.* New York: McGraw Hill.

Nero v. Kansas State University, 861 P.2d 768 (Kan. 1993).

Onondaga-Cortland-Madison Board of Cooperative Educational Services. (1997). *Violence in our schools.* OCM-Boces Update Extra. Syracuse, NY: Author.

Survey Research Center. (1995). *Monitoring the future, 1994, 1995.* Ann Arbor, MI: University of Michigan.

U.S. Department of Justice. (1994). *Firearms and crimes of violence: Selected findings from National Statistical Series* (Report No. MCJ-146844). Washington, DC: U.S. Department of Justice.

Van Der Werf, M. (1995, February 28). Bid for ASU cuts strikes sparks on political ax. *The Arizona Republic,* pp. A1, A4.

World Health Organization. (1990). *World health statistics annual.* New York: Author.

Yozwiak, S. (1993, August 15). Parole uproar called GOP politics. *The Arizona Republic,* pp. B1, B4.

CHAPTER 4

Weapon Carrying on Campus

Randal W. Summers and Allan M. Hoffman

INTRODUCTION

Each day millions of travelers pass through metal detectors. Some are subjected to additional handheld scanners, and many are asked to open bags or turn on computers. All of this is done with the intent of detecting weapons or explosive devices. This scenario has become commonplace, with very few travelers complaining about the delays and inconvenience. In addition to preventing violent incidents, the security procedures increase passengers' sense of "feeling safe."

Recent surveys suggest that students on high school and college campuses are carrying weapons. Is it that same sense of "feeling safe" that is motivating these individuals to carry or possess weapons? What are the implications of this weapon carrying for violent crime? This chapter reviews the extent of weapon carrying and explores the underlying dynamics of doing so. It also explores the controversial relationship between violent crime and weapons (specifically handguns).

WEAPONS BORNE BY STUDENTS

The Centers for Disease Control and Prevention (CDC) has conducted a number of nationwide surveys (CDC, 1993) and found that weapon carrying has increased. For example, in 1990 4% of students indicated they carried a firearm; this increased to 6% in 1991 and 8% (1 out of every 13 students) in 1993. Because 42% of the high school students in the survey reported that they were in a fight in the last 12 months, the researchers hypothesized that fighting behavior may be related to weapon carrying. Students carry weapons to ward off attack or to protect themselves in a fight. Jennifer Friday reported that 20% of victimizations in schools involved a weapon (Friday, 1996). These statistics are quite alarming, but according to the CDC survey, weapon carrying and fighting decrease with age and grade level (CDC, 1993). Therefore, one might surmise that the extent of weapon carrying on the college campus should be considerably less than on the high school campus. However, this does not seem to be the case.

Anecdotal evidence suggests that weapons are involved in violence on campus. For example, several people were shot at a fraternity party ("Florida A&M Students Shot," 1997); and an arsenal of weapons including dynamite, hand grenades, a grenade launcher, and an M-16 rifle were found in a campus apartment ("Wright State U. Finds," 1997).

According to a recent national survey of 26,000 college students on 61 campuses ("Nearly 1 Million," 1997), 7% of students carried a gun or knife in the previous 30 days. The study indicated 11% of the men and 4% of the women surveyed carried weapons. Extrapolated, this means that approximately 1 million (980,000) students carry weapons on campus.

Is this part of a trend toward increasing crime on the college campus? The answer is no. According to a *Chronicle of Higher Education* report on campus crime, campus drug arrests increased 18%, while other crimes such as robbery, aggravated assault, burglary, vehicle theft, and violations of weapons laws declined in 1995 ("Campus Drug Arrests," 1997). Arrests for liquor-law violations increased from 15,097 in 1994 to 15,208 in 1995. Drug law violations increased from 5,764 in 1994 to 6,797 in 1995. Weapon violations decreased from 1,233 in 1994 to 1,002 in 1995. Reported incidents of forcible sex offenses increased from 955 in 1994 to 973 in 1995 and nonforcible sex offenses (incest and statutory rape; some include lewdness and stalking) increased from 86 to 87. The data were reported for 4-year colleges with enrollments of 5,000 or more. Data from community colleges were not reported; these institutions tend to have low crime rates, which has been attributed to students' living and socializing off campus. Slightly more than 1,000 arrests were made for crimes involving weapons. These arrests were for violation of the weapons law which prohibits the manufacture, sale, purchase, transportation, possession, concealment, or use of firearms, knives, explosives, or other deadly weapons. Table 4–1 lists those colleges with more than 15 weapons violations in the years 1994 and 1995.

It is interesting to note that the University of California-Los Angeles campus, with a population of 35,110, had only two weapon violations in 1995 and five in 1994. Similarly, the University of California-Riverside, with a campus population of 8,590, had nine violations in 1995 and seven in 1994. The surrounding areas of both these campuses have been associated with relatively high crime statistics, yet the weapons violations on campus are quite low. At face value, it would seem that weapon carrying (7% of students) does not correlate with the extent of weapons violations (less than one third of 1%) or the number of murders. However, caution is advised since there has been considerable debate about the interpretation of

Table 4–1 Campuses with 15 or More Weapon Violations, 1994 or 1995

Campus	1995	1994	Campus Population	Murders
Alabama State University	0	32	5037	0
Arizona State University	29	23	42189	1 (1995)
Duke University	18	24	11352	0
East Carolina University	17	7	18076	0
California State University-Fresno	7	16	17293	0
Jackson State University (MS)	15	13	6224	0
Michigan State University	20	21	40254	0
North Carolina A & T University	20	22	8136	0
North Carolina State University	26	11	28223	0
San Jose State University	12	15	26299	1 (1994)
Southern University (LA)	15	15	9904	1 (1994)
University of Alabama-Birmingham	33	23	15362	0
University of California, Berkeley	38	32	29634	0
University of California, Irvine	29	12	17073	0
University of Maryland (College Park)	8	15	32493	0
University of Carolina-Greensboro	9	28	12658	0
University of Nevada-Las Vegas	9	18	18954	0
Virginia Commonwealth University	21	18	21523	0

Source: Data from Many Campus Murders Involve Neither Students Nor Faculty, *The Chronicle of Higher Education,* p. A46, March 21, 1997.

campus crime data. For example, a higher number of incidents of crime on a campus does not mean that a campus is less safe. It could reflect the rigor of law enforcement on a campus. Declines in violent and property crimes at colleges reflect national trends. The Federal Bureau of Investigation (FBI) reported that the rate of crimes per capita fell in 1995 to the lowest level since 1985 ("Campus Drug Arrests," 1997). There was an average of 1% decrease in property crimes including burglary and vehicle theft.

It should also be noted that murders that occur on campus frequently do not involve students or faculty members. According to one report ("Many Campus Murders," 1997), 8 of the 15 murders reported by colleges and universities "just happened to occur on college property" (p. A46). Nevertheless, such incidents are included in the crime statistics reported by the institutions where the incident occurred.

WEAPONS IN SOCIETY—ARMED AND DANGEROUS?

It is important to review campus crime levels and weapon carrying in the context of society. It is apparent that campus crime reflects national trends, but what about weapons carrying? According to Steven Barkan (1997), 25% to 30% of all households have a handgun and another 25% have a rifle or shotgun. The primary motive for 40% of them is self-protection and secondarily recreation. It has been estimated that there are 225 million firearms in the United States. Of these firearms, 75 million are handguns. In addition, 2 million handguns come on the market each year and another 340,000 are stolen each year. Geographically, the South has the highest ownership rate and the Northeast the lowest. In the inner cities, more than 20% of male high school students reported owning a handgun. More Caucasians own guns than African Americans.

Joseph Sheley and James Wright studied selected juvenile samples from high crime areas (Sheley & Wright, 1993). They found that 12% of juveniles carried guns all the time. They also found that 5% to 6 % of the students used drugs. When asked where they could obtain a gun, students indicated they could borrow one from a friend or relative, or get it off the street. Although those involved in selling drugs have higher levels of gun ownership, drug use is only moderately related to gun activity. According to Steven Donziger in *The Real War on Crime* (1996), this extensive gun ownership has contributed to 22,000 homicides per year in the United States—10 times the per capita rate of European countries, which have more restrictive weapon policies.

VIOLENCE IN THE UNITED STATES IN PERSPECTIVE

Because of firearms, the United States leads the world in the proportion of violent crime that results in injury. For assault with force, Canada and Australia have higher rates than the United States. The percentage of the population victimized in the United States is 2.2; in Canada it is 2.3 and in Australia it is 2.8. In respect to robbery, the United States has a 1.7% rate of victimization; Spain's rate is 2.9%. For car theft, the United States has a 2.3 % rate of victimization; Australia's rate is 2.7% and England's 2.8% (Donziger, 1996). When crime rates in industrialized countries were compared, the United States had the highest crime rate in only 1 category (homicide) of the 14 that were measured (Donziger, 1996). A closer look at the U.S. homicide statistics reveals that homicide represents only 0.2% of all violent and nonviolent crime (U.S. Department of Justice, 1994a). In fact, violent offenses only make up 13.6% of all crime. Yet the prison system is mostly filled with nonviolent offenders. The rate of incarceration in the United States for homicide is 9 per 100,000. This is quite high relative to other countries. For example, Canada has an incarceration rate of 2.2 per 100,000 for homicide, and Australia's rate is 1.9 per 100,000. Some of the reasons postulated for lower incarceration rates in other countries include social safety nets to protect children from poverty, severe restrictions on the availability of firearms, shorter prison terms for nonviolent crime, and the focus on rehabilitation in the prison system.

It appears that the homicide rate in the United States is unique and may be related to the availability of deadly firearms. However, on closer analysis, one can be far more fearful of other events than homicide by strangers. Jay Livingston in *Crime and Criminology* (1996) indicates that more people die from events such as accidents that are neither violent nor criminal than from homicides. Home accidents and faulty consumer products injure more people than assault and robbery crimes. Job-related diseases and on-the-job accidents cause more deaths or injuries than murders, rapes, and robberies. People have much less to fear from being murdered by an unknown person than from dying in an auto accident. For example, in 1992 there were 9,300 murders committed by unknown persons compared with 22,000 who died in accidents at work or home and another 40,000 who died on the highways. Livingston also points out the myth associated with the cost of street crime. It was estimated in 1992, for example, that street crime cost society $16 billion. He points out that this is pale in comparison with the cost of white-collar crime and tax cheating, which costs untold billions. The other side of crime—which involves law enforcement, the operation of the court system, and jails and prisons—is estimated to cost $60 billion annually.

THE PERCEPTION OF CRIME

Beginning in the 1980s, teenage boys in all racial and ethnic groups were more likely to be shot to death than to die from all natural causes combined. There has been a 154% increase in the rate of homicide for males aged 15 to 19 years. The homicide rate is eight times higher for African-American males than Caucasian males (Donziger, 1996). There is no doubt that violent crime is a major problem in the inner city. However, because inner-city crimes are communicated through the media, people in low-crime areas perceive that they are unsafe. In reality, they are more safe than people were in the 1970s. Most neighborhoods outside of some highly dangerous areas are relatively safe. Although people's greatest fear is of being murdered by a stranger, this is an extremely unlikely event; most violent crime is committed by friends or family. A government study reviewed 8,000 homicides in urban areas in 1994 and found that 8 out of 10 murder victims were killed not by a stranger but by a family member or friend (U.S. Department of Justice, 1994a). The study reported that women were more likely to be assaulted by their spouse or boyfriend; children are more likely to be molested by family or friends. Although the perception is that elderly people are vulnerable and frequently victimized, they are at less risk than young people. A person aged 16 to 19 is four times more likely to be victimized than a person aged 35 to 49.

Women also have a perception that they are at great risk. However, the risk of murder, rape, robbery, or assault for a white woman aged 65 is 1/60 that of an African-American, male teenager (U.S. Department of Justice, 1994b). Generally speaking, males not only commit more crime than females but they are victimized more than females. An interesting component of the perception of crime is that the more a victim is removed or the more indirect the link between the criminal and the victim, the weaker the public reaction. That is why the perception of street crime has more psychological impact on people than other forms of less visible criminal activity, such as white-collar crime. So despite the lack of dramatic changes in the overall crime rate, Americans still react as though they are in imminent danger. By 1977, spending on private security had passed public spending. This trend to self-protection has not abated.

In the 1980s, 2% to 5% of residences had burglar alarms. By 1990, this number had reached 10% (Donziger, 1996). In 1996 government spending on law enforcement reached $30 billion, but spending on private security hit $52 billion. During the past 30 years, the perception of crime not only affected the extent of spending on law enforcement and private security but also affected policy. Each political campaign focused on crime and a "Get Tough Policy." Between 1968 and 1995, Congress passed six anticrime bills. Many of these bills were controversial in that Congress threatened to withhold federal money to get states to adopt certain policies on crime.

Crime and politics became inseparable partners. Conservatives believed that crime was caused by deviant individuals. Their approach was to give government more power to punish these individuals with more law enforcement and the building of more prisons. Liberals, on the other hand, viewed crime as stemming from harsh socioeconomic conditions. They were reluctant to give government any more power. They believed that public officials were building more prisons and filling them up with nonviolent offenders. They believed that the government funded $10 billion for prisons by underfunding drug treatment programs, educational programs, and violence prevention programs. The most recent federal legislation (enacted in 1995) has been referred to as the "three-strikes policy"; three-time offenders are sentenced to life in prison. This policy has serious cost implications. The average cost of maintaining an inmate is $22,000 per year. As the inmate ages, the cost can reach $69,000 per year. The Rand Corporation estimated that it will cost $4.5 to $6.5 billion dollars every year to implement the three-strikes law (Greenwood, 1994). Zimbardo (1994) estimated that the cost of a life term for an average California prisoner is $1.5 million. Critics of the three-strikes law charge that the expected increase in prison population by the year 2020 will cost hundreds of billions of dollars. This spending is especially controversial, because funds for higher education were being slashed at the same time the law was implemented in California.

Not only have politics and crime become entangled, but crime has became synonymous with race. When people speak out against crime, they often speak about their hidden belief that crime is associated with race. The public defenders office reports that nonwhites are being charged at 17 times the rate of Caucasian offenders with the same criminal history (Barkan, 1997).

Since the 1980s, the perception of crime has given rise to the feeling of "I don't feel safe!" This public perception has resulted in law enforcement expenditures quadrupling and prison and jail populations tripling. However, these efforts have been to no avail; polls still show that most of Americans don't feel safe. The question is, will we continue to increase enforcement, lengthen sentences, and build more prisons?

The 1980s also were a time when neighborhood crime watches began. College campuses also began establishing these watches. One story in particular highlights the point that perception can lead to peculiar responses. Ronald Troyer, in an article in the *Justice Quarterly* (1988), described an anticrime patrol that was established at Drake University in Des Moines, Iowa. Apparently, there was a rape on campus that had alarmed the student body. At about the same time, the student newspaper began increasing its coverage of campus crime. This was attributed not to the new crime wave, but to a new editor who once worked as a police reporter and enjoyed stories about crime. To make matters worse, the student body president took a full page ad out entitled, "Are You Safe? A Message on Crime in the Drake Community." Although the crime rate in the surrounding area and campus crime (largely vandalism) had decreased, the campus was very fearful. This led to the formation of a campus patrol. Besides trying to do something about crime on campus, students joined the patrol for community service credits (13% of them) or to improve the image of their fraternity. They wore yellow jackets and carried flashlights and two-way radios. Most of the activity of the patrol involved escorting female students, helping to fix flat tires, and giving directions to visitors. The crime-stopping expectation of the patrol waned, crime stories in the newspaper subsided, and crime was no longer an issue in student elections; by the fourth year, only six students were still patrolling. This story illustrates how the perception of crime can initially heighten fear and concerns about crime and then diminish, even though the actual crime statistics never changed.

HANDGUNS AND HOMICIDE

According to the FBI's *Uniform Crime Reports,* there are 275 million guns in the United States (FBI, 1994). One third of these guns are handguns. They account for 68.5 % of the murders that are done with firearms. The murder rate increased from 32 per 1 million population in 1987 to 51 per 1 million in 1993. This increase is attributable to handguns. Some have argued that the number of incidents of shooting has not changed, but the sophisticated design and caliber of handgun technology has made guns more lethal. Another factor that may reduce the homicide rate is the availability of medical help. Aggravated assault also relates to murder. An aggravated assault is defined as an attack by one person upon another, in which the offender uses or displays a weapon in a threatening manner or the victim suffers severe injury ("Campus Drug Arrests," 1997). What begins as an aggravated assault often ends in murder. According to the National Crime Victimization Survey, there is a continuing increase in the use of handguns in violent crime (Rand, 1994). In 1994 handguns were involved in 13,000 homicides, 20,000 rapes, 280,000 robberies, and 580,000 assaults (Greenwood, 1994). For people aged 16 to 19, African-American males have the highest victimization rate (39.7 per 1,000); the rate is 9.5 for Caucasian males, 9.1 for African-American females, and 3.6 for Caucasian females. Michael Rand in *Guns and Crime* (1994) indicated that more teenagers are killed by guns than from all diseases combined. From 1979 to 1990, guns were involved in the death of 50,000 children and teenagers.

As pointed out earlier, the murder rate is higher in the United States than anywhere else. But what about gun ownership? Within the United States, the South has the highest rate of gun ownership (Barkan, 1997). Gastil (1971) suggests that ownership itself is not a cause of murder but rather a symptom of an underlying attitude toward deadly violence. This attitude appears to be part of the southern culture. Even after eliminating the murders committed with guns, the murder rate in the U.S. South is still twice as high as in European countries (Gastil, 1971).

There is also little doubt that handguns and automatic weapons are a significant factor in crime. What is surprising about handguns is that they are more of a danger to the owner and the owner's family and friends than to strangers or attackers. Kellerman (1993) reported that a gun at home is 43 times more likely to be used to kill a family member than to kill an intruder and 5 times more likely to be used to commit suicide (Donziger, 1996). The argument for the use of handguns for protecting one's home is weak at best. Another surprising fact is the high cost of health care associated with gun injuries. The estimated cost for treating a gunshot wound ranges from $15,000 to $20,000. If the victim requires intensive care, the cost usually reaches $150,000. The total medical spending for firearm victims is astronomical. In 1993, for example, medical spending was $3 billion (Donziger, 1996).

WHY CARRY GUNS?

Why do students carry guns? They carry guns mainly out of fear. However, when the facts are considered, we see that this fear is irrational.

The latest survey indicates that the extent of weapon carrying on campus is 7% or approximately 1 million students. What does this actually mean? Murder on campus is not a common occurrence. In 1997 there were only 15 murders among 489 college campuses, and none occurred where weapon arrests on campus are the highest. So, the gun "toters" are not killing people. Nor are armed robberies common on campus. However, aggravated assaults are high relative to other campus crimes. It is possible that students carry guns because they have been assaulted or know someone who has been assaulted. Gun carrying in high school has been associated with gaining respect and status. This might be a component of gun carrying on the college campus. A survey of weapon carrying on campus found that weapon carriers had been harassed more than noncarriers and they felt unsafe ("Nearly 1 Million," 1997).

Because the college campus reflects society, there is a strong possibility that fear motivates gun carrying. Survey after survey indicates Americans don't feel safe. Students in high school and on the college campus may feel unsafe as well. The perception of fear is real, but the underlying data do not warrant this fear except in highly dangerous areas. For example, death by violent crime is far less common than death caused by noncrime events. The cost of medical care for shooting victims is high, but the cost of all violent crime is less than the cost of white-collar crime. Because Americans are safer than ever before, our fear is irrational. Our response to our fear of crime has been to arm ourselves and to build more prisons. Ironically, most prisons are filled with nonviolent offenders and guns kill more family members and friends than strangers.

Not all fear of crime is irrational. It is important to distinguish between nonviolent and violent crime. Violent crimes are crimes committed against people and include murder, rape, robbery, kidnapping, and assault. Victims of these crimes have an increase in subsequent fear. Nonviolent crimes are crimes against property and, with a few exceptions, do not elicit as much fear. For example, burglary, defined as breaking into a dwelling, has a heavy fear component because of the psychological sense of intrusion associated with violent crime (Donziger, 1996). Drug offenses and car jacking, although they technically involve property, may involve or lead to violence. There is considerable debate about the influence of the media and television on our perception of crime. Some believe that television may desensitize us to violent crime or, in some instances, create a false impression that violent crime is running rampant. In reality, most crime is not violent crime. One in 10 arrests is for a violent crime, and only 3 in 100 crimes result in injury. The majority of prison inmates are incarcerated for nonviolent property and drug offenses. One of the consequences of fear is that it can affect social life by causing people to change their behavior. Researchers have investigated how fear changes the way people live (Friedberg, 1983; Bennack, 1989). The extent of public fear of crime has varied over the years as measured by the response to the question, "Are there any places within a mile of where you live where you'd be afraid to walk alone at night?" In 1967, 31% said yes. By 1982 this percentage rose to 48%, but has declined in the last decade to just over 40%. Although the crime rate was the same for 1993 and 1981 (Federal Bureau of Investigation, 1994), and the levels of fear were also the same, more people took self-protective measures in 1993 than in 1981 (Maguire & Pastore, 1995). There was an increase in the sale of guns, private security services, and burglar alarms. Fear changed the way people behaved. Every measure for 1993 increased over those in 1981: the percentage of people who used special locks increased by 30%, people who walked only with others increased by 20%, people who kept a dog increased by 18%, people who bought a gun increased by 14%, people who carried a weapon increased by 15%, people who installed a burglar alarm increased by 14%, and people who carried a whistle increased by 5% (Maguire & Pastore, 1995).

UNDERLYING DYNAMICS OF FEAR

Younger African-American and Hispanic males living in urban areas are more likely than other Americans to carry weapons (U.S. Department of Justice, 1994a). Of the students in central-city public schools, 17% admitted to carrying weapons to school (Harris, 1993). When asked why, their responses included protection, showing off to impress friends, feeling important, and emulating friends. Nationwide, 4% of high school students carry a weapon (Harris, 1993). When teachers and police were asked why students carry weapons, they indicated that students carry weapons for the following reasons:

- for protection on the way to school
- to impress friends
- for self-esteem
- for protection while at school

Personal characteristics such as having been a victim, gender, and age contribute to feelings of vulnerability and fear. Everyone is more fearful after having been victimized. Women are twice as likely to indicate they feel unsafe, but they are less likely to be victimized. Older people feel vulnerable and are more fearful but are less likely to be victimized. In a study of 10,000 households in eight cities, only 5.7% of the total sample had been victimized (Garogalo, 1979).

The rate of victimization was very low, even for young males (12.2 %). Only 1.7% had received any kind of injury requiring medical attention. Despite the low victimization rate, 50% said they felt unsafe. According to Skogan and Maxfield in *Coping with Crime* (1981), it is difficult methodologically to link crime with fear of crime because there are strange aspects to victimization. For example, they found that women had the same increased fear from purse snatching as from rape.

Generally speaking, the higher the rate of violent crime in a neighborhood, the more fear—despite the fact that most violent crimes are assaults such as hitting, slapping, or verbal threats. Only about 8% of victims of violent crime require hospital emergency room treatment, and most are released the same day. Only 1% of crime victims require a hospital stay of more than 1 day. Crime, in and of itself, is not the most important factor contributing to fear. When crime is linked with disorder, people become more isolated in their communities and more fearful. Thus, fear is more of a reaction to one's immediate environment than to the crime itself. People typically feel safer in their own neighborhood, regardless of the crime rate. So, familiarity becomes a significant variable in fearfulness.

Wesley Skogan pointed out in *Disorder and Decline: Crime and the Spiral of Decay in American Neighborhoods* (1990) that people do not normally know whether crime is increasing or decreasing in their area. They perceive a sense of safety or fear based on signs of disorder and incivility (homeless people, vandalism, graffiti, abandoned buildings, teenagers in the street). It is the socioeconomic condition of the people and the appearance of the neighborhood that is important in raising the level of fear.

Even the presence of police in the neighborhood can alter the perception of crime. The Newark, New Jersey, experiment, which later became known as the community-based policing model, put police back on foot patrol (Skogan, 1990). Although the actual crime rate did not go down, the people felt safer, felt better about the police, and felt that crime went down.

Visibility of disorder and incivility are keys to people feeling threatened. Unfamiliarity or "something or someone being different" can also cause fear and apprehension. A presidential commission concluded that the basis of a fear of violent crime is basically a fear of strangers (CDC, 1990). People are more fearful of others when they are less predictable. This means that anyone with different behavior, language, or physical attributes could elicit fear. This fear often involves race issues. In 1992, in response to the verdict in the Rodney King case (the acquittal of four Los Angeles police officers who were accused of beating a black man), one of the worst riots of the century erupted. The riots culminated in more than 50 fatalities and injuries and over a half billion dollars in property damage from the burning and looting. This riot encapsulated the points about differentness, less predictability, and visibility of disorder. Televising every minute of the riot resulted in increased fear among television viewers. People living 60 miles away, in low-crime neighborhoods, became fearful of the civil unrest and began buying guns and signing up for gun classes for self-protection.

ROLE OF THE MEDIA

The influence of the media on crime has been a topic of controversy for a long time. It is not necessarily the volume of crime news that affects us but rather the way it is portrayed. The news editor chooses to cover a select number of events among thousands of events. The media tend to focus on unusual crimes or those that resemble crimes depicted in fictional television or movies. Crime is presented as isolated individual incidents that relate to insanity or rage. The public does not really learn about the most frequent crimes or about those that are likely to directly affect us. Because we do not see the justice system in action, the media can cause a perception of a "crime wave" (which is nonexistent) and an increase of fear in the public. Crime waves are created by a number of media covering the same crime stories. News editors also watch television and read other media and are influenced by what they see or read. The image spread by the media usually has no correlation with the actual statistics.

Popular prime-time television news programs often select and overdramatize incidents of crime and murder, which again leaves the public with an unrealistic perspective and considerable fear. Even more distortion occurs when the police select certain information for release to the media. The police can also discourage the media from reporting on certain unpopular themes. For example, when the media attempted to run a series of articles on crime in the New York subway system, the transit authorities said there was no crime wave in the subways, and that ended that story. Usually crime stories have a limited media life span. A dramatic incident occurs and receives media attention. Then the various media continue reporting similar crimes. Public officials present a well-publicized response that everything possible is being done to solve the problem, and the media and the public lose interest.

Another controversial aspect of crime and the media is the influence of television on violent behavior. Unfortunately, there is no agreement as to which causes which. There are many variables associated with this question, including sex, age, and socioeconomic status. However, people who watch a lot of television tend to overestimate the levels of violent crime, and they are more fearful of crime than infrequent watchers. Is it the television that is influencing the person or is it simply fearful people staying home and watching more television? There is agreement on the fact that the media have little direct and lasting effect on fear. In the short run, the media can influence fear with crime-wave stories of car jackings and rapes;

but, in the long run, it is the victimization and environmental cues that affect our perception of crime and form our fear.

POTENTIAL SOLUTIONS

Given the statistics presented in this chapter on weapons, crime, and fear, it is appropriate to address various solutions to these problems. However, the fundamental disagreements about the nature of these problems in our society could fill a whole text. Each of the approaches to gun control, crime legislation, and crime reduction is highly controversial. For example, with respect to the media and violence, some advocate governmental regulation and censorship. Others feel this would be unconstitutional; instead, they support the notion of maintaining basic freedoms and propose self-regulation.

Another highly controversial approach to curbing violence is gun control. Advocates suggest that, to be effective, gun control must be legislated at the highest jurisdictional level to prevent importation from other states. Because millions of guns are already in place, gun control would probably have to focus on the availability of ammunition to have any hope of succeeding. Interestingly enough, the 1990s has witnessed an increase in drive-by shootings and gang violence (Norman, 1993). This has prompted efforts to restrict firearms. However, only one third of the people were in favor of a complete ban, and more than 70% indicated they wanted less restrictions. The pro-gun lobbyists suggest that, because only one sixth of felons obtained their guns legally, gun control would not reduce crime or gun availability. Anti-gun lobbyists support 7-day, mandatory waiting periods. They are advocates for putting a tax on guns and ammunition to deter their purchase and to pay for medical care and crime prevention. Other, more moderate interest groups suggest implementing "safe technology." This refers to fingerprinted weapons, which can be fired only by the legal owner.

There are also those who favor the reduction of crime by improving economic deprivation and inner-city conditions, eliminating racial discrimination, and developing programs to eliminate inadequate and abusive parenting, which are all viewed as contributing factors to crime. In educational institutions, the outright ban and zero-tolerance policies seem to be the most obvious approach to weapon carrying at school or on the college campus. However, we also need to focus on reducing the psychological fear of crime and developing alternative programs and methods for enhancing self-esteem. Will weapon-carrying on the college campus increase? College campuses, in many ways, reflect our society in general. If weapon-carrying becomes the norm in our society, then we can expect to see more weapons on campus. Regardless of the extent of weapons on campus, college officials must be cognizant of the fact that the potential for weapons creeping into the classroom and dormitories will likely increase in the future. As weapon-carrying increases, so does the risk of campus violence.

REFERENCES

Barkan, S.E. (1997). *Criminology: A sociological understanding.* Englewood Cliffs, NJ: Prentice Hall.

Bennack, Jr., F.A. (1989). *The American public's hopes and fears for the decade of the 1990's.* New York: Hearst.

Campus drug arrests increased 18 per cent in 1995; Reports of other crimes fell. (1997, March 21). *The Chronicle of Higher Education,* p. 1.

Centers for Disease Control and Prevention. (1990). *Weapon-carrying among high school students-United States.* Atlanta, GA: Author.

Centers for Disease Control and Prevention. (1993). Youth Risk Behavior Survey on 16,296 students grade 9–12 nationwide. In *Violence prevention: School's newest challenge, school safety.* Atlanta, GA: Author.

Donziger, S.R. (Ed.). (1996). *The real war on crime: The report of the national criminal justice commission.* New York: Harper Perennial.

Federal Bureau of Investigation. (1994). *Crime in the United States, 1993: The uniform crime reports.* Washington, DC: Author.

Florida A&M students shot at fraternity party. (1997, May 9). *The Chronicle of Higher Education,* p. A6.

Friday, J. (1996). Weapon carrying in school. In A.M. Hoffman (Ed.), *Schools, violence, and society* (pp. 21–32). Westport, CT: Praeger.

Friedberg, A. (1983). *America afraid: How fear of crime changes the way we live* (The Figgie Report). New York: New American Library.

Garogalo, J. (1979). Victimization and the fear of crime. *Journal of Research in Crime and Delinquency, 16*(1), 80–97.

Gastil, R.D. (1971). Homicide and a regional culture of violence. *American Sociological Review, 36,* 412–427.

Greenwood, P.W. (1994). *Three strikes and you're out—Estimated benefits and costs of California's new mandatory sentencing law.* Santa Monica, CA: Rand Corporation.

Harris, L. (1993). *A survey of experience, perceptions, and apprehensions about guns among young people in America.* Chicago: LH Research for the Joyce Foundation.

Kellerman, A.L. (1993). Gun ownership as a risk factor for homicide in the home. *New England Journal of Medicine, 329,* 1084–1092.

Livingston, J. (1996). *Crime and criminology* (2nd ed.). Englewood Cliffs, NJ: Prentice Hall.

Maguire, K., & Pastore, A.L. (Eds.). (1995). *Sourcebook of criminal justice statistics, 1994.* Washington, DC: U.S. Department of Justice, Bureau of Justice Statistics, U.S. Government Printing Office.

Many campus murders involve neither students nor faculty. (1997, March 21). *The Chronicle of Higher Education,* p. A46.

Nearly 1 million college students may be armed, survey finds. (1997, July 3). *The Chronicle of Higher Education,* p. A6.

Norman, J. (1993, December 30). Since 1930s, support for reform. *USA Today*, p. 4A.

Rand, M.R. (1994). *Guns and crime*. Washington, DC: National Academy Press.

Sheley, J.F., & Wright, J.D. (1993). *Gun acquisition and possession in selected juvenile samples*. Washington, DC: National Institute of Justice.

Skogan, W.G. (1990). *Disorder and decline: Crime and the spiral of decay in American neighborhoods*. New York: Free Press.

Skogan, W.G., & Maxfield, M.G. (1981). *Coping with crime*. Beverly Hills, CA: Sage.

Troyer, R.L. (1988). The urban anti-crime patrol phenomenon: A case study of a student effort. *Justice Quarterly, 5*(3), 397–419.

U.S. Department of Justice, Bureau of Statistics. (1994a). *Murder in families*. Washington, DC: U.S. Government Printing Office.

U.S. Department of Justice, Bureau of Statistics. (1994b). *Criminal victimization in the United States*. Washington, DC: U.S. Government Printing Office.

Wright State U. finds a cache of weapons in a campus apartment. (1997, June 20). *The Chronicle of Higher Education*, p. A8.

Zimbardo, P. (1994). *Transforming California's prisons into expensive old age homes for felons: Enormous hidden costs and consequences for California's taxpayers*. San Francisco: Center on Juvenile and Criminal Justice.

CHAPTER 5

Matters of Civility on Campus

John H. Schuh

INTRODUCTION

One of the most challenging problems faced by many colleges and universities is how to maintain an atmosphere of civility on campus while protecting the free speech rights of students, faculty, and staff. With increasing frequency in the past decade, students and others have demanded a nonoffensive environment and protection for the rights of victims of hate speech and violent crimes. Administrators have been challenged to ensure that free-speech rights are preserved in circumstances where the very nature of the free speech is offensive. Consider the following scenarios.

- A student calls other students names they find offensive. They claim that he has engaged in "hate speech" and they should have redress of their grievance. The student calling other students names asserts that he is merely exercising his right of free speech. What does the law say? How does the college respond?
- Members of a student organization wear a tee shirt that features a caricature of a member of a racial minority group that members of the minority group find offensive. The members of the organization who prepared the tee shirt simply assert that they mean no harm and were having fun with the caricature. Members of the minority group demand action on the part of the university.
- A group of students, all of whom are members of a racial minority group, invite a speaker to campus who has vigorously criticized a different minority group in speeches at other campuses. Is the speaker allowed to present his views on campus, or is access denied? Administrators fear that violence

might break out if the speaker is permitted access to the campus but are uncertain of the legality of denying him access to the campus.

These scenarios underscore the challenges faced on the contemporary campus. Coupled with these challenges are the ever-increasing number of lawsuits filed against colleges and universities. According to Gose (1994b), "Lawyers at several colleges say they are seeing a greater number of what they consider frivolous claims and lawsuits every year. Many believe that the unique relationship between students and universities invites litigation. Parents may feel that as tuition rises ever higher, so does the duty of universities to insure the safety of their children" (p. A27). As a consequence, campus officials are pressed to maintain a civil atmosphere while staying true to the principles of free speech and intellectual dialogue.

One approach to dealing with issues of civility has been introduced at the University of Pittsburgh. Pitt has announced that all new students are to take a pledge "that calls for them to pursue civility, academic honesty, mutual respect and compassion for others" ("U of Pittsburgh," 1997, p. A37). Even this modest attempt has been met with criticism. One faculty member called the pledge meaningless and claimed that it "might hinder open debate" ("U of Pittsburgh," 1997, p. A37). Still, maintaining decorum and civility is an increasing problem on many campuses and faculty are challenged to foster an atmosphere of civility in the classroom (Schneider, 1998). "Some scholars argue that academe has never been above a good slugfest. But close encounters of the uncivil kind are leaving many professors stunned, even shaken" (p. A12).

The purpose of this chapter is to identify and discuss some of the issues that are related to the general topic of maintaining civility on the contemporary campus. The issues identified are by no means a complete list. Rather they have been selected because they illustrate difficult challenges faced by campus administrators that often have no easy solution. In many of these situations, the campuses identified did nothing to invite the problems they encountered; rather, they were simply the victims of bad fortune in that the incident or situation just happened to occur at the specific institution rather than some place else on the higher education landscape. The reader should not infer legal advice from this chapter. Legal counsel should be sought regarding how specific situations should be addressed.

STUDENT PUBLICATIONS

The First Amendment of the U.S. Constitution ensures freedom of the press. According to Kaplin and Lee (1995), "The chief concern of the First Amendment's free press guarantee is censorship. Thus, whenever a public institution seeks to control or coercively influence the content of a student publication, it will have a legal problem on its hands" (p. 539). "As perhaps the most staunchly guarded of all First Amendment rights, the right to a free press protects student publications from virtually all encroachments on their editorial prerogatives by public institutions. In a series of forceful cases, courts have implemented this student press freedom, using First Amendment principles akin to those that would apply to a big-city daily published by a private corporation" (pp. 369–370). It would seem that the student press should be able to publish what it chooses without restriction, other than by laws that protect against libel, and the freedom of the press should go unchallenged. But that is not always the case.

It is not uncommon for some student newspapers or other publications to exercise bad judgment or bad taste. For example, student publications at Bryn Mawr, Duke, Oberlin, and the University of Michigan "have been at the center of flaps over supposed slurs" (Shea, 1994b, p. A39). The student newspaper at the University of Minnesota at Duluth published an April Fool's Day issue with a number of gag stories in addition to "an article about a fictional gay bar that included dozens of slurs against gay and lesbians" ("April Fool's Issue," 1996, p. A41). In these instances, the editors were contrite about their exercise of poor taste and apologized for offending other members of the campus community. But other instances related to freedom of the press have not been resolved so amiably.

A Pennsylvania State University student newspaper that allegedly was antifeminist in its editorial content had all 4,000 copies of an issue stolen one night, and several hundred copies were burned in front of the building that housed a conservative trustee's law office. The university's response was to arrest two journalism majors and charge them with theft (Shea, 1993d). Some students and others concluded that the theft was justified; others indicated that the paper's freedom of speech was abridged by the theft. The result, in addition to the destruction of the newspapers, was that vigorous controversy was fueled on campus, and anger was engendered on both sides of the issue.

A similar incident occurred at the University of Pennsylvania. Several African-American students were arrested after they were apprehended trying to remove all 14,000 copies of the April 15, 1993, edition of *The Daily Pennsylvanian* (Shea, 1993b, p. A32; Shea, 1993e, p. A27). The paper had been critical of administrators' responses to criticisms from African-American students. A report issued after the incident criticized the police for treating the attempted removal of the newspapers as a criminal act rather than a form of protest. Police also were criticized for using excessive force in apprehending one of the students.

The university police chief was quoted as saying "How do you handle the situation, given that this is an academic community?" According to Shea, "That question is one that administrators at a number of institutions have had to answer, as the tactic of protesting the content of newspapers by taking copies of them spreads. The responses from college officials have ranged from a shrug of the shoulders to filing charges against the students involved" (Shea, 1993e, p. A27).

The two Pennsylvania cases illustrate how violence, or the potential for violence, can arise over a difference of opinion about the content of newspapers and how far the First Amendment extends in such cases. On the one hand, the newspaper is protected by its First Amendment right to publish content that is not libelous. On the other hand, a group of students believes that their right not to be offended supersedes the newspaper's First Amendment right. The result in one case was that many copies of the paper were burned and, in the other, students attempted to remove all the copies of the paper. In each case, some students were arrested. Administrators, as is often the case, attempted to take both sides of the issue with the result that students on both sides were upset with the university's response. Although the violence that resulted was relatively minor, seeds of ill will were sown that had the potential to bloom for years to come.

OUTSIDE SPEAKERS

Speakers invited by student groups or other members of a campus community can have the effect of forcing freedom of speech to its absolute limit. Consider the attention generated by speeches delivered by Khalid Abdul Muhammad in 1994. According to Shea, "Mr. Muhammad blends exhortations about the need for black self-reliance with mockery of whites and condemnation of Jews" (Shea, 1994a, p. A32). He has been characterized as "an anti-Semite and hate-monger" (Mercer & Lederman, 1994, p. A31) as a result of speeches he has delivered on several college campuses.

Complaints about speakers do not always result from situations as dramatic as the previous example. A commencement speech by Michael Eric Dyson at the University of North Carolina resulted in some people walking out. *The Chronicle of Higher Education* reported that "Many said they were offended by the language of the speech" ("Some U. of N.C. Graduates," 1997, p. A8). "The chancellor later said he regretted the tone of the speech, and that it offended people but added that he had no right to censure it. Mr. Dyson accused the chancellor of failing to defend free speech" (p. A8).

Robert M. O'Neil (1994), founding director of the Thomas Jefferson Center for the Protection of Free Expression and Professor of Law at the University of Virginia concluded that "a college campus, of all places in society, must be open to the full range of viewpoints" (p. A52). O'Neil (1994) recommends several steps that institutions might take if a highly volatile speaker is invited to campus.

1. Describe the implications and consequences of bringing inflammatory messages to campus.
2. Unfettered press coverage could be a condition of the event.
3. The institution should provide security for the event.

4. The speaker should be required to answer questions.
5. The campus community should be ready to condemn abhorrent views (p. A52).

O'Neil (1994) concludes with this observation: "The college campus has long been, and probably will remain, a forum for hateful messages as well as ennobling ones" (p. A52).

Controversial speakers have been a part of the life on college campuses for decades. The Free Speech Movement, for example, marked the beginning of the protest movement more than three decades ago, and that certainly was not the first time that controversial speech found its way to colleges and universities (Lucas, 1994). Although campuses can take the steps recommended by O'Neil to mitigate the effects of the controversy, eliminating the potential for violence associated with controversial speakers is impossible.

What is particularly interesting about the case of Khalid Abdul Muhammad is that the speech delivered by a person from one minority group was deemed offensive by people of another minority group. Do the free speech rights of one minority group prevail over the rights of another minority group not to be the subject of a hateful, at least in their eyes, diatribe? The answer appears to be that there is no way, at least for a public institution, to stop hateful speech regardless of how hurtful and offensive it might be, unless there is evidence to suggest that the speech will incite a riot.

HATE SPEECH AMONG MEMBERS OF THE CAMPUS COMMUNITY

According to Kaplin and Lee (1995), "Hate speech is an imprecise catch-all term that generally includes verbal and written words and symbolic acts that convey a grossly negative assessment of particular persons or groups based on their race, gender, ethnicity, religion, sexual orientation or disability. Hate speech thus is derogatory and degrading, and the language typically is coarse" (p. 509). "The purpose of the speech is more to humiliate or wound than it is to communicate ideas or information. Common vehicles for such speech include epithets, slurs, insults, taunts and threats" (Kaplin & Lee, 1997, p. 384). Hate speech may include printing on tee shirts, posters, classroom bulletin boards, flyers, leaflets, phone calls, letters, or electronic mail messages (Kaplin & Lee, 1995).

Issues related to hate speech have been the focus of controversy when campuses have tried to regulate student speech. An example of this is a controversy at the University of Pennsylvania, where one student allegedly referred to a group of women students of color in terms they found offensive ("Penn Student Disputes," May 5, 1993). The situation hinged on the one student's right to express himself compared with the other students' right not to be the subject of harassment. Ultimately the offended students dropped the charge (Shea, 1993c), but that did not stop the controversy of speech codes on campuses. In this case, the student, after graduating, sued the "university for breach of contract and reckless, intentional infliction of emotional distress" ("Student at Center," April 19, 1996). Moreover, the university dropped its racial harassment policy and replaced it with a "code that does not punish students who use derogatory or insulting speech" (Gose, 1994a, p. A30).

Campuses have adopted speech codes for a variety of reasons, among them to ensure civility in discourse among members of their communities. These efforts have not always been successful. According to *The Chronicle of Higher Education,* "Hate speech codes at the Universities of Michigan and Wisconsin were struck down by federal judges after the ACLU sued" (Dodge, 1992, p. A35). Stanford University, a private institution, even had its antiharassment policy overturned. "A California Superior Court struck down a Stanford University antiharassment policy last week, saying it restricted the free-speech rights of students.... California lawmakers opened the door for the students' suit in 1992 by requiring that private universities grant students the same constitutional protections as public institutions provide" (Shea, 1995a, p. A32). Still, according to one study of 384 public institutions, "36 percent of the universities forbid verbal abuse based on race, sex or other personal attributes" ("Speech Codes in Place," 1994, p. A4).

Kaplin and Lee (1995) conclude that public institutions face "exceeding difficulty . . . in attempting to promulgate hate speech regulations that would survive First Amendment scrutiny" (p. 513). That is not to suggest that the problem of hateful speech and acts are not problems on college campuses. According to Levin and McDevitt, "The Prejudice Institute of the Center for the Applied Study of Ethnoviolence, at Towson State University, reported an upsurge during the late 1980s in racial and anti-Semitic episodes on college campuses. It estimated that 20 percent of all students from racial and ethnic minority groups were either physically or verbally harassed during their college years" (1995, p. B1).

How is hateful speech operationalized? In one case, a campus humor magazine published a parody on an article that had appeared in a magazine edited by students of color. The editors of the parody thought they had done nothing wrong, while the students toward whom the parody was directed "argued that the publication of the parody just before a final-exam period created a hostile educational environment for [the campus's] minority students" (Shea, 1993f, p. A37). On the other side of the argument is the following recommendation: "And we must remind our colleagues that the most effective means for countering any speech with which we disagree is also the means that is true to the First Amendment: more speech, rather than suppressions" (Strossen, 1993, p. B1).

But the matter of regulating speech is extraordinarily complex, as Lawrence (1989) clearly points out. "There can be no meaningful discussion of how we should reconcile our commitment to equality and our commitment to free speech until it is acknowledged that there is real harm inflicted by racist speech and that this harm is far from trivial" (p. B1). He adds, "We must look for ways to offer assistance and support to students whose speech and political participation are chilled in a climate of racial harassment" (p. B1).

Kaplin and Lee (1995) suggest that, instead of trying to regulate speech, institutions should consider nonregulatory approaches for dealing with hate speech. "Nonregulatory initiatives may also be more in harmony with higher education's mission to foster critical examination and dialogue in the search for truth" (p. 514). They also suggest that colleges and universities should consider adopting regulations that deal with hate "*conduct or behavior*" (p. 514). In this way, the institution may regulate certain actions that are associated with speech, but not the speech itself. Kaplin and Lee make several suggestions along these lines, which taken together, can provide the framework for an excellent institutional response to matters related to hateful speech.

1. When hateful speech is combined with hateful behavior, the behavior can be regulated. Defacing the campus in the name of free speech does not have to be tolerated, so the painting of a swastika on the Hillel Center could result in disciplinary action against the perpetrators. Free speech, in this instance, does not extend to destroying property.
2. The time and place at which hate speech is uttered or the manner in which it is uttered, "as long as they use neutral regulations that do not focus on the content or viewpoint of the speech" (Kaplin & Lee, 1995, p. 515) can be regulated. This would mean that screaming epithets in a classroom building in the middle of the day could be regulated the same way a student musical group may not sing songs that disrupt students studying in the campus library.
3. "Institutions may regulate hate speech in the form of threats or intimidation aimed at particular individuals and creating in them a realistic fear for their physical safety or the security or their property" (Kaplin & Lee, 1995, p. 515). Consequently, personal attacks on specific individuals (for example, "Sally Jones should be horse whipped for her political opinions and that's what all of us should plan on doing later today") rather than broad-ranging, vague threats against entire classes of people (for example, a speaker who asserts "all men should be violently punished for their sins"") can be addressed and regulated.
4. Personal areas within the campus, such as residence hall rooms, might be a place where hateful speech can be regulated. Thus, the occupant of a residence hall room should not be subject to the kind of speech that might be acceptable in a public park. "Accordingly, under established free-speech tenets, students should have the right to avoid being exposed to others'

expressions by seeking refuge in their rooms" (White, 1994, p. POV). "For first amendment purposes, such private areas are not considered 'public forums' open to public dialogue; and the persons occupying such places may be 'captive audiences' who cannot guard their privacy by avoiding hate speech" (Kaplin & Lee, 1995, p. 515).

Clearly, the area of hate speech is difficult to address and manage on the college campus. The ideas above provide some approaches to dealing with this issue, but as Kaplin and Lee (1995) conclude, "Often the harassment or abuse involved in campus incidents or covered by conduct rules has been conveyed by the spoken or written word or by symbolic conduct—thus raising difficult issues concerning students' free speech and press rights and having important implications for both academic freedom and equal educational opportunity for students" (p. 509).

As it is difficult for colleges and universities to control hateful speech by those not associated with the institution, it is also difficult to control hateful speech by members of the college. As a consequence, the strategy that seems to work best is to deal with behaviors associated with hateful speech (such as dealing with someone who spray paints another person's car window), which can be regulated under a code of conduct. One of the strengths of higher education, free and open discourse between individuals who are members of the institution, also is a weakness. That is, it is extremely difficult for the campus to stop one person from engaging in odious, disgusting speech towards another. The result, of course, is the potential for violence towards either the speaker or the recipient of the message. The challenge, quite obviously, is to protect every person's rights in such discourse and to minimize the potential for violence.

ELECTRONIC MAIL AND THE INTERNET

An interesting phenomenon that has emerged during the past several years has been the use of the Internet for messages that some regard as intimidating, hateful, or threatening. In one case, a former student was prosecuted for sending electronic hate mail in which it was alleged that he tried to intimidate 10 individuals because of their race ("Former Student Indicted," 1996); in another case, a student who was accused of sending stories over the Internet had his case dismissed. In the latter case, the student was suspended, but his position was sustained by a federal judge ("Judge Dismisses Case," 1995) and a federal appeals court (Young, 1997a).

The use of the Internet for transmitting offensive, objectionable material has been debated widely. Congress waded in on the issue (Wilson, 1996), and a variety of opinions were expressed on the issue. O'Neil (1995) observed, "Apparently the most dangerous desperado on an American campus earlier this year was armed not with a gun or a knife or a bomb but with a keyboard" (p. A68). The student's "offense was writing a bizarre and twisted tale in which he fantasized the rape and torture of a female classmate, whom he named. Had his story appeared in an underground magazine, little more would have been heard of it. What made his story different was that he had posted it on the Internet" (p. A68). Lemisch (1995) offered additional thoughts about limiting speech over the Internet: "Legally, it is an open question whether existing laws regulating free speech—including bans on certain types of expression such as slander, libel and obscenity—apply to the Internet. The system operates in a gray area between what is considered public speech and personal correspondence" (p. 156).

Attempts at regulating use of the Internet for speech that some regard as offensive, disgusting, or threatening have been unsuccessful. At the time of this writing, the Communications Decency Act is being reviewed by the U.S. Supreme Court after it was struck down by a federal appeals court (Young, 1997b, p. A33). Pending the outcome of the court's deliberations, it is likely that colleges and universities may be best served by doing nothing to try to regulate what goes out over the Internet. "Public institutions . . . must keep a watchful eye on the First Amendment when drafting and enforcing computer use policies" (Kaplin & Lee, 1997, p. 392). "Private institutions . . . may voluntarily protect student free expression through student codes or bills of rights, or computer use policies themselves, or through campus custom—or may be bound by state statutes or regulations to protect free expression" (Kaplin & Lee, 1997, p. 393).

RAPE VICTIMS

Another dimension of violence on the contemporary college campus is dealing with the victims of violent crimes, including rape. All too frequently, the institution is dragged into the fray because of the allegation that the institution did not prevent the incident or that it handled the incident poorly after it occurred. Following are examples of each.

At Carleton College, four women who reported being raped sued the college because they contended "that the college bungled their cases by advising them not to go to the police and then failing to punish the men adequately through its own judicial proceedings" (Collison, 1991a, p. A29). According to the women, the alleged perpetrators had been reported for previous sexual assaults ("Victims of Alleged Rapes," 1991). Ultimately the suit was settled out of court ("Carleton Settles Suit," 1991).

Two other institutions were sued because, in one case, campus police officers allegedly were negligent in handling the investigation and put the victim through a "brutal, insensitive and humiliating interrogation" ("George Mason University,"

1992, p. A4) and, in the other case, the institution, by taking inadequate measures, contributed to the attack ("Award in USC Rape Trial," 1993). In the latter case, the award that had been granted at trial was reversed. In yet another case, a student sued her institution by alleging that "she had been raped during a party in her coeducational dormitory and argued that the university had been negligent for housing females on the same floor with male students" ("Court Rules University," 1991, p. A2). In this case, the award by the trial court was reversed on appeal.

And yet another institution experienced protest because a rape victim alleged that the members of the campus hospital did not treat her well. "A protest and continuing pressure from angry students has led officials at the Ohio State University to promise more sensitive treatment for rape victims" (Shea, 1993a, p. A32). "A rape victim complained that staff members at the campus hospital had been cold to her and left her unattended in the emergency room for several hours" (Shea, 1993a, p. A32).

Rape charges also can engender campus protest. In one instance, students protested the way the campus handled a charge of rape for failing to take certain actions against sexual assault and harassment complaints ("Alleged Rape Prompts Protest," 1996); in another instance, protesters campaigned to defend the alleged perpetrator of the incident ("Students Protest Amid Charges," 1995). Although these two incidents took place virtually at opposite ends of the country (North Carolina and Montana), they demonstrate how campuses are placed under a microscope after an incident occurs. In still another incident, students protested that one institution had not moved quickly enough in investigating an alleged rape. The accused rapist finally was barred from class but then was allowed to return a week later (Strosnider, 1996). Finally, a student who was accused of rape, and punished as a result of the act, sued his accuser and the university (Gose, 1996; Gose, 1997). In this very complex case, the accuser was sued for libel while the university was sued for breach of contract and negligence.

Rape is a very serious crime, and always needs to be treated as such. In the cases cited above, a variety of issues arise. In several situations, the institution was charged with not protecting students from rape, either because of allegedly insufficient action against known perpetrators (i.e., students who had been charged with offenses but were allowed to remain in school) or by not providing safe enough premises so as not to contribute to the conditions that allowed the rape to occur. In other cases, an institution experienced protests because either it did not act vigorously after the act to punish the perpetrator, or it acted too severely and punished the perpetrator too quickly after the act. And, finally, there is the case in which the institution and the victim were sued by the alleged perpetrator. It would appear that too often the violent crime of rape leaves an institution in a situation where the institution cannot win. If it deals with the incident slowly and deliberately, it can be criticized for moving too slowly. If it moves swiftly, the result can be criticism for moving recklessly. In short, the institution appears to be forced to take one side or the other, with the result potentially being protest or even lawsuits. And just to make the situation even more complex, consider this 1993 case, as reported by Gose: "In a 1993 case in which one student raped another at Kansas State University, the State Supreme Court decided the university should have warned students about the rapist because he was accused of a separate sexual assault a few weeks earlier. The Court's decision . . . would put universities in a Catch-22: Label and punish an accused student who has not yet been charged, or face liability for failing to protect students from danger" (1994b, p. A27).

There is no best way for an institution to handle a rape case. In addition to the fact that the alleged rape is ordinarily a violation of the campus code of conduct, it is also a serious crime with the implication of a prison sentence for a person who is found guilty. A carefully coordinated approach to handling the crime involving law enforcement officers, campus officials, and advocates for both the victim and perpetrator may provide a bit of comfort for those charged with managing the aftermath of such an incident. Even then, there are no guarantees that a wide range of problems will not result as the case is sorted out.

HAZING

Hazing is a crime in some states (Gehring, 1987) and a thoroughly despicable practice. Yet examples of hazing are all too common on the contemporary campus. For example, *The Chronicle of Higher Education* published a series of photographs that depicted hazing at unidentified institutions, including one photo that "depicts a pledge who believes he is waiting to be branded with a fraternity symbol. Another shows blindfolded students drenched with vomit and molasses being led away from the camera" ("Photographs Prove Persistence," 1996). Kappa Alpha Psi and several members of a chapter agreed to pay $2.25 million to settle a wrongful death suit ("Fraternity Pays Millions," 1997). A fraternity at Texas A&M and one of its members faced "criminal charges after a pledge was seriously injured by a wedgie" ("Fraternity Indicted," 1997).

Hazing is not the exclusive territory of fraternities. Other examples of hazing can be found at colleges and universities. Two women quit their college after asserting that they had been victims of hazing. The two women charged "that cadets doused them with a flammable liquid and set their clothing on fire" (Guernsey, 1997, p. A32). Two men who were accused of hazing dropped out of the college ("Two Citadel Men Quit," 1997). And, another institution paid more than $2 million as the settlement in a hazing case. The student's lawyer claimed that "his client had been traumatized by around-the-clock hazing during the two weeks he was enrolled at Norwich" ("Norwich U. Ordered," 1997, p. A6.)

Hazing is violence directed at specific individuals as part of the initiation process to join a student organization or become a member of the college at large. It includes physical and/or emotional harassment of prospective members. The consequence, quite obviously, is that the student has to endure punishment if he or she wants to become a member. Hazing supposedly is not tolerated on college campuses. But it still happens all too frequently. Students who are hazed have their lives disrupted or worse. Too often they have little choice in dealing with the hazing other than to drop out, and perhaps have their dream of membership shattered.

FRATERNITIES

In addition to the hazing activities described above, fraternities sometimes engage in other activities that contribute to an atmosphere of incivility—or violence—on campus. A fraternity ritual at one campus that included cross-burning and the wearing of Confederate uniforms and flags resulted in spirited debate ("U. of Nebraska Fraternity's," 1997). On another campus, a fraternity's suspension for sponsoring an "ugly woman contest" was overturned by a U.S. District Court judge. According to Collison (1991b), "The university had suspended the fraternity for two years because administrators said the contest was racist and had offended black students and women. The fraternity members filed suit, saying the suspension violated their rights to free speech" (p. A45).

Another example is the charge by women at a college that "fraternity members ran to a sorority singing a song about the mutilation of women's genitalia with a chain saw, and that the men then pulled down their pants" (Jaschik, 1993, p. 28). This case led to the college, a private institution, being charged with violation of a state law it did not know existed; after complying with state law, the college was accused of violating federal law (Jaschik, 1993). None of this would have happened had the fraternity members exercised better judgment and not been involved in the offensive behavior in the first place.

Moving from the territory of poor taste to violence, a fraternity at Indiana University managed to have a lawsuit that resulted from the assault of a freshman at a party thrown out. "'The bottom line is that the mere drinking of alcohol in a fraternity house does not make sexual assault foreseeable,' says Michael Rabinowitch, a lawyer for Delta Tau Delta" (Shea, 1995b, p. A39). But in another venue, a federal jury "concluded that both a fraternity, Tau Kappa Epsilon, and the University of Rhode Island had a responsibility to stop a party before things got out of hand and freshman Michelle Eckman was assaulted in a private room" (Shea, 1995b, p. A39). And, fraternity activities can lead to death. A student was found dead in the upstairs room of a fraternity house after members of the fraternity had engaged in getting pledges to "throw up and to see who they could make throw up first" ("12 Charged in Death," 1997, p. A10).

Some fraternities continue to behave in ways that are antithetical to the norms of contemporary society and the regulations of their college. Such behavior involves incredibly poor judgment and use of the First Amendment as a shield to offend members of groups that have no means to redress their grievances. To be sure, many student organizations have adopted risk-management policies (Dunkel & Schuh, 1998) and often publicly decry the kinds of activities identified above. But, examples of violence toward students resulting from terribly poor judgment exhibited by members still occur. Clearly more work needs to be done to stamp out this form of campus violence.

OTHER ISSUES

Other examples easily can be identified that illustrate the dilemmas facing colleges and universities as they attempt to create an atmosphere of civility and good taste on campus. Certainly the definitions of "good taste" and "civility" depend to a great extent on whose ox is being gored. But consider the following illustrations. "A student has sued the University of South Florida for stopping her from showing on the campus a videotape depicting lesbian sex. She said the campus police violated her free speech by confiscating the video, which she was projecting onto the side of a building" ("Student Sues U. of South Florida," 1997). As the case developed, a federal judge "barred the University of South Florida from disciplining a student who showed a sexually explicit videotape outdoors on the campus" ("U. of South Florida," 1997).

The movie *Showgirls* was shown on the University of Alabama campus over the protests of the state governor and 44 state lawmakers. The movie depicted "violence, obscenity, and nudity" and the "graphical portrayal of a gang rape" ("U. of Alabama," 1996). Once again, the university was put in the position of trying to find middle ground between a group of students who wanted to exercise their First Amendment right to show the movie and another group of students who claimed to be offended and insulted by the showing of the film, which included violence.

The National Association for Biomedical Research and the National Institutes of Health issued security advice to colleges and universities to protect laboratories from vandalism by animal-rights activities (Mangan, 1990). Among the tips were ways for researchers to protect themselves if they began receiving harassing telephone calls or hate mail.

And there are other examples of poor taste or worse, including the founding of a pagan group that, according to the president, "hopes to end perception that pagans 'eat babies and kill virgins'" ("Pagan Group Wins," 1997); the founding of a pro-heterosexual group ("Pro-Heterosexual Group," 1997); and the painting of anti-gay slogans in a campus tunnel, which resulted in the withdrawal of charges for free-speech reasons ("Free-Speech Tunnel," 1994).

All of these examples point to the potential for violence that exists on campus related to one group of people exercising its right of free speech and consequently offending members of another group. Again the institution stands in the middle, having to ensure the rights of one group while trying to assure the other group that they will not experience violence, even if they are called vile names or worse.

Clearly, students and others feel passionately about their points of view, and institutions are caught in a conundrum as to which group to support when quite obviously no middle ground exists. Litigation often is the result and, most often, the group exercising its free-speech rights is the one whose position is sustained. But that rarely means that the controversy, or the potential for violence, is over. In fact, as can be seen in many of these examples, such controversy all too often results in physical or emotional violence.

REFERENCES

Alleged rape prompts protest at Appalachian State U. (1996, December 6). *The Chronicle of Higher Education,* p. A10.

April Fool's issue of student newspaper sparks outcry. (1996, April 12). *The Chronicle of Higher Education,* p. A41.

Award in USC rape trial is reversed by court. (1993, July 7). *The Chronicle of Higher Education,* p. A6.

Carleton settles suit brought by women. (1991, November 6). *The Chronicle of Higher Education,* p. A4.

Collison, M.N-K. (1991a, May 15). 4 women sue Carleton College for not protecting them from sexual assault. *The Chronicle of Higher Education,* p. A29.

Collison, M.N-K. (1991b, September 4). Judge cites First Amendment protection in overturning suspension of fraternity. *The Chronicle of Higher Education,* p. A45.

Court rules university not negligent in rape. (1991, March 27). *The Chronicle of Higher Education,* p. A2.

Dodge, S. (1992). Campus codes that ban hate speech are rarely used to penalize students. *The Chronicle of Higher Education,* p. A35.

Dunkel, N.W., & Schuh, J.H. (1998). *Advising student groups and organizations.* San Francisco: Jossey-Bass.

Former student indicted for electronic hate mail at Irvine. (1996, November 22). *The Chronicle of Higher Education,* p. A6.

Fraternity indicted after alleged hazing at Texas A&M. (1997, January 10). *The Chronicle of Higher Education,* p. A8.

Fraternity pays millions to settle hazing suit. (1997, January 24). *The Chronicle of Higher Education,* p. A6.

Free-speech tunnel put to test at N.C. State. (1994, March 23). *The Chronicle of Higher Education,* p. A4.

Gehring, D.D. (1987). Legal rights and responsibilities of campus student groups and advisers. In J.H. Schuh (Ed.), *A handbook for student group advisers* (2nd ed., pp. 115–149). Alexandria, VA: American College Personnel Association.

George Mason University is sued for "botching" rape case. (1992, December 9). *The Chronicle of Higher Education*, p. A4.

Gose, B. (1994a, June 29). Penn to replace controversial speech code. *The Chronicle of Higher Education*, p. A30.

Gose, B. (1994b, August 17). Lawsuit "feeding frenzy." *The Chronicle of Higher Education*, p. A27.

Gose, B. (1996, October 11). Handling of date-rape case leaves Brown U. policies in dispute. *The Chronicle of Higher Education*, p. A53.

Gose, B. (1997, February 21). Student who was accused of rape sues Brown U. and his accuser. *The Chronicle of Higher Education*, p. A38.

Guernsey, L. (1997, January 24). 2 women leave the Citadel, blaming hazing incidents on college. *The Chronicle of Higher Education*, p. A32.

Jaschik, S. (1993, July 7). Free speech and "grotesque" behavior. *The Chronicle of Higher Education*, p. A28.

Judge dismisses case against former Michigan student. (1995, June 30). *The Chronicle of Higher Education*, p. A17.

Kaplin, W.A., & Lee, B.A. (1995). *The law of higher education* (3rd ed.). San Francisco: Jossey-Bass.

Kaplin, W.A., & Lee, B.A. (1997). *A legal guide for student affairs professionals.* San Francisco: Jossey-Bass.

Lawrence, III, C.R. (1989, October 25). The debates over placing limits on racist speech must not ignore the damage it does to its victims. *The Chronicle of Higher Education*, p. B1.

Lemisch, J. (1995, January 20). The First Amendment is under attack in cyberspace. *The Chronicle of Higher Education*, p. A56.

Levin, J., & McDevitt, J. (1995, August 4). The research needed to understand hate crime. *The Chronicle of Higher Education*, p. B1.

Lucas, C.J. (1994). *American higher education: A history.* New York: St. Martin's Griffin.

Mangan, K.S. (1990, September 19). Lab-security tips from 2 groups. *The Chronicle of Higher Education*, p. A19.

Mercer, J., & Lederman, D. (1994, March 16). Aftermath of a fiery message. *The Chronicle of Higher Education*, p. A31.

Norwich U. ordered to pay $2.2 million in hazing case. (1997, January 31). *The Chronicle of Higher Education*, p. A6.

O'Neil, R.M. (1994, February 16). Hateful messages that force free speech to the limit. *The Chronicle of Higher Education*, p. A52.

O'Neil, R.M. (1995, November 3). Free speech on the electronic frontier. *The Chronicle of Higher Education*, p. A68.

Pagan group wins recognition at U. of Tennessee at Chattanooga. (1997, January 31). *The Chronicle of Higher Education*, p. A27.

Penn student disputes accusation of racial harassment. (1993, May 5). *The Chronicle of Higher Education*, p. A39.

Photographs prove persistence of extreme hazing rituals. (1996, November 15). *The Chronicle of Higher Education*, p. A49.

Pro-heterosexual group at Penn State is refused recognition. (1997, February 28). *The Chronicle of Higher Education*, p. A43.

Schneider, A. (1998, March 27). Insubordination and intimidation signal the end of decorum in many classrooms. *The Chronicle of Higher Education,* pp. A12–A14.

Shea, C. (1993a, January 13). College looks at how it treats rape victims. *The Chronicle of Higher Education,* p. A32.

Shea, C. (1993b, April 28). Campus paper angers black students. *The Chronicle of Higher Education,* p. A32.

Shea, C. (1993c, June 2). Outcome at Penn leaves no one satisfied. *The Chronicle of Higher Education,* p. A24.

Shea, C. (1993d, July 28). Charges against women roil Penn State. *The Chronicle of Higher Education,* p. A30.

Shea, C. (1993e, August 4). Police criticized at Penn. *The Chronicle of Higher Education,* p. A27.

Shea, C. (1993f, December 1). The limits of free speech. *The Chronicle of Higher Education,* p. A37.

Shea, C. (1994a, March 16). Dealing with virulent speakers. *The Chronicle of Higher Education,* p. A32.

Shea, C. (1994b, November 30). No laughing matter. *The Chronicle of Higher Education,* p. A39.

Shea, C. (1995a, March 10). Anti-harassment policy is struck down. *The Chronicle of Higher Education,* p. A32.

Shea, C. (1995b, July 28). Courts disagree over liability for fraternity-house assaults. *The Chronicle of Higher Education,* p. A39.

Some U. of N.C. graduates walk out on commencement speech. (1997, January 10). *The Chronicle of Higher Education,* p. A8.

Speech codes in place despite court's ruling. (1994, February 2). *The Chronicle of Higher Education,* p. A4.

Strosnider, K. (1996, December 13). Sit-in spurs U. of Penn. to bar accused rapist from class. *The Chronicle of Higher Education,* p. A42.

Strossen, N. (1993, July 7). Threats to free speech. *The Chronicle of Higher Education,* p. B1.

Student at center of "water buffalo" incident sues U. of Penn. (1996, April 19). *The Chronicle of Higher Education,* p. A6.

Student sues U. of South Florida over lesbian video. (1997, January 24). *The Chronicle of Higher Education,* p. A6.

Students protest amid charges of date rape. (1995, March 17). *The Chronicle of Higher Education,* p. A6.

Twelve charged in death of fraternity pledge at Clarkson U. (1997, March 7). *The Chronicle of Higher Education,* p. A10.

Two Citadel men quit; 24 women accepted for next fall's class. (1997, February 14). *The Chronicle of Higher Education,* p. A8.

U. of Alabama campus to see "Showgirls" movie despite protests. (1996, January 19). *The Chronicle of Higher Education,* p. A25.

U. of Nebraska fraternity's cross-burning ritual stirs debate. (1997, February 7, 1997). *The Chronicle of Higher Education,* p. A8.

U. of Pittsburgh to require pledge of civility from students. (1997, June 13). *The Chronicle of Higher Education,* p. A37.

U. of S. Florida student wins round in suit over sex videotape. (1997, February 7). *The Chronicle of Higher Education,* p. A8.

Victims of alleged rapes sue Carleton College. (1991, April 10). *The Chronicle of Higher Education*, p. A2.

White, L. (1994, May 25). Hate-speech codes that will pass constitutional muster. *The Chronicle of Higher Education*, p. POV.

Wilson, D.L. (1996, February 9). Congress passes bill banning on-line "indecency." *The Chronicle of Higher Education*, p. A23.

Young, J.R. (1997a, February 7). Student's on-line rape fantasy protected by First Amendment. *The Chronicle of Higher Education*, p. A27.

Young, J.R. (1997b, March 28). High court hears challenge to law on Internet "indecency." *The Chronicle of Higher Education*, p. A33.

CHAPTER 6

Violent Crime in the College and University Workplace

Allan M. Hoffman, Randal W. Summers, and Ira Schoenwald

INTRODUCTION

What would be your reaction if your family was asked to relocate to a town that reported it had experienced, last year alone, 1 murder, 2 rapes, 13 robberies, 20 aggravated assaults, 190 burglaries, 66 car thefts, 167 arrests for liquor violations, 158 arrests for drug violations, and 32 arrests for weapons violations? Would you be concerned? Would you ask how those statistics compare to other places in America? This place is not a town, it is a university. The statistics are self-reported, in compliance with the Student Right-to-Know and Campus Security Act of 1990 ("Fact File," 1993). The academic campus, once thought to be a sanctuary, is no longer safe.

Serious concern has been expressed in a variety of quarters about violence on college and university campuses. The campus is unique because it is an educational environment comprised of students and faculty, it is a workplace for administrators and staff, and it is also a target of outsiders (an extension of societal violence). Therefore, prevention and intervention strategies need to encompass hiring, firing, management and leadership practices, and staff and employee training. Administrators must be cognizant of the legal costs in liabilities and the responsibilities toward all members of this diverse community. They must embrace the principle of the "student's right to know" with accurate reporting of violent incidents. They must address policies and educational programs that are directed at safe practices related to student dating, drugs, alcohol, and personal safety. Prevention and intervention strategies must also encompass the institution's responsibility in protecting assets with security measures; providing supportive systems such as counseling and crisis centers, preventive education programs, management/leadership training; and developing workplace violence action plans. The administration must develop contingency plans to deal with the aftermath of violent incidents and provide crisis leadership that admits problems, creates violence mission statements, and plans. Perhaps, over time, new and more effective prevention and intervention methods will be developed as more research is done by groups such as the Center for the Study of Campus Violence at Towson State University, in Maryland, and the American College Personnel Association. However, violence is increasing on college and university campuses. Even with the development of progressive legislative programs and improved methods of dealing with violent situations, campuses must pay increased attention and resources to this growing problem. More needs to be done in research, prevention, and intervention strategies. There is little doubt that violence on campuses will become an increasing concern for colleges and universities. Professionals need to develop an arsenal of research, skills, and practical knowledge about violence on college and university campuses as an important part of their professional knowledge.

This chapter examines issues surrounding violence on college and university campuses with a special emphasis on the campus as workplace. It explores the relative uniqueness of the campus in that a college is an employer with employees who may be susceptible to the recent epidemic of violence in the workplace. The college campus is also comprised of students, who may be perpetrators of violence or victims of other students. In addition, the property of employees, faculty, students, and the campus itself may be easy targets for outsiders (perpetrators from the surrounding community). After presenting a definition of violence, national statistics on violence, and information about the cost of such violence, the chapter reviews selected prevention and intervention strategies.

THE CAMPUS AS AN ACADEMIC SANCTUARY: MYTH OR REALITY?

It is believed that colleges are far safer than the communities that surround them. However, the Federal Bureau of Investigation (FBI) reported that between 1992 and 1993 there was a 1.5% decrease in violent crime and a 3.4% decrease in property crime in the nation (Lederman, 1995). During this same period there was an *increase* in crime on college campuses. It seems that the college campus as a sanctuary is more myth than reality. A certain amount of crime has existed on college campuses throughout history. For example, two crime reports of different time periods were compared at Pennsylvania State University. A 1972 report indicated a total of 1,032 crimes (including murder, rape, robbery, aggravated assault, burglary, larceny theft, motor-vehicle theft, and arson), while in 1992 the university reported a total of 1,018 crimes (Collison, 1993).

Data about the extent of campus crime have only recently been compiled and publicized. In 1986, the murder of a student at Lehigh University resulted in an intense lobbying campaign. This activity culminated in the adoption of the Student Right-to-Know and Campus Security Act (November 1990), which requires every postsecondary institution that receives federal aid to provide its faculty and students with a report about its crime statistics and the institution's crime-prevention policies. In 1992, more than 42,500 crimes were reported by colleges and universities ("Fact File," 1993). The statistics included 30 murders, nearly 1,000 rapes, 1,800 robberies from persons, 32,127 burglaries, and 8,981 motor-vehicle thefts.

In a national sample of college women in 1989, one in four respondents reported experiencing rape or attempted rape (Koss & Dinero, 1989). In 1993 the Hate Crimes Statistics Act was passed; this law requires colleges and universities to determine if crimes committed on campus were racially or ethnically motivated. In a survey of 796 campuses (each with enrollments over 5,000) reported in *The Chronicle of Higher Education* (Lederman, 1995), 20% identified hate crimes in their statistics. The contemporary campus as an academic sanctuary seems to be more myth than reality.

THE CAMPUS AS A WORKPLACE: VIOLENCE IN THE WORKPLACE

The campus has always been perceived as a place for scholarly activities. However, it is also a workplace for thousands of employees. Unfortunately, the violence that is manifest in our society has now crept into the workplace. Between 1980 and 1988, homicide was the third leading cause of occupational death among all workers in the United States, accounting for about 12% of job-related deaths (Jenkins, Layne, & Kesner, 1992). Between 1980 and 1985, homicide was the leading cause of occupational death for women (Levin, Hewitt, & Misner, 1992). In 1991, homicide accounted for 20% of on-the-job deaths in California and 70% in New York City (Mantell, 1994). In 1992, according to the National Safe Workplace Institute, there were 110,000 serious workplace violence incidents including 750 employees killed by disgruntled coworkers (Mantell, 1994). Marvin Runyon, Postmaster General and Chief Executive Officer of the U.S. Postal Service, observed, "Recent statistics show that what was practically unheard of a quarter of a century ago, violence in the workplace, is now one of the fastest-growing crimes in the United States. Last year, 25 million people were victimized by fear and violence in the workplace. No company or organization should think it is immune to having a tragedy in its workplace. It can happen" (Mantell, 1994, p. vii). It seems that the workplace may be hazardous to our health. The campus as a workplace is at risk.

THE CAMPUS AS A TARGET: AN EXTENSION OF SOCIETAL VIOLENCE

Colleges and universities can be characterized as having a trilogy of violence. The campus is at risk from certain individuals in the student body. It is also at risk as a workplace in that disgruntled workers and inept, frustrated supervisors may act out their frustrations. Moreover, the campus can be an appealing target for villains from the communities outside of the campus. For example, Johns Hopkins University (East Baltimore campus) is located in a very dangerous neighborhood. Security measures were stepped up after a physician was abducted in the parking lot at gunpoint and left for dead in the trunk of his car. Two months later, a medical student on the way to class was beaten and raped. Our societal crime wave has washed upon the shores of our colleges and universities.

Over the past two decades, the amount of violence in our society has increased. It is estimated that almost one third of all Americans are victimized by crime each year (Mantell, 1994). Although the amount of campus crime is alarming, campus crime is pale when compared with national crime statistics. For example, in 1992 there were 17 murders on college and university campuses and 23,000 homicides nationally (Lederman, 1995). Today, it is estimated that a crime occurs every 17 seconds. The situation has been characterized as "toxic" (Mantell, 1994).

Ever-increasing levels of child abuse, domestic violence, marital breakup, unemployment, weapons, drug and alcohol abuse, and media-glorified violence are seen as contributing factors in the increase in violence in society. For example, children watching 3 hours of television per day will have seen, by age 12, approximately 8,000 murders and 100,000 acts of violence (Mantell, 1994). The effect of television on violence is sometimes debated. However, a meta-analysis of 217 studies from 1957 to 1990 indicated a relationship between exposure to television violence and antisocial and aggressive behavior (Paik & Comstock, 1994). Violence, which has multicausal origins, can occur as acts of aggression and crime in the home, in the school, in the workplace, and on the college campus.

DEFINITION OF VIOLENCE: THE VIOLENCE CONTINUUM

People often think of violence mainly as assault or murder. However, the definition of violence can be much broader, including any antisocial, destructive, or dangerous behavior. According to *Webster's College Dictionary* (1992), the definition of *violence* is "(1) swift and intense force; (2) rough or injurious physical force, action or treatment." The definition of *violent* is "acting with or characterized by uncontrolled, strong, rough force."

Violent acts can range from covert, disruptive activities aimed at objects (small-scale vandalism) to overt acts that can result in bodily harm to people and extreme damage to property. For example, violence includes vandalizing campus property; vandalizing vehicles in parking lots; implanting a computer virus; sabotaging laboratory experiments; tampering with computers (hacking); stealing personal and campus property; intimidating with threat of assault through letters and facsimiles; fighting and disorderly conduct; punching a supervisor, professor, or administrator; bomb threats; arson; campus and date rape; stabbing or shooting students, faculty, or administrators; and suicide on or near campus. People, too, fall on a violence continuum—from normal individuals who do not commit violent acts to armed and dangerous psychotics who commit multiple homicides.

NATIONAL STATISTICS FOR VIOLENCE ON CAMPUS

College campuses have a history of violence with some periods of unrest worse than others. For example, a generation ago, the American Council on Education, in a survey of 382 U.S. colleges and universities, found that 6.2% had experienced violent protest and 16.2% had experienced nonviolent disruption in the 1968–1969 academic year (Bayer & Astin, 1969). According to the FBI, the 1968–1969 academic year had been the most violent ever. There were 61 reports of arson and bombing and 4,000 arrests for student disturbances ("Campus Revolts," 1970). In 1970, U.S. campuses were threatened by terrorist bombings. Leaders of the Weathermen faction of the Students for a Democratic Society had intended bombing Columbia University. Instead, the bombs accidentally detonated and destroyed a New York townhouse ("The House," 1970).

By the 1980s, it was not student protest but incidents of sexual assault on college campuses that alarmed college administrators, faculty, students, and parents. In a large-scale national survey (1987), the incidence of sexual assault on college campuses became well established (Koss, Gidycz, & Wisniewski, 1987). Although sexual aggression from strangers is very uncommon among college students, coercive sex among acquaintances, or date rape, has become a serious problem on campus (Miller & Marshall, 1987). More recent surveys suggest that alcohol consumption and affiliation with certain groups are linked to being a perpetrator or victim of a violent act.

For example, a nationwide study of 60,000 college students reported that, among drinking students, fraternity men were three times more likely to commit violent acts and athletes five times more likely than their student counterparts (Lenihan & Rawlins, 1994). Membership in a fraternal group and alcohol abuse have also been linked to gang rape. Women in sororities were found to be at greater risk for sexual assault than nonsorority women (Sanday, 1990). Various research surveys over the years have provided information on the nature and extent of certain types of violence. However, until very recently, no systematic or consistent method of gathering national data on campus violence was available.

As mentioned earlier in this chapter, the first statistics required by law for violence on campus were reported for the year 1992. *The Chronicle of Higher Education* conducted a survey of 796 reporting colleges with enrollments greater than 5,000 (Lederman, 1995) and compared the crimes reported for 1992 and 1993. Drug law violations increased by 34.3%, weapon violations increased by 11.2%, forcible sex offenses increased by 3.1%, aggravated assaults increased by 2.7%, and robbery increased by 2.2%. The number of murders declined from 17 to 15 (17.6%), rape declined from 458 to 367 (19.9%), nonforcible sex declined from 90 to 83 (7.8%), burglary declined from 21,079 to 20,123 (4.5%), and motor-vehicle theft declined from 7,270 to 7,032 (3.3%). The new federal law provided the impetus and format for recording national statistics on campus violence. However, concerns have been expressed about the comprehensiveness and meaningfulness or interpretation of the data.

PROBLEMS WITH REPORTING AND INTERPRETING VIOLENCE STATISTICS

The problems associated with the campus crime statistics vary from the omission of recording some types of crime to the incorrect interpretation of the data.

Omission

College safety officials report that 75% of the crime committed on campus is larceny (theft), vandalism, and fire alarms (Lenihan, 1994). No requirements exist to report them. In addition, the categories of liquor, drug, and weapon violations include only the number of people arrested. What is omitted is the number of incidents reported to campus police that do not result in arrest and are handled through campus disciplinary procedures. Such incidents are not included in the campus violence statistics.

Another problem of omission is the category of rape. One college, for example, reported no rapes in the formal section of its crime report but indicated elsewhere in the report that four date rapes had occurred. Rapes are seldom reported to campus police, and when a rape victim seeks help at the campus health center, the center is not compelled to report the rape in the crime report. Coercive sex referred to as "date rape" is seldom reported. In one study, 20% of female respondents at Kent State University reported that they had been raped (by someone they knew), but only 8% of them reported it to the police. There were similar findings at the University of South Dakota, where 20% of 247 women interviewed said they had experienced "date rape." This phenomenon of rape victims not reporting the incident is not unique to the college campus; Miller and Marshall (1987) estimated that only 40% to 50% of all rapes are reported to the police, despite the severe emotional and physical trauma it causes the victims.

Another area of omission in crime reporting on campus includes a variety of sexual offenses against women (obscene phone calls, sexual molestation, exhibitionism, stalking, and attempted rape). In a study of 103 university women, Herold, Mantle, and Zemitis (1979) found that 61% of the respondents experienced an obscene phone call, 44% sexual molestation, 27% exhibitionism, 24% being followed, 16% attempted rape, and 1% rape. Although most victims of these offenses discussed the incident with friends, very few reported the incident to parents, relatives, police, or health or social work professionals. These findings support the conclusion that many offenses are not reported.

Additional confusion exists in regard to offenses that overlap two categories. For example, separate categories are found for sexual offenses and state liquor law violations, but alcohol plays a significant role in most sexual offenses. In the Miller and Marshall study (1987), 70% of the men accused of sexual offenses indicated that they had been under the influence of drugs or alcohol when they were involved in the incident. Lenihan and Rawlins (1994) also pointed out that alcohol was a contributing factor in 75% of sexual assaults among college students.

As mentioned earlier, certain offenses are omitted from reporting when a college or university deals with them internally. For example, the University of Nevada at Reno in 1993 reported 200 cases of disciplinary action for violations of their alcohol policies, but they reported no arrests for state liquor law violations. In addition, certain definitions of campus crime may result in some statistics being omitted. For example, if a student is abducted and taken off campus, the crime is classified as a community crime, not a campus crime. Crimes committed in the surrounding community, including adjacent streets, such as car theft or rape, can be left out of the statistics as well. Because of the natural tendency of competing campuses to compare statistics, campuses are likely to underreport crime statistics.

Interpretation

There are also problems associated with the interpretation of campus crime statistics. The intent of campus crime reports is to disclose the information and provide an impetus for prevention and intervention strategies. Parents and students might take the crime statistics into account when choosing a college or university. However, the interpretation of crime reports can be misleading.

One campus might report ten assaults and two rapes, while another reports five assaults and one rape. The second campus may appear to be safer until the size of the campus is considered. The first campus may have 20,000 students and therefore a be a relatively safe campus, whereas the second campus with only 2,000 students would be less safe, despite the lower crime statistics.

The type of campus may be another factor that can affect interpretation of the statistics. For example, a campus with no on-campus housing would have fewer incidents than one with on-campus housing.

Other factors such as the policies of the campus police department and presence of special treatment programs may influence the campus crime statistics. A campus may not experience any more crime in absolute terms, but the reported statistics may increase as a result of increased aggressiveness/responsiveness of campus security. Similarly, the numbers could increase with the introduction of special programs for treating victims such as a rape crisis center where counselors encourage the victim to report the incident to the police. Such was the case at the University of South Dakota; when the university introduced a rape crisis team, the reported rapes increased from zero in 1990–1991 to seven in 1991–1992. The increase was attributed to an increased willingness on the part of the victims to report rather than an actual increase in crime (Lederman, 1993).

THE COST OF VIOLENCE

The costs associated with crime and violence in society are staggering. Mandel and Magnusson (1993) reported that crime and violence cost Americans $425 billion a year, and the entire criminal justice system costs an additional $90 billion per year. According to Ronet Bachman (1994), crime victimization in the workplace cost 500,000 employees 1,751,100 lost days of work each year. This missed work costs $55 million per year in lost wages, not including sick or vacation pay (Bachman, 1994). Mantell (1994) estimates that each episode of violence in the workplace costs $250,000 dollars in lost work, employee medical benefits, or legal expenses.

The cost of violence on campus also is considerable. Tampering or sabotaging computers or lab and office equipment, vandalizing or stealing institutional and personal property, and arson can add up to millions. The aftermath of any violence on campus usually results in indirect costs such as increased insurance rates and increased costs of security (personnel, devices, and controls). A campus may incur additional costs with the establishment of special prevention and intervention programs such as education programs and crisis centers.

Human Costs

If the campus is viewed as a workplace, the human costs of violence are significant. When violence erupts, the campus environment can become permeated with fear. Soon the number of sick days for employees increases. The efficiency, morale, and productivity of staff declines. In an environment of fear, one's sense of well-being suffers, job satisfaction diminishes, and absenteeism and employee turnover increase. An environment of fear and violence also gives rise to employees' filing stress claims. Mantell (1994) estimates that the average stress claim is between $15,000 and $20,000; in addition, employees with stress injuries stay away from work longer than those with physical injuries.

The cost of violence not only affects the campus as a workplace but also affects the credibility of the campus as a center of education. A negative reputation can make it difficult to attract new faculty and students. A campus that is plagued with violence has many psychological and social problems. When students are the victims they become depressed, their grades decline, and they may even drop out of school. The emotional scarring caused by a violent incident can last months and even years.

Legal Costs: Liabilities and Obligations

If faculty or staff commit a violent act, the college or university can be held liable. Various types of liability include negligent hiring, negligent retention, failure to warn, respondent superior, and negligent entrustment. Other local and state laws also may be violated, thereby placing the campus at legal risk.

Anfuso (1994) reports that if an employer knows or should have known that a person is at risk for committing violence but hires or retains the person anyway, the employer is responsible for any harm the employee causes. Hiring a person who is unfit for the job is termed *negligent hiring*. An employer who knows an employee is unfit but keeps the individual employed can be charged with *negligent retention*. Feliu (1994–1995) indicated that increasingly courts are holding employers liable for violent incidents based on violation of the doctrine of "duty of care." Duty of care is breached when employers fail to exercise reasonable care to ensure that employees and customers are free from the risk of harm. The Colorado Supreme Court, for example, along with 14 other states has recognized the tort of negligent hiring. Negligent hiring can result in additional costs to the institution, if an employee is found responsible for committing acts that placed others at risk or resulted in costs that must ultimately be assigned to the institution.

Under workers' compensation statutes, injuries in the workplace can result in liability (but not when nonemployees are injured). However, these statutes define the limit of an employer's liability and can actually block or prevent legal action to sue for workplace violence. For example, a woman attempted to sue for negligence when a coworker assaulted her. However, a Maryland court held that she was barred from suing under workers' compensation laws. In another case, a Minnesota court held the opposite. An employer had retained an employee with a criminal record. This employee threatened another employee and later shot and killed the person. The employer was sued for wrongful death on the basis of negligence. Workers' compensation law could not block the suit in this case.

Elliott and Jarrett (1994) report that the only piece of legislation today concerning homicide in the workplace is the Occupational Health and Safety Act (OSHA) of 1970. They point out that the act has no specific standards, but it does ensure that every worker works under safe and healthful working conditions.

The *Stevens OSHA Reporter* ("Warning Signs," 1995) indicated that OSHA could assign liability under the general duty clause. An employer is considered to be on notice of the risk of workplace violence when a violent incident has occurred or when an employer has become aware of threats or intimidation in the workplace. Once on notice, the employer must initiate a preventive program. Failing to do so could result in OSHA issuing a citation with monetary penalties and government-ordered abatement requirements. Violent acts normally are considered to be outside the scope of employment, so an employer would not be responsible for violent acts of its employees. This concept is referred to as *respondeat superior*. However, this rule has been modified, and employers can be held liable for incidents that occur outside the scope of employment. In other words, the employer is responsible for the behavior of its employees.

Another type of liability is when an employer has knowledge that an employee is not experienced or is incompetent in the use of a weapon. This concept would apply in employment situations where individuals carry firearms (such as armed security guards). This type of liability is called *negligent entrustment* and would seem to occur only in rare instances. However, on closer examination, wherever armed security is used, the liability is quite real if not quite probable. Several reports indicated that almost all alarm personnel and proprietary security employees who carry firearms had discharged them on duty (Clark & DeGeneste, 1995). This fact is quite alarming considering that the same research indicates that the average amount of firearms training for security officers is less than 8 hours and that training does not focus on "tactical savvy" or the handling of dangerous situations but on the mechanics of firing the gun. Henry I. DeGeneste, senior vice president of corporate security for Prudential Securities, Inc., in New York, emphasizes the potential for liability: "Security officers generally do not have the training and preparation for armed public contact assignments" (Clark & DeGeneste, 1995).

The Student Right-to-Know and Campus Security Act requires universities and colleges to report any occurrence of hate crimes. A hate crime could also have legal implications under antidiscrimination laws such as the Equal Employment Opportunity laws (such as Title VII). An employer would be held liable if it failed to protect employees from racial or sexual harassment.

Still another legal issue could arise when employees turn to their unions for help with workplace violence. For example, the Service Employees International Union has developed contract language to ensure workplace safety. It wants employers to provide specific policies to prevent on-the-job assault. These policies must include prevention, management of violent situations, providing legal counsel, installation of security devices, employee training, employer assistance in prosecuting offenders, and post-traumatic support to employees (victims).

Stone and Hayes (1995) indicated that organizations must be sensitive to various legal issues related to a safe workplace. These issues have direct bearing on their obligation to provide a safe and healthful workplace and their potential liability. In summary, the direct and indirect costs and legal liabilities and obligations associated with campus violence can be extensive, making the costs of prevention pale in comparison and completely justified.

PREVENTION AND INTERVENTION STRATEGIES RELATED TO EMPLOYEES

The following strategies for intervention and prevention of campus violence focus upon the employee and the institution. Although these approaches contain some strategies that have been used for a while—such as focused hiring practices, counseling, and training programs—the focus on violence prevention and intervention is relatively new to the college and university campus.

Employee Hiring Practices

The modern philosophy of recruiting and hiring is to find a match between prospective employees and the organization. The goal is to find a person who has not only the right skills for the job, but also values, attitudes, and behaviors that are compatible with the organization's values, mission statement, and social milieu. This philosophy and practice is especially applicable to the prevention of violence in the workplace. Preemployment drug screening, appropriate (legally defensible) vocational testing, background checks, thorough application review, and effective selection interviewing (preferably multiple interviews) will help to avoid many problems.

A manager at a company in California placed a poster on the wall next to the human resources office where people picked up job applications. The poster announced the company's drug-screening policy for all applicants. The manager believed that the poster, which was put up for legal reasons, was effective in discouraging drug "users" or alcohol "abusers" from even submitting an application. Another practice that is becoming commonplace in large organizations is background investigations. Some organizations hire a security company to check the prospective employees' work history, military history, criminal history, driving record, and sometimes even their credit history. The investigation is done before the organization makes any offer to an applicant.

A thorough review of the person's application or resume at times can help eliminate potential problem employees. Mantell (1994) suggests that gaps in a resume or sloppy applications should prompt employers to ask some probing questions. If a person makes any comments in the interview about having been treated unfairly by previous employers, this is a cause for concern. Most acts of violence committed by disgruntled workers seem to have been motivated by their perception of having been treated unfairly. Mantell (1994) suggests that applicants should be asked the following seven questions:

1. When do you feel you were treated unfairly in your life?
2. What complaints have you had about your supervisor in the past?
3. What could a supervisor do to make you angry?
4. What did you do about it?
5. What would you have liked to do about it?
6. Why did you feel you were treated unfairly?
7. What has your supervisor done in the past to make you really angry?

Safe hiring practices are equivalent to the proverbial "ounce of prevention." They are definitely worth the time and expense because a disgruntled worker can cause a myriad of problems. The signs of a potentially dangerous disgruntled worker include the following:

- Is disgruntled; has disruptive attitude and behavior.
- Is socially isolated (a loner).
- Has poor self-esteem.
- May seek help in appropriate or inappropriate ways (a cry for help).
- Has a fascination with the military.
- Is a gun or weapon collector.
- Has temper-control difficulties.
- Has made threats on the job.
- Has few if any outlets for the rage inside.
- Has a preoccupation or keen interest in media reports of violence.
- Has an unstable family life.
- Other employees have voiced concern about this person.
- Is frequently involved in labor-management disputes.
- Has unresolved physical- or emotional-injury claims.
- Frequently complains about working conditions.
- Complains of stress at work.
- Is a 30-to-40-year-old male.
- Has a history of drug and alcohol abuse.
- Has a psychiatric impairment.

A report in the *Los Angeles Times* illustrates what can happen if an employee has a psychiatric impairment (Lynch, 1996). In 1993, a disgruntled former postal worker in Dana Point, California, went on a deadly rampage. He stabbed his mother to death, killed her pet dog, and went to the post office where he had worked. He shot two coworkers, killing one of them. The accused had been previously hospitalized for manic depression and bizarre behavior (going to work carrying a bag of rabbits and wearing his undershorts on his head or on the outside of his clothing). Just after the murders, he told police he had to kill his mother to protect her from a "catastrophe" on Mother's Day. While incarcerated, awaiting trial, he attempted suicide by jumping off a second-story walkway in the jail. Although thorough screening and background checks may not be foolproof, safe hiring practices will significantly reduce the risk of hiring a "problem."

Positive Management Practices

Some individuals who have various types of personality disorder (such as paranoid, antisocial, or borderline) may not pose a problem until they are subjected to poor supervision or a negative organizational climate. This is a formula for disaster. A "toxic" boss or stress carrier with poor people skills can lower morale and productivity and push the borderline employee "over the edge." Some supervisors still practice an autocratic management style today ("Do it my way or suffer the consequences"), although this approach is not effective in the long run and can often result in anger and potential aggression. Even well-adjusted workers suffer when exposed to supervisors who are rude, intimidating, or abusive. Progressive management practices value employees and treat them with respect.

In a positive work climate, teamwork thrives and employees feel a sense of pride of ownership in their work. E.W. Deming (1992), one of the originators of total quality management (TQM), expounded 14 management principles. One of them emphasized getting rid of fear in the workplace. Some companies avoid violence in the company by fostering a violence-free culture ("Creating a Violence-Free Company Culture," 1995). For example, Wainwright Industries provides all of its employees with interpersonal relations training, which emphasizes treating people with respect and dignity. Managers are committed to responding to employees' suggestions within 24 hours. Wainwright is one of only six small businesses to have won the Malcolm Baldridge National Quality Award in 1994.

Too many negative side effects result from managing by fear and intimidation. This point is especially relevant to the area of discipline and termination.

Discipline Not Punishment

The violence in the workplace scenario, too often played out, is one of a suspended or fired worker returning to the job with a gun. Safe discipline and termination procedures may help prevent a situation from escalating into disaster and also avoid wrongful termination suits. Grote and Harvey (1993) provide a helpful model in dealing with discipline. They suggest that most issues can be categorized as problems with attendance, conduct (following work rules), or performance (how well they do the job). Grote and Harvey's model includes five steps in the employee discipline process:

1. Identify what is the desired behavior (performance) and what is the actual behavior (current undesirable or inappropriate behavior). A supervisor should not focus on the employee's personality traits but rather on the specific aspect of the employee's performance.
2. Analyze the seriousness or impact of the problem, decide what consequences the employee could face and determine the next action step with the employee.
3. Discuss the problem with the employee in a respectful manner. Obtain the employee's commitment to change. Discuss some alternative solutions to the problem, and agree on the action the employee will take to improve.
4. Document everything. Describe the problem, its history, and the counseling or discussion sessions that have been held to correct the problem. Documentation becomes invaluable in the event of a wrongful-discharge lawsuit.
5. Follow-up. The ongoing monitoring of the situation will reveal whether the problem is being solved. Praise any improvement in the desired direction, and take any required action if there is no change.

Successful managers communicate with their employees and provide specific feedback on job performance. Problems provide opportunities for discussion, getting a commitment from the employee to improve, and coaching toward collaboration where all parties are satisfied that the issue is resolved.

Safe Termination Practices

Organizations must have clear guidelines for terminating employees. If a marginal or disgruntled employee is terminated without due care, there is a risk that he or she may return to the work site. In one instance, a terminated employee went out to the car for a gun, came back in immediately, and shot the supervisor. In other instances, the disgruntled worker returned years later. It is difficult to predict these situations, so fair, legal, and compassionate handling of terminations is necessary. The best managers coach their employees to success. They consider termination only as a very last alternative. They use progressive discipline with compassion. This way, the employee has ample opportunity to think matters through and improve as a result of oral warnings, written warnings, and suspension. There are exceptions where progressive discipline is not followed. In some cases, the behavior of the employee might warrant taking swift action—termination—such as when the employee violates a zero-tolerance policy regarding theft or aggressive behavior.

An appointment is arranged for the terminated employee to come back to pick up personal belongings, and he or she is escorted off the premises by security. If there are warning signs that the person is psychologically impaired and termination is unwise, then the person is referred to an employee-assistance program or outside agency for help. Thousands of workers are laid off or terminated each year, but only a small number return for revenge. The way an employee is terminated can make a difference in whether the severance is peaceful or results in a lawsuit or violence.

Even though an employer may believe that termination is justified, subtle pressure from unions or the fear of upsetting other workers may cause the organization to retain a marginal employee. This organization is at legal risk of negligent retention. The employer may knowingly retain a potentially harmful employee. If the employee acts out and hurts someone, the organization may be held liable.

Workplace Violence Action Plan

In 1993, Northwest National Life Insurance reported that more than 6,000 workers were killed on the job (Crabb, 1995). Of these, 1,000 were homicides. Of these homicides, 82% were associated with robberies, 6% were associated with police in the line of duty, and 4% were associated with personal disputes among employees (Crabb, 1995). Although 4% might seem like a relatively small number of murders on the job, what these statistics do not reveal is that 2 million people were physically attacked at work and another 16 million were harassed on the job. In view of these statistics a violence action plan makes perfect sense.

Virginia Gibson (1992) observed that as overall incidents of violent crime rise in the country, the likelihood of workers' experiencing an incident on the job also increases. She asserted that the perpetrators of workplace violence are sometimes people from outside the organization. However, the perpetrator is a more often a co-worker who abuses drugs or alcohol, is angry over being fired or disciplined, or is emotionally unstable and vents frustrations in the workplace. Gibson maintains that organizations should develop a workplace violence action plan. The plan or policy development should include the human resources department, safety department, and the training department (Gibson, 1992).

Johnson and Indvik (1994) suggest that organizational policy on workplace violence should contain a description of the characteristics of potentially violent behavior, an outline of who should be contacted if violence is suspected, and the basics of how to conduct an investigation. Mattman (1995) suggested that organizations should create an executive committee to establish a workplace violence program. The committee would establish policies, implement, and administer the program. He asserted that the success of the program depends on the committee's initial development work and how well they oversee the program. Following a potential violent situation, the Good Samaritan Health System used a similar process to develop a workplace violence policy. Their policy development included the following steps (Mantell, 1994):

- a self-assessment
- a climate survey
- briefing of senior management
- identification of the potential perpetrator
- training

Freidman (1995) points out that companies should offer training programs that teach supervisors and employees what to look for to prevent domestic violence from entering the workplace. The above policies and programs are aimed at *preventing* violence, but Henderson-Loney (1995) encourages organizations to create a trauma plan to provide guidelines for employees to follow when violence does occur (the aftermath of trauma). A violent incident or death of a co-worker can upset employees and greatly disrupt performance or cause lost time. A trauma plan would involve a quick response by the organization in dealing with employees' feelings and reactions to the violence and help return work activities to normal.

CAL-OSHA: Model Injury and Illness Prevention Program for Workplace Security

On March 30, 1994, the California Department of Industrial Relations, Division of Occupational Safety and Health (DOSH) adopted guidelines for workplace security.* In California, assaults and violent acts compose the highest category of occupational fatalities. DOSH devised a set of guidelines to help employers meet this rising problem. They divided workplace violence into three categories based on the relationship of the violent offender or perpetrator to the business or workplace.

---

*DOSH is the agency responsible for enforcement of the California OHSA regulations. *California Code,* sec. 142, at 6307–6308. These regulations can be adopted from the California Occupational Health and Safety Code.

1. A type I event is caused by a person with no legitimate relationship to the business or workplace. An example would be a robbery or other criminal act resulting in violence toward a worker. Type I violent crimes account for most common types of workplace violence occurring usually in late evening hours in convenience stores, against night security guards, and against employees of bars and motels.
2. A type II violent event is committed by a recipient of services (such as a client, customer, or student or former student). In type II events, faculty, staff, and managers would be likely victims of violence from a student, former student, or relative of a student.
3. A type III event is committed by a perpetrator who has an employment-related relationship with the institution. In this scenario, the violence is committed by an employee of the college or a relative of an employee.

The DOSH model is available from the California Department of Occupational Safety and Health Administration. It can be a useful guide for colleges and universities looking for a plan or set of guidelines to assess themselves and establish a consistent program of violence prevention.*

Bomb Awareness

The university campus and industrial workplace has also been confronted with bomb threats. This potential danger became manifest in the Oklahoma City bombing. In addition to threats and actual bombing of work sites or campuses, colleges and universities have recently had to deal with a new type of peril—the letter bomb. The Unabomer allegedly chose a university professor as one of his victims of a mail bomb. Michael Gips (1995) suggests that organizations can improve their bomb-detection capabilities with simple policies and basic training on bomb awareness. He reports that the U.S. Treasury Department's Bureau of Alcohol, Tobacco, and Firearms has a pamphlet with bomb-detection and safety tips. For example, a package or letter with lumps, bulges, or protrusions; one that is wrapped in string; or one that has no return address or a nonsensical return address might indicate that it contains a bomb.

---

*The materials can be obtained through the Department of Industrial Relations, Division of Occupational Safety and Health, 455 Golden Gate Avenue, Room 5202, San Francisco, CA, under "Cal/OSHA Guidelines for Workplace Security" and "Cal/OSHA Model Injury and Illness Prevention Program for Workplace Security, March, 1995."

Security Measures

The concept of maintaining security at a facility or institution is not a new phenomenon. What is new are the proactive and innovative approaches that many organizations are taking to protect assets. Many high schools, for example, are now installing closed-circuit video cameras to monitor valuable assets such as computers and audio-visual equipment. Some schools are even installing cameras in hallways to limit vandalism and monitor student conduct.

To protect hospital employees from abusive patients or their families, the National Rehabilitation Hospital in Washington, D.C., uses a group approach in its treatment. Its security program involves a careful sign-in procedure, and the hospital education programs help employees to recognize and handle potentially dangerous or abusive individuals (Overman, 1995).

The security measures used on the college and university campus also have evolved. In 1971, Richard Post asserted that campus security had outgrown its night watchman origins. He advocated the development of a comprehensive plan that includes policies related to law enforcement, crime and violence protection, and general campus police or security services. A total program would use qualified and trained security staff. Detection equipment, automatic sensing equipment, and alarm equipment would be installed in administrative and academic buildings. A central station must be identified for the annunciation of manual alarms. Identification would be required for exit and entry of buildings after the close of a normal day. Electronically controlled doors and an identification card/building pass system would be monitored by security. Closed-circuit television would be used to monitor areas requiring a high degree of security. The program would include the development and on-going evaluation of security and safety policies.

Although law enforcement activities are necessary for arresting violators, prevention should be the focus. It makes good economic sense. Workers' compensation evaluates employers in respect to an Experience Modification Factor, which is calculated by reviewing the claims and the activities/programs put in place to prevent claims over the three previous years. A factor lower than 1.00 means the organization's claims are less than industry norms.

Policy and Procedure for Strikes

A strike, walk-out, or staff "sick day" can be debilitating on a college campus. In addition, student sit-ins can render the campus helpless and draw unwarranted media coverage. Although these forms of protest have not been a frequent occurrence on contemporary college campuses, institutions should have a policy for dealing with such situations and a method to communicate it.

Supportive Institutional Systems

An essential determinant of a successful prevention and intervention strategy is the presence of supportive institutional systems. These include counseling programs; crisis centers; action teams made up of students, faculty, and employees; and preventive education programs for employees and students.

Earlier in this chapter, statistics on sexual assault (rape, date rape, and so forth) on the college campus were highlighted. The victims of such crimes need access to special supportive systems. Howard Doweiko (1981) indicated that victims of sexual assault need to regain a sense of control over their life, personal autonomy, and an understanding of their response to the assault. The average age of sexual assault victims is 24 (college age). Therefore, college counselors need to know sexual assault counseling to help victims adjust (Doweiko, 1981). According to the Minnesota Program for Victims of Sexual Assault (1978), a victim may go through various stages of adjustment following assault. In stage 1, the victim must adjust to the situation involving immediate medical care. In stage 2, the victim must accept that a sexual assault has happened. Stage 3 requires the victim to accept that he or she is now in a victim role (Minnesota Program for Victims of Sexual Assault, 1978). Crum (1976) reports that a sexual assault constitutes a major life crisis, and a person's adjustment depends on the uniqueness of the assault and the victim's characteristic coping mechanisms.

Sutherland and Scherl (1970) described three phases of adjustment that a victim goes through in reacting to sexual assault. Phase 1 (acute reaction) is characterized by a wide range of emotions including shock, disbelief, anxiety, fear, humiliation, loss of control, anger, restlessness, tears, agitation, tenseness, inappropriate laughter, crying, and sobbing. In this phase, victims deal with medical attention, legal and police matters, notification of family and friends, and housing and transportation matters. In phase 2 (outward adjustment), defense mechanisms of denial, suppression, and rationalization create an impression of adjustment, but that is an outward facade. In phase 3 (resolution), the victim becomes depressed; feels guilty, soiled, or damaged; and engages in self-blame. Baker and Peterson (1977) found that the greater the victim's physical injury, the less self-blame occurs. In phase 3, victims have to integrate a new view of self and resolve their feelings about the assailant. Evans (1978) found that one half of the victims attempt to repress memories of the assault.

Training as a Preventive Strategy

Training is an effective and versatile strategy in preventing violent incidents. Management and leadership training can help to prevent or counteract "toxic" organizational climates. Training programs for effective coaching, selection interviewing, progressive discipline, and safe termination will help to avoid many problems and also support progressive management practices. Johnson and Indvik (1994) suggest that companies who offer training on sexual harassment could easily include a component on workplace violence. Employees could also be trained in the use of security equipment, how to recognize suspicious activity, who to call in an emergency, and to be aware that employee assistance programs are available as a helpful resource.

Contingency Plans for Crisis Management

Maintaining support services, public safety, and campus security are important. Equally important is anticipating and minimizing the effects of major disruptions on campus. Contingency planning involves the process of asking "what if?" For example, what if there were an electrical shutdown, a water main break, a natural disaster, a fire, or violent acts threatening the safety of staff and students? If any of these situations occurred and was managed poorly, it could prove to be very costly. The purpose of a contingency plan is to minimize the effect on the educational environment and reduce the human and financial costs.

A contingency plan should address steps for before, during, and after (recovery steps) a crisis. It includes the following components (Mantell, 1994; Young & Smith, 1971):

1. A definition of a *critical situation* (situations beyond the ordinary that affect the college such as fires, floods, power failures, explosions, or riots).
2. A process for getting people to think about why contingency plans are necessary. This helps people become aware of their responsibility in a crisis, which is as important as the written plan.
3. Direction on who should be involved in developing the plan. Those who would implement the plan should be involved in developing it. One person from each vice president's office should be appointed to coordinate the plan. The coordinator and supervisor should develop the framework for the plan. The supervisor along with his or her staff should develop the specific details of the plan. The plan is then put into an overall or campus-wide contingency plan.
4. Direction on a variety of issues. For example, how are people notified of a critical situation? What regulations limit alternative actions? How is the reimbursement of funds addressed if the college closes? How can academic work be completed? How are administrative records secured? What are the arrangements between local fire and police services and campus security? What are the implications for student housing? How are paychecks disbursed? What special equipment or facilities will be needed? How is fund-

ing arranged for overtime or other nonbudgeted personnel services? What system can be developed for communication, should normal services be unavailable?
5. Direction on how to ensure the plans are operable. For example, the plans should be evaluated through simulations. Plans need to be updated as exigencies change and should be reviewed for their effectiveness after the crisis is over.
6. After an incident occurs, a number of activities need to be undertaken. Among them are the following:
- notification of law enforcement and rescue personnel
- protection of the scene
- on-site crisis counseling for students and employees
- carrying out of media-relations plan
- notification of families of victims
- assistance with grief and trauma recovery for victims, their families, and witnesses
- employee assistance programs
- designation of a clean-up crew
- provision of ongoing psychological assistance
- notification of human resources office so that appropriate actions can be taken concerning any employee who is a victim of the incident
- notification of the dean of students' office so that appropriate actions can be taken related to the records of any student who may have been a victim of the incident
- memorials for any loss

All of the above actions must be accomplished in an environment of care, compassion, and concern. Anything less will cause additional problems for the institution.

Crisis Leadership

When an organization experiences a violent incident, the people involved go through stages of shock. They first experience shock (numbness), disbelief, and denial. Then they experience emotions and then, in the last stage, there is reconstruction of equilibrium. It is vitally important that the leadership of the organization restore confidence as quickly as possible by conveying a message of hope (such as, "We are here to stay. We will rebuild even better than before.") This response will help to avoid prolonged grief reactions and help return things back to normal. Crisis leadership also involves admitting problems, creating violence mission statements, and revamping contingency plans where they may have failed. Evaluation of violence prevention and intervention strategies should focus on whether they protect people, preserve the physical environment, and ultimately maintain a positive educational environment.

REFERENCES

Anfuso, D. (1994). Violence-prevention strategies limit legal liabilities. *Personnel Journal, 73*(10), 72.

Bachman, R. (1994, July). Violence and theft in the workplace. *Crime Data Brief Bureau of Justice Statistics.* Washington, DC: U.S. Department of Justice.

Baker, A.L., & Peterson, C. (1977). Self-blame by rape victims as a function of the rape's consequences: An attributional analysis. *Crisis Intervention, 8,* 92–104.

Bayer, A.E., & Astin, A.W. (1969). Violence and disruption on the U.S. campus, 1968–1969. *Educational Record, 50,* 337–350.

Campus revolts over? (1970, February 16). *U.S. News & World Report, 68,* 53–55.

Clark, B., & DeGeneste, H.I. (1995, October). A call to arms—A disarming question. *Security Management, 2,* 48–51.

Collison, M. (1993, January 20). Law may push colleges to spend more money on security and campus-police training. *The Chronicle of Higher Education,* p. 32–33.

Crabb, S. (1995, July 27). Violence at work: The brutal truths. *People Management, 1,* 25–27.

Creating a violence-free company culture. (1995, February). *Nation's Business, 83,* 22–24.

Crum, B.S. (1976). Counseling rape victims. *Journal of Pastoral Care, 28,* 112–121.

Deming, E.W. (1992). *Out of crisis.* Cambridge, MA: Massachusetts Institute of Technology Press.

Doweiko, H. (1981). Counseling the victims of sexual assault. *Journal of College Student Personnel, 22,* 41–45.

Elliott, R., & Jarrett, D. (1994). Violence in the workplace: The role of human resource management. *Personnel Management, 23*(2), 287–299.

Evans, H.I. (1978). Psychotherapy for the rape victim: Some treatment models. *Hospital and Community Psychiatry, 29,* 309–312.

Fact file: Crime data from 2,400 colleges and universities. (1993, January 20). *The Chronicle of Higher Education,* pp. 34–43.

Feliu, A. (1994-1995). Workplace violence and the duty of care: The scope of an employer's obligation to protect against the violent employee. *Employee Relations Law Journal, 20,* 381–406.

Freidman, S. (1995, August 7). Fear of workplace violence prompts safety initiatives. *National Underwriter, 99,* 1.

Gibson, V. (1992, September). Safety training benefits employees on and off the job. *HR Focus, 69,* 19.

Gips, M. (1995, May). Blasting impression. *Security Management, 39,* 11.

Grote, R., & Harvey, E.L. (1993). *Discipline without punishment.* New York: McGraw-Hill.

Henderson-Loney, J. (1995, August). The crying game. *Training and Development, 49,* 54–56.

Herold, E.S., Mantle, D., & Zemitis, O. (1979). A study of sexual offenses against females. *Adolescence, 14,* 65–72.

The house on 11th street. (1970, March 23). *Newsweek, 75,* 30.

Jenkins, L.E., Layne, L., & Kesner, S. (1992). Homicides in the workplace. *Journal of the American Association of Occupational Health Nurses, 40*(5), 215–218.

Johnson, P.R., & Indvik, J. (1994, Winter). Workplace violence: An issue of the nineties. *Public Personnel Management, 23,* 515–523.

Koss, M.P., & Dinero, T.E. (1989). Discriminant analysis for risk factors of sexual victimization among a national sample of college women. *Journal of Counseling and Clinical Psychology, 55,* 242–250.

Koss, M.P., Gidycz, C.A., & Wisniewski, N. (1987). The scope of rape: Incident and prevalence of sexual assault and victimization in a national sample of higher education students. *Journal of Counseling and Clinical Psychology, 55,* 162–170.

Lederman, D. (1993, January 20). Experts say disclosure law allows colleges to omit categories that could give clearer picture of crime. *The Chronicle of Higher Education,* p. A33.

Lederman, D. (1995, February 3). Colleges report rise in violent crime, crime data from 796 colleges and universities. *The Chronicle of Higher Education,* pp. 31–42.

Lenihan, G.O., & Rawlins, M.E. (1994). Rape supportive attitudes among Greek students before and after a date rape prevention program. *Journal of College Student Development, 35,* 450–455.

Levin, P.F., Hewitt, J., & Misner, S. (1992). Female workplace homicides. *Journal of the American Association of Occupational Health Nurses, 4,* 229.

Lynch, R. (1996, June 17). Foundation of Hibun rests on sanity. *Los Angeles Times,* Orange County Edition, p. B7.

Mandel, M., & Magnusson, P. (1993, December 13). The economics of crime. *Business Week,* 72–85.

Mantell, M. (1994). *Ticking time bombs: Defusing violence in the workplace.* New York: Irwin.

Mattman, J.W. (1995). What's growing in the corporate culture? *Security Management, 39,* 42–46.

Miller, B., & Marshall, J.C. (1987). Coercive sex on the university campus. *Journal of College Student Personnel, 28,* 38–47.

Minnesota Program for Victims of Sexual Assault. (1978). *Sexual assault: A statewide problem.* St Paul: Author.

Overman, S. (1995, March). Preventing violence against health-care workers. *HR Magazine, 40,* 51–53.

Paik, H., & Comstock, G. (1994). The effects of television violence on antisocial behavior: A meta-analysis. *Communication Research, 21,* 516–546.

Post, R. (1971, August). Campus security. *College and University Business, 51*(2), 33–34.

Sanday, P.R. (1990). *Fraternity gang rape: Sex, brotherhood and privilege on campus.* New York: New York University Press.

Stone, R.A., & Hayes, R. (1995, Autumn). Developing policies addressing workplace violence. *Employee Relations Today, 22,* 25–57.

Sutherland, S., & Scherl, D.J. (1970). Patterns of response among victims of rape. *American Journal of Orthopsychiatry, 40,* 503–511.

Warning signs, prevention of workplace violence subject of new report. (1995, November 6). *Stevens OSHA Reporter, 6*(42), 1.

Webster's College Dictionary. (1992). New York: Random House.

Young, J.H., & Smith, S.A. (1971, August). What can be done if disaster strikes? With contingency planning, you already know. *College and University Business, 51*(2), 35–37.

CHAPTER 7

Violence at Home on Campus

Carolyn J. Palmer

INTRODUCTION

It is well known that most violent acts are committed close to home. "Relatives attack relatives and neighbors attack neighbors more often than one stranger attacks another stranger from a different part of town" (Grayson, 1993, pp. 140–141).

Many campuses may be considered "home" to students who live in residence halls, fraternity and sorority houses, apartment buildings, and other residential facilities that are owned and operated by colleges and universities. Although some students may live in campus apartments with their spouses and/or children, most residential students are single, traditional-aged undergraduates whose biological families live at some distance from the campus. Often substituting as "extended family" during the college years are friends, lovers, classmates, roommates, neighbors, and staff members—particularly those who live in the same housing units as the students. An equivalent to "domestic violence" involving spouses, children, and other family members is "relationship violence" involving individuals who know one another as a result of any number of different relationships within the campus community.

Towson State University campus violence studies indicated that "most violence on college campuses is student-to-student violence" (Sherrill & Siegel, 1989, p. 1). Although random acts of violence are occasionally committed by strangers or "outsiders" who invade student residences without permission, most of the perpetrators of violence and other forms of victimization occurring in campus housing units are residents of those units and their invited guests.

For example, in a study conducted by Parsonage et al. (1992), 182 residence hall staff members described the perpetrators of the "single most serious" incidents that had victimized them during their careers. Of these perpetrators, 145

(80%) were identified as residents, 11 (6%) were guests, 15 (8%) were unknown/ unidentified (and thus may or may not have been residents or guests), and only 11 (6%) were described as unknown visitors or other persons. In another study, resident assistants (RAs) were asked to describe in their own words the "most serious" incidents they had encountered in their work. In 90 of the descriptions of violence, the resident and/or student status of the perpetrators was identified: 54 (60%) were residence hall students, 24 (27%) were students with residence unspecified, and 12 (13%) were nonstudents (Palmer, 1996).

WHY VIOLENCE OCCURS IN CAMPUS RESIDENCES

According to Rickgarn (1989), "Clearly violence in residence halls is not new. While it may be ignored or denied, violence has existed since the beginning of residential units on the campuses of this country" (p. 29). Nevertheless, Smith (1989) emphasized that the frequency and seriousness of violent incidents occurring on today's campuses are unprecedented in the history of American higher education. On residential campuses, many (perhaps most) of the violent incidents occur where students live. Why is this so? An intuitively reasonable, although perhaps simplistic, explanation is that campus residences are where students are most likely to be at night and on weekends, particularly after consuming alcohol, when violence is most likely to occur.

The Role of Alcohol and Other Drugs

Concerns regarding students' use of illicit drugs should never be minimized. Clearly, drug possession, use, and trafficking have been associated with some of the most serious and tragic incidents of violence in student housing. Witness the recent murder of a live-in staff member who had contacted police to report a drug violation, followed by the suicide of the student perpetrator in their Purdue University residence hall. The "drug of choice" among most college students and the drug most often linked to campus violence is alcohol.

In spite of laws and institutional policies regarding alcohol, many students, including those who are underage, drink in campus residences or return to their residences after drinking off campus (Palmer, 1996). And while under the influence of alcohol, students are at greater risk of becoming not only the *perpetrators,* but also the *victims* of violence. In a study conducted at a large university, Ellison, O'Shaughnessey, and Palmer (1991) found that 71% of the women and 81% of the men had been drinking when sexual assaults occurred. Based on the findings from his survey of more than 17,000 students on 140 campuses, Weschler (1996) concluded that "when women abuse alcohol, they increase their risk of becoming victimized by unwanted or unprotected sex" (p. 23). Weschler also noted risks associated with "secondhand effects" resulting from someone else's drinking. One example of a secondhand effect is the finding reported by Ellison et al. (1991) that 62% of the nonsexual batteries associated with "courtship violence" occurred when the perpetrators were under the influence of alcohol.

The Challenges of Living in "Concentrated Proximity" with Others

Even under the best of circumstances (such as within loving relationships and under ideal environmental conditions), people who live together occasionally get on one another's nerves. According to Amada (1994), "When large numbers of persons reside together in concentrated proximity, it is inevitable that interpersonal tensions, misunderstandings, incivilities, and disharmonies will arise, at times reaching serious proportions" (p. 39). The concentrated proximity within which many students live is often perceived as "overcrowding," particularly by students who, during their precollege years, had their own room and complete dominion over the personal space, possessions, activities, and sense of privacy associated with that room.

Those who have never had roommates with whom they have had to share, compromise, negotiate, and so forth may find it difficult to temper their sense of territoriality and control over what they identify as their own personal space in a campus residence. Some students appear to be oblivious to the needs and rights of their roommates and neighbors, yet pursue their own needs and exert what they believe are their own rights with considerable vehemence. Clearly, many students do not like institutional policies that regulate noise, prohibit underage drinking, and otherwise specify what students can and cannot do—particularly within the privacy of their own rooms. Perhaps the most common complaint in student residences, particularly since walls are generally not soundproofed, is that someone else is making too much noise. Consider the following two items, which were included on a residence hall environmental assessment form at a university that provides on-campus housing for approximately 10,000 students:

1. I consider noise to be a major problem in my residence hall.
2. To be honest, I have to admit that I contribute my fair share to the noise in my residence hall.

Most students agreed or strongly agreed with the first statement, but almost none agreed or strongly agreed with the second statement. Because what is "music" to one person's ears is "noise" to someone else, students do not acknowledge their own contributions to noise levels perceived as excessive within the living community. When alcohol is involved, noise levels increase, tempers are more likely to flare, and inappropriate responses (including violence) tend to predominate. Consider the following incidents described by RAs in Palmer's 1996 study:

- Two residents had been drinking somewhere off campus. When they came into the building, they got documented for being noisy. One blamed the other and they got into a fist fight. The police were called in to break it up. There was one broken wrist, a bloody nose, and many bruises.
- Two drunken residents were involved in a conflict that led to a fight and resulted in injuries—just because of noise!
- Two students had an altercation over noise. At first they just exchanged words. Then the noisy resident punched the complainant in the mouth—in front of several witnesses.
- A student beat up two guys who lived on my floor because he thought they told on him for being loud.
- Harassing phone calls (5 am, five times in a row type of thing) from the guys upstairs to a particular room on my floor because the residents had the guts to complain about noise.
- I approached a noise violation in a room at 2:00 am. The resident and her boyfriend were totally smashed. I asked him to leave and he got extremely upset and threatened violence. I called security for backup. As they escorted the man out, he shoved me up against the wall and called me a _____.
- Residents were engaged in a heated verbal argument over noise and were about to physically harm each other. When I tried to intervene, one of them got all up in my face threatening me, but I stood my ground and talked calmly, and he finally settled down.
- A student threatened to hit me because I asked him and his guests to keep quiet at 3 AM.
- While enforcing quiet hours a student called me a ___ and threatened to kill me in the hallway.
- After a music complaint that I had handled, a resident phoned me and said that the music had not ceased and that he would kick my _____ if it didn't.

Acceptance of Violence as a Conflict-Resolution Strategy

Noise problems may be compounded by the fact that many students underutilize common areas (such as study lounges, television rooms, and recreation facilities) located in their campus residences and try instead to use their rooms for quiet activities (such as studying and sleeping) as well as noisy activities (such as playing music, watching television, and socializing with friends). However, disputes concerning noise and many other problems may result in violence, vandalism, and other destructive or disruptive behaviors primarily because many of today's students have not developed effective skills in using words to settle arguments or solve conflicts.

Students who enter college during the next few years may experience even greater difficulty in using positive means to address what they believe are negative situations. The March 27, 1995, edition of the *Synfax Weekly Report* summarized the results of selected national studies of attitudes and behaviors related to violence as a problem-solving technique used by many students at the elementary and secondary school levels. The report concluded: "It's tempting to see reports of school violence as exaggerated, or inapplicable to the college setting. That's probably a mistake. We can't dismiss out of hand the growing number of independent surveys with consistent data about a spreading culture of violence, . . . especially as we try to expand educational opportunities for larger numbers of students" (p. 343).

Transitions from Homogeneous to Diverse Communities

As the student population becomes not only larger, but also more diverse, housing officers must meet the challenge of helping students who differ from one another. Housing officers must encourage students in any number of ways to live together in harmony, with civility, and with an appreciation for the very diversity that may otherwise foster intolerance and yield intolerant behaviors. Amada (1994) noted that many students have spent most of their precollege lives in relatively homogeneous communities and that, in their college residences, they are expected for the first time to live with those of different races, religions, nationalities, sexual orientations, social and economic backgrounds, lifestyles, value systems, and sensibilities. Indeed, students who live on campus often must live with strangers they might not choose as their closest neighbors or roommates.

Prejudice and bigotry often result in the victimization of "different" others. In a study conducted by Palmer (1993), housing officers at 49 institutions (28 public and 21 private) located in 30 states indicated that a total of 1,078 "officially reported" incidents had victimized racial/ethnic minority, gay/lesbian, and Jewish students in their residence halls during the previous 2 years. These included 210 incidents of violence (against persons), 426 incidents of vandalism (affecting property), and 442 incidents of verbal harassment (face-to-face, by telephone, by computer, in written notes, and so on). Descriptions of the "most common" and "most serious" incidents included examples of attacks, beatings, and other physical assaults; vandalism of cars, clothing, and other personal property; racist, homophobic, and anti-Semitic graffiti; demeaning, intimidating, and threatening messages (often anonymous); racial slurs resulting in fist fights and a "near riot"; and continuous harassment described as taunting and tormenting behaviors that ultimately forced victimized students to leave residence halls or withdraw from college.

SPECIAL PROBLEMS IN FAMILY HOUSING AND GREEK HOUSING

Individual institutions undoubtedly keep records of violent and other incidents reported in family housing units (on-campus apartments for students who are married and/or have children). However, there are no known studies that have synthesized this information across institutions. Personal communication with colleagues who supervise family housing units suggests that violence in these units differs from violence in residence halls in at least two important ways. First, children and adolescents who live with their student parents sometimes assault or fight with one another on playgrounds or in other common areas located within the family housing complex. Indeed, policies regarding the behaviors of spouses, children, and other nonstudent residents and their guests must be established and enforced. Second, family housing staff must address concerns related to domestic violence, particularly among international students whose home cultures are more accepting of physical restraint and battering as strategies for controlling or disciplining one's spouse (usually one's wife) and children.

Violence in on-campus fraternity and sorority houses (or portions of residence halls reserved for members of Greek letter organizations) may be similar in many ways to the violence that occurs in residence halls. However, it should be emphasized that most fraternity and sorority houses are located off campus and are not owned, operated, or staffed by the colleges and universities with which they are associated. Consequently, one must use caution in generalizing research on fraternities and sororities to those that actually have their houses on campus. For example, hazing incidents may be more common in off-campus houses than in residential units where institutional staff are responsible for monitoring fraternity and sorority activities.

Because students choose to join—and are chosen by the members of—specific fraternities and sororities, their groups are generally rather homogeneous in terms of race, religion, socioeconomic status, and other demographic characteristics. Although living only with very similar others during the college years may not be conducive to reducing prejudice, developing multicultural appreciation, and learning to work effectively within diverse groups, within-group violence related to cultural differences may be minimal in Greek housing.

Because most campus violence is alcohol-related, it should be noted that Wechsler (1996) reported that "the single strongest predictor of binge drinking was found to be fraternity or sorority residence or membership" (p. 22). Binge drinkers included 62% of sorority members, 35% of other women, 75% of fraternity members, and 45% of other men. Also noteworthy was the finding that 80% of the women and 86% of the men who actually lived in sorority and fraternity houses, respectively, were binge drinkers. Neuberger and Hanson (1997) cited literature concerning "the troubling association between Greek affiliation and alcohol abuse . . . [and the] contention that Greek affiliation reinforces traditional male-dominant, female submissive attitudes" (p. 92).

These factors may be related to "widespread knowledge that fraternity members are frequently involved in the sexual assaults of women" (Martin & Hummer, 1989, p. 457). However, Koss and Cleveland (1996) emphasize that sexual aggression and coercive sexuality among fraternity members is commonly associated with alcohol, particularly when large quantities of alcohol are available and binge drinking is encouraged at parties and other events that are unsupervised. If one assumes that activities are better supervised in fraternity houses that are located on campus, one might also speculate that the prevalence of binge drinking and sexual assault in on-campus fraternity houses may be lower than in off-campus fraternity houses.

THE MOST SERIOUS AND MOST COMMON FORMS OF VIOLENCE

Violent Deaths and Weapons

Serious violent incidents occur within student residences. Indeed, some students and staff have lost their lives via homicide or suicide, along with arson or pranks that unintentionally led to someone's death. *Campuses Respond to Violent Tragedy* (Siegel, 1994) and "Homicide in the University Residence Halls: One Counseling Center's Response" (Waldo, Harman, & O'Malley, 1993) provide detailed accounts of such incidents and institutional responses to them.

Concerns regarding increasing access to weapons in our society and on our campuses are particularly acute in residential facilities, in part because students who have weapons, or objects they might use as weapons, are most likely to keep them where they live. Numbers and percentages of students who possess weapons on their campuses are, and will most likely remain, unknown. However, of the 287 "most serious" incidents described by RAs who participated in Palmer's (1996) study of violence and other forms of victimization in residence halls, 20 involved guns (including pellet guns and BB guns), 5 involved knives or razor blades, and 6 involved bats or other objects used to strike people or property. Objects thrown at people ranged from a cosmetic case to a beer bottle, and objects thrown at property (most often staff cars or the doors or windows of staff rooms) ranged from spaghetti to a cement block. The most serious incidents also included five fires (arsons), one smoke bomb, and one case involving the release of pepper gas (p. 273).

"Hidden" Violence

Palmer (1996) reported that 374 RAs from 12 institutions identified a total of 5,472 incidents that had victimized them and students on their floors during their tenure as RAs (average 2.0 semesters). This yields an average of 14.6 incidents per RA per academic year. More specifically, these incidents included 775 acts of violence (2.1 per RA), 1,881 cases of vandalism (5.0 per RA), and 2,816 incidents of verbal harassment (7.5 per RA). Of the violent incidents RAs identified in this study, only 61.0% had been officially reported to their supervisors and 34.3% had been reported to police.

Of the violent incidents described as "most serious" (in cases where genders were identified), 90% of the perpetrators were men, whereas 75% of the victims were women. This finding supports Whitaker and Pollard's (1993) identification of male college students as "the most frequent immediate perpetrators of campus violence" (p. xiii) and female college students as "the principal victims of campus violence currently" (p. xi).

In Palmer's (1996) study, the most serious incidents of men's violence against women generally took the form of sexual violence (such as rape) or courtship violence (such as battery), involved perpetrators and victims who knew each other, occurred in privacy "behind closed doors," and were not reported to institutional officials or to police. Indeed, much of the relationship violence that occurs at home in student residences may be described as a premarital version of the domestic violence that occurs at home in our larger society. Because most of this violence remains "hidden" from law enforcement personnel, it is not reflected in official crime statistics.

ADDRESSING THE NONREPORTING OF VIOLENT CRIMES

One of the more tragic realities associated with much of the violence that occurs in student residences is that professionals who are in positions to address the problem are not given an opportunity to do so. Because of the reluctance on the part of victims, witnesses, student staff, and others to report relationship violence, housing officers, police officers, judicial officers, counselors, health care professionals, senior administrators, and others are not able to identify and then assist those involved. In the absence of accurate and complete information regarding the extent and nature of problems, it is difficult to plan and implement effective prevention, intervention, and follow-up programs.

Reasons for Nonreporting

Why are so many violent incidents not reported? Many of the reasons are associated with the common signs and symptoms of *primary victimization* (experienced by directly targeted victims) and *secondary victimization* (experienced by witnesses, friends, staff, and caring others who are aware of incidents) (Roark, 1993). Reactions to victimization include shock, fear, outrage, confusion, disorientation, self-blame, rationalization, cynicism, helplessness, and despair. Some victims may feel betrayed by and yet remain loyal to the very people who harmed them and even make excuses for the perpetrators. For example, "I know he didn't really mean to hurt me; he was just drunk and out of control," and "It's my own fault [that he hit me and broke my jaw]; I shouldn't have argued with him and made him mad" are quotations from student victims of relationship violence.

In focus group interviews of residence hall staff at 10 institutions, Palmer (1996) found that the two most common reasons for not reporting violence were fear of retaliation and lack of faith in "the system" (referring to both the criminal justice system and the institutional discipline system). Following are examples of comments pertaining to these reasons:

- A resident grabbed me by the neck after disciplinary paperwork was delivered. Upset over the outcome.
- A male resident told on another male resident, and the second resident beat the other resident badly.
- A guy on my floor was slapping his girlfriend around, and when I tried to break it up, he threatened to beat my_____. Truthfully, it scared the _____ out of me.
- One of my students was raped. Friends took her to the hospital, but she didn't want the police involved. She knew that even if the rapist was arrested, he'd be out on bail until his case went to court, and she was afraid of what he would do to her if she talked to the police.
- Maybe the police can put him in jail tonight, but I know he'll be back tomorrow and then he'll be even more _____ at me because I got police involved.
- He'll just deny everything and it will come down to my word against his. I have no real proof, no evidence, no witnesses, so what do you expect me to do? Even in serious cases, I tend to believe there's no way the police can do anything about it anyway.
- I think the administration does care about what's happening, but there's really nothing they can do about this stuff either.

Encouraging (versus Forcing) Reporting

As this book is being written, many college and university administrators are challenging proposed federal legislation (H.R. 715) entitled the "Accuracy in Campus Crime Reporting Act of 1997" (ACCRA). According to the March 17, 1997, *Synfax Weekly Report,* this law, as currently proposed, would require not only campus housing officials, but also counselors, disciplinary officers, and others to report crimes that come to their attention, even when victims ask that the information be held in confidence. In addition, the law would open disciplinary records and proceedings alleging criminal misconduct to the media and to public inspection. Although the sponsors of this bill may indeed have positive intentions, the bill's implementation could negatively affect hidden violence in student residences in two important ways.

1. Staff may be forced to choose between what the law requires and what their own ethical and moral standards of care for the victims of violence require. This may be particularly true in cases where victims of sexual assault want and need counseling, but would be victimized further if the intimate details of such assaults were described to the police, the media, and the public.
2. Victims of rape and other sexual offenses, courtship violence, and other forms of relationship violence might be forced even further into hiding. That is, even fewer victims may approach staff for counseling and other assistance if they fear that confidentiality will be jeopardized.

The outcome of the controversy regarding the proposed ACCRA may not be determined for some time. But even now, the two issues cited above create dilemmas for residential staff and students, as most staff are required to report any information pertaining to violations of laws and policies. Palmer (1996) emphasized that reporting duties and the consequences of nonreporting should be clarified for all staff. However, she also noted that addressing the issues contributing to nonreporting may be more effective than merely threatening to terminate employment for nonreporting. Suggested strategies include: "creating support networks for victims, witnesses, and RAs who report incidents; ensuring that response systems will not revictimize those who have already been harmed; treating all parties with sensitivity, compassion, and dignity; fostering positive working relationships among RAs, supervisors, counselors, campus police, and judicial officers; and discussing with RAs the ways in which reporting incidents can help those involved and can prevent further incidents" (pp. 276–277).

VIOLENCE REDUCTION MEASURES: SIGNS OF SUCCESS

Grimm (1996), in summarizing the Federal Bureau of Investigation's (FBI's) Campus Crime Statistics, noted declines in reports of murder, rape (sex offense), aggravated assault, armed robbery, arson, burglary, larceny-theft, and motor-vehicle theft between 1992 and 1994. Of course, campus crimes reported to the FBI do not include crimes that victims, witnesses, and others did not report to institutional or law enforcement officials in the first place. However, it is conceivable that—as a result of legislative, political, professional, and media attention given to campus crime in the late 1980s and early 1990s—administrators have taken actions that have been effective in reducing the amount of criminal activity occurring on their campuses. The following represent examples of the many measures that have been taken by student housing officers:

- 24-hour desk service
- full-time security guards and student night patrols
- entrance door card access
- residence hall education and awareness programs
- safety and security audits
- campuswide safety weeks
- crime report bulletin boards
- residence hall neighborhood watch
- safety office operation ID
- date rape awareness programs
- self-defense classes
- vegetation surveying and trimming
- lighting surveys
- building signage
- student room door peepholes
- blue light phone systems
- safety and security workshops and seminars available through ACUHO-I (The Association of College and University Housing Officers—International) and other national and regional organizations (Grimm, 1996, p. 10)

Because colleges and universities vary greatly in terms of their institutional missions, housing systems, student populations, and so forth, no single strategy or set of strategies for addressing violence in student residences will be effective at all institutions. Rather, each institution is encouraged to involve students, staff, faculty, and others (as appropriate) in assessing needs, establishing goals, and developing a comprehensive plan to address violence "at home" on campus now and in the future.

REFERENCES

Amada, G. (1994). *Coping with the disruptive college student: A practical model.* Asheville, NC: College Administration Publications.

Ellison, H., O'Shaughnessey, M., & Palmer, C. (1991, April). *Sexual and other violence on campus: Developing an institutional response.* Paper presented at the annual conference of the National Association of Student Personnel Administrators, Washington, DC.

Grayson, P. (1993). Keeping their antennas up: Violence and the urban college student. In L. Whitaker & J. Pollard (Eds.), *Campus violence: Kinds, causes, and cures* (pp. 139–150). Binghamton, NY: Haworth Press.

Grimm, J. (1996). The campus crime rate: Some light at the end of the tunnel. *Journal of College and University Student Housing, 26*(1), 7–10.

Koss, M., & Cleveland, H. (1996). Athletic participation, fraternity membership, and date rape. *Violence Against Women, 2*(2), 180–190.

Martin, P., & Hummer, R. (1989). Fraternities and rape on campus. *Gender & Society, 3*(4), 457–473.

Neuberger, C., & Hanson, G. (1997). The Greek life self-study: A powerful process for change on campus. *NASPA Journal, 34*(2), 91–100.

Palmer, C. (1993). *Violent crimes and other forms of victimization in residence halls.* Asheville, NC: College Administration Publications.

Palmer, C. (1996). Violence and other forms of victimization in residence halls: Perspectives of resident assistants. *Journal of College Student Development, 37*(3), 268–278.

Parsonage, W., Barbiero, M., Bartoo, K., Febbraro, A., Hoffman, R., Ichter, L., Lawrence, T., Rivera, S., Sculimbrene, P., Terry, E., Traub, M., & Wernovsky, H. (1992). *The victimization of Pennsylvania State University resident assistants and coordinators in the line of duty: A research report.* State College, PA: The Pennsylvania State University, Department of Administration of Justice.

Rickgarn, R. (1989). Violence in residence halls: Campus domestic violence. In J. Sherrill & D. Siegel (Eds.), *Responding to violence on campus* (pp. 29–40) (New Directions for Student Services Sourcebook No. 47). San Francisco: Jossey-Bass.

Roark, M.L. (1993). Conceptualizing campus violence: Definitions, underlying factors, and effects. In L. Whitaker & J. Pollard (Eds.), *Campus violence: Kinds, causes, and cures* (pp. 1–27). Binghamton, NY: Haworth Press.

Sherrill, J., & Siegel, D. (1989). *Responding to violence on campus* (New Directions for Student Services Sourcebook No. 47). San Francisco: Jossey-Bass.

Siegel, D. (1994). *Campuses respond to violent tragedy.* Phoenix, AZ: Oryx Press.

Smith, M. (1989). The ancestry of campus violence. In J. Sherrill & D. Siegel (Eds.), *Responding to violence on campus* (pp. 5–15) (New Directions for Student Services Sourcebook No. 47). San Francisco: Jossey-Bass.

Synfax Weekly Report. (1995, March 27). Crofton, MD: Author.

Synfax Weekly Report. (1997, March 17). Crofton, MD: Author.

Waldo, M., Harman, M., & O'Malley, K. (1993). Homicide in the university residence halls: One counseling center's response. In L. Whitaker & J. Pollard (Eds.), *Campus violence: Kinds, causes, and cures* (pp. 273–284). Binghamton, NY: Haworth Press.

Wechsler, H. (1996). Alcohol and the American college campus: A report from the Harvard School of Public Health. *Change, 28*(4), 20–25, 60.

Whitaker, L., & Pollard, J. (Eds.). (1993). *Campus violence: Kinds, causes, and cures.* Binghamton, NY: Haworth Press.

CHAPTER 8

Reducing Racial and Ethnic Hate Crimes on Campus: The Need for Community

Robert H. Fenske and Leonard Gordon

INTRODUCTION

Why is a hate crime so deeply disturbing to society and so traumatic to the victim? One reason is that although the violence is directed impersonally—sometimes even randomly—it also has highly personal, long-lasting effects on the victim. By definition, a hate crime occurs when the perpetrator selects victims because of their membership in or affiliation with a hated group or class of people. The crime, therefore, threatens by association all members of the target group or class. The perpetrator's hate exists before the opportunity for the commission of the crime presents itself. In some instances the perpetrator may be acquainted with the victims, but more often victims are targeted solely because they represent a hated group, not because of their personal characteristics. When a hate crime occurs on a campus, it must be reported as a prejudicial act. Lewis and Greene (1997) report that "According to the Campus Security Act, postsecondary institutions are required to report statistics concerning the occurrence of certain criminal offenses that 'manifest evidence of prejudice based on race, religion, sexual orientation, or ethnicity, as prescribed by the Hate Crimes Statistics Act (28 V.C.C. 534)'" (p. 17).

Increased violence on college campuses is a function of multiple domestic and global social forces in an economically and socially changing American society. Within this context are conflicts involving interracial and cross-ethnic relations in collegiate settings. This chapter begins by examining the nature of hate crimes in our society and describing the laws enacted to control hate crimes. The chapter then discusses the incidence of hate crimes and racial tensions on college campuses, especially campuses that have experienced increasing racial and ethnic student diversity. Following this discussion, theoretical explanations of ethnoviolence are reviewed. Finally, the chapter explores community-building approaches aimed at improving conflict resolution and campus cultural diversity, while addressing crucial issues of political correctness and academic freedom. The conflicts are deep and consequential, and for an increasingly pluralistic society, ultimately unavoidable. Although conflict is inevitable, compromises and solutions can be sought that can benefit a society that is becoming more diverse with each passing year. Therefore, the challenge to higher education is to address racial and ethnic conflicts in such a manner as to advance the legitimate needs and interests of all students and in the process reduce the tensions that have led to increased violence.

THE NATURE OF HATE CRIMES

This chapter focuses on one type and locality of hate crime—physical and verbal abuse directed toward racial and ethnic minorities occurring on a college or university campus. Because the title of this book is *Violence on Campus,* hate crimes are discussed under the rubric of "campus ethnoviolence" as defined by Howard Ehrlich, a leading authority in the field. He suggests that campus ethnoviolence reflects relationships among groups in the larger society and is based on prejudice and discrimination (Ehrlich, 1995, pp. 3–4). Although ethnoviolence often is directed randomly at persons having membership in or affiliation with a group, the group itself is specifically differentiated and targeted apart from other groups. Typically, there is consensus or at least strong agreement among potential perpetrators "that the group is an acceptable target. For example, Jews and Muslims are acceptable targets; Methodists and Presbyterians are not. It is the relationship of groups in society that determines the likelihood, the content, and the intensity of prejudice towards a particular group" (p. 4).

Obviously, there are instances where persons who are members of a majority group can be targets of ethnoviolence. Protestants may find themselves threatened in a Catholic, Jewish, or Muslim neighborhood. Television viewers throughout the world watched in horror as Reginald Denny, a white truck driver, was dragged from his vehicle and beaten nearly to death by a black mob during the 1990 Los Angeles race riot. The riot had been ignited by the acquittal of the white policemen who had been filmed beating Rodney King, a black motorist—an event that previously had been viewed with equal horror by a worldwide television audience. On college campuses, minority ethnoviolence directed toward majority persons occurs much less often than ethnoviolence directed toward minorities. Nonetheless, one must not minimize the seriousness of such crimes or their deeply disturbing impact on majority victims.

As indicated earlier, prejudice and discrimination are the bases of ethnoviolence. According to Ehrlich (1995), "Prejudice is an attitude toward a category of people.... To say that a person is prejudiced against some group is to say that the person holds a set of beliefs about that group, has an emotional reaction to that group, and is motivated to behave in a certain way toward that group" (p. 3). Ehrlich describes discrimination as "the behavioral corollary of prejudice. It constitutes actions that deny equal treatment to a category of people. The result is to restrict the opportunities or social rewards available to others, while maintaining those opportunities and rewards for one's own membership group" (p. 4).

Hate crimes in general, regardless of whether they occur on campus, vary widely along two dimensions: (1) the way they are carried out and (2) their effects on victims. As Levin and McDevitt point out, "Many hate crimes consist of not only nasty talk or mischievous pranks, but also violent assaults and even murder" (1993, p. x). Although the hate and violence are usually impersonal, the victim is deeply and personally traumatized, not only by the pain of the verbal or physical assault but also because the victim was targeted for religious affiliation or an immutable attribute such as race or gender, rather than personal words or actions. Of course, victims of other crimes such as mugging are also generally innocent of provoking an attack. But a repetition of the attack can usually be avoided by extra precaution and prudent behavior. Such is not the case with hate crimes. In the case of race, a minority person, or a majority person in a hostile minority environment, is always a potential victim of the perpetrator of a hate crime. Aside from adopting a disguise or always avoiding a potentially threatening situation (obviously not always possible or foreseeable), the threat is ever present. Furthermore, the threat is generalized beyond the victim to all members of his or her group. In 1993, on the large main campus of Arizona State University, an Asian student was severely beaten by two men—one black and the other white. The attack was entirely unprovoked and was seen as an act of ethnoviolence. The entire campus community of Asians and Asian Americans reacted with expressions of concern and fear and with demands for greater protection (Hogan, 1993). According to Levin and McDevitt, "In effect, an attack inspired by bigotry says in unequivocal terms to each and every member of the victim's group that 'the same thing could happen to you'" (1993, p. x).

Acts of ethnoviolence are often situational and opportunistic. In 1989, a group of young black men in Kenosha, Wisconsin, viewed the film *Mississippi Burning* and, incensed by the film's depiction of injustice toward blacks, discussed "getting even" with whites (*Wisconsin v. Mitchell,* 1993). They saw a lone, young, white male across the street, surrounded him and beat him severely. Their viewing of and reaction to the film created the situation, and the appearance of a lone member of the target group presented the opportunity for an act of ethnoviolence. The fact that this particular hate crime was perpetrated on a majority victim by members of a racial minority group underscores the ubiquity of prejudice and discrimination as the basic root of hate crimes. It also underscores the "causal chain" effect of ethnoviolence. *Mississippi Burning* graphically depicts the notorious 1960s murder of three civil rights workers, one of whom was black and the other two Jewish, by white supremacists. The young black movie viewers very likely had been targets of prejudice and discrimination in their lives, and the movie triggered their act of ethnoviolence. It is not difficult to imagine the feelings of the victim of their brutal, unprovoked attack. Conventional views of human nature suggest that his victimization may very well predispose him, and perhaps his friends and relatives, to retaliatory ethnoviolence if and when the opportunity presents itself. This chain of actual and hypothetical events indicates that ethnoviolence is not only sequential, it is also multiplicative, involving a widening circle of people over time unless the process is checked and, if possible, reversed.

Levin and McDevitt (1993) identify three additional characteristics of hate crimes based on findings of recent research. "First, hate crimes tend to be *excessively brutal*. . . . The *hatred* in such crimes gets expressed when force is exercised beyond what may be necessary to subdue victims [emphasis in original]" (p. 11). These authors reported that, although only 7% of all criminal physical assaults require medical treatment, "many victims of hate crimes—fully 30 percent—wind up requiring treatment at a hospital because of the severity of their injuries" (p. 11).

A second characteristic of hate crimes is that they are frequently perpetrated at random on total strangers (p. 12). This characteristic is the basis for the deep societal fear of hate crimes that almost places them beyond reach of prediction and prevention. Ethnoviolence evokes a type of primeval fear akin to that of the violent, destructive forces in nature. Persons who have lived through a major earthquake may be left with a permanent morbid fear of being indoors; persons who have experienced a violent tornado or hurricane may fear the outdoors to some extent for the rest of their lives. Ethnoviolence can generate a similar irrational, pervasive fear. Levin and McDevitt cite the example of news reports of violence resulting from gangs disputing territory in which to deal drugs. The typical reaction might be one of little concern to someone not involved in using or dealing drugs and who furthermore resolves to avoid the area disputed by the gangs. In contrast, an unprovoked, violent hate crime in a public place elicits a basic fear that transcends the time and place of the crime itself. The reaction might be "if it happened to her there, it could happen to me or my loved ones any time, any place." That is why Levin and McDevitt point out that "unlike attacks motivated by greed, jealousy and lust, hate offenses can be regarded as acts of domestic terrorism" (p. ix). Ethnoviolent crimes committed on campuses have a similar chilling effect.

The third characteristic of hate crimes identified by Levin and McDevitt is that they are usually perpetrated by multiple offenders (p. 16). Based upon a review of research, these authors found that nationally "only about 25 percent of all crimes of violence are committed by more than one offender. By contrast, 64 percent of hate crimes reported to the police involve two or more perpetrators" (p. 16). They point out that group criminal activity provides safety in numbers, anonymity, and mutual psychological support that can escalate from internal group discussion to verbal and then physical attack on the victim (p. 17).

LEGAL BASES AND REPORTING REQUIREMENTS FOR HATE CRIMES

This section briefly describes the various federal, state, and campus laws and reporting requirements for hate crimes, all of which affect campus ethnoviolence. These and other legal matters are discussed in greater detail elsewhere in this book.

Several federal statutes deal directly with defining hate crimes and the penalties related to them. These laws were enacted beginning in the late 1960s to codify punishment of violations of the Civil Rights Act of 1964; they deal mainly with conspiracy to prevent or limit enjoyment of such civil rights as free choice of housing, schooling, and political activism (Levin & McDevitt, 1993, pp. 180–186). Other federal statutes relating to hate crimes deal with gathering data and reporting them to the federal government. These are discussed later in this chapter.

Nearly all states have some form of hate crime legislation. The earliest of these laws, dating back to the 19th century in the former slave states, proscribed the wearing of masks and hoods associated with racial terrorism. They were in direct response to the activities of the Ku Klux Klan and other racist organizations (National Victim Center, 1995, p. 1). Another wave of state legislation in the late 1960s and early 1970s mirrored the federal statutes of the same era.

According to the National Victim Center, "[Federal and state] anti-bias crime legislation serves three distinct purposes. First, such laws protect the right of individuals to be free from random violence against their person or property. Second, these laws serve to deter hate crime—to discourage people from acting on their biases—and to punish violent behavior. Finally, hate crimes statutes serve as a statement of abhorrence of all violence motivated by prejudice" (1995, p. 2). State statutes are basically of two types: (1) 17 states use existing laws and provide for enhanced penalties when a crime can also be proven to include bias as a motivation; and (2) 29 states have enacted laws directed at personal physical injury, property destruction, or intimidating threats motivated by a specific bias. These laws define prejudicial behavior as criminal, thus creating a special category of actionable crimes based on prejudicial motivation. The difficulty of determining motivation and concern for protecting First Amendment rights have limited the applicability of both types of laws. Where such offenses have been successfully prosecuted, the federal Hate Crimes Statistics Act of 1990 requires that they be reported in order to form a national statistical base to be developed by the Federal

Bureau of Investigation's (FBI's) Uniform Crime Reporting (UCR) program. Because of the difficulty in determining an offender's subjective motivation, UCR asked law enforcement agencies to report bias "only if the investigation revealed sufficient objective facts to lead a reasonable and prudent person to conclude that bias motivated, in whole or in part, the offenders' actions" (FBI, 1992, p. 2).

The problem of forcing all purported hate crimes into one format to be applied across all 50 states, more than 16,000 local enforcement agencies, and more than 7,000 postsecondary campuses has proven very difficult. Federal statutes, state laws, local ordinances, and campus security regulations can, and often do, conflict with one another in definition of crimes and penalties that may be imposed. State laws tend to have precedence in the area of hate crimes. As of 1992, 14 states required reporting of hate crimes by local enforcement agencies to a designated state agency. As might be expected, the 14 states have definitions of hate crimes that are not always consistent with one another and with the UCR definition (National Victim Center, 1995, pp. 2–3).

The legal and reporting basis for campus ethnoviolence rests mainly on Title II of Public Law 101-542, the Student Right-To-Know and Campus Security Act of 1990. This law is usually referred to as the Campus Security Act, and is sometimes called the "Clery Bill" in recognition of the events that precipitated its development and passage. Jeanne Clery, an 18-year-old freshman at Lehigh University, was brutally raped and murdered in April 1986 in her residence hall room. Her parents, appalled at the lax security and failure of the university to inform students about potential danger in the halls, resolved to attack these problems on a national basis. Their efforts culminated in the Campus Security Act. The law has been amended several times, notably in 1992 and in 1995, to strengthen reporting requirements and procedures. Problems relating to accurate, comprehensive reporting of campus crimes, including "hate crimes" to be reported pursuant to the Hate Crimes Statistics Act (28 U.S.C. 534) of 1990, have persisted. Congress attempted to codify and improve hate crime reporting through introduction of the Open Police Logs Bill of 1995. Hearings on the bill continued through 1996, and the bill was replaced early in 1997 by H.R. 715, the Accuracy in Campus Crime Reporting Act of 1997. The Senate followed suit in November 1997 by introducing S. 1493, the Campus Hate Crimes Right-To-Know Act. Both of these bills seek to remedy what is perceived as a massive underreporting of hate crimes and other crimes under the principal law relating to campus crime, the Campus Security Act of 1990. The protracted hearings on the Open Police Logs Bill revealed the frustration and concern of various constituencies on college and university campuses. Final regulations for the 1990 Campus Security Act were not approved until 1994 (Lively, 1996). Even then, the reporting requirements were seen as lax and resulting in substantial underreporting (Burd, 1996; Schmidt, 1996).

A basic and as yet unresolved problem is that such reporting can be seen as running counter to the provisions of the so-called Buckley Amendment (the Family Education Rights and Privacy Act, or FERPA) of 1974 as interpreted by colleges and universities. Campuses are caught in the dilemma of, on the one hand, protecting the privacy rights of students who are charged with crimes (including hate crimes) and, on the other hand, protecting the security and right to know of other students who may be potential victims of such crimes.

By the end of 1997, the Clinton administration began to figure prominently in the concern about what was perceived as a "rising tide" of hate crimes in the United States. President Clinton began speaking to the issue in a radio address on June 7, 1997, in which he referred to campus hate crimes and a renewed commitment by the U.S. Department of Justice to begin "a thorough review of the laws concerning hate crimes and the ways in which the federal government can make a difference to help us build a more vigorous plan of action" (Office of the Press Secretary, 1997a, p. 2). At the first-ever White House Conference on Hate Crimes on November 10, 1997, the president identified specific new federal initiatives. These included establishment of a national network of work groups of U.S. attorneys to "develop enforcement strategies, share best practices, and educate the public about hate crimes" (Office of the Press Secretary, 1997b, p. 2). He also stated that "over 50" additional FBI agents would be assigned to work on hate crimes enforcement, that "the National Crime Victimization Survey used by the Justice Department will finally include questions about hate crimes" and that "the Education and Justice Departments will distribute to every school district in the country a hate crimes resource guide . . . and also the Justice Department is launching a Web site where younger students can learn about prejudice and the harm it causes" (pp. 3–4).

INCIDENCE OF HATE CRIMES

The most recent national figures available were cited in a November 11, 1997, *Washington Post* article reporting on the White House Conference on Hate Crimes the previous day. "There were 8,759 hate crimes reported to the federal government in 1996, a 10 percent increase over the previous year, when 7,947 were reported. Federal officials say race was a factor in 63 percent of the reported hate crimes" (Fletcher, 1997, p. A8). The data were from the Criminal Justice Information Services of the FBI. Such data are reported to the FBI by legal authorities; however, participating local agencies only cover 75% of the population (FBI, 1996). According to the 1995 figures, the most common targets were blacks (2,988) and whites (1,226); and the most common crimes were intimidation (4,048), simple assault (1,796), and aggravated assault (1,268) (FBI, 1996).

There are contrasting views of the rise in incidence of hate crimes. On the one hand, the rapid rise in recent years can be correlated to the expansion of reporting resources. The 10% increase from 1995 to 1996 (from 7,947 to 8,759) can be compared to an even greater proportionate increase the previous year. The total of reported hate crimes in 1994 was 5,852, so the 1995 total of 7,947 represented a 30% increase. However, in 1994, only 7,200 local police agencies reported to the FBI. These agencies covered 58% of the U.S. population. In 1995 the 9,500 police agencies that reported covered 75% of the U.S. population (Sniffen, 1996).

Underreporting remains a problem as long as the coverage is significantly less than the total population. Furthermore, underreporting is inherently more likely than overreporting, because of the difficulty of determining the perpetrator's racial or ethnic bias. Such motivation is often difficult to determine from the mere facts of the incident, such as in cases of assault. It is not enough for the victim to claim that the attack was motivated by racial or ethnic hatred. The perpetrator must admit such motivation or it must be proven in a court of law. Because hate crimes carry greater penalties than the same crime absent this motivation, the avoidance of such charges and the difficulty of proving them suggest that many hate crimes go unreported. The FBI is keenly aware of the problem. "Federal officials warned that their statistics capture just a small portion of the problem, because they believe hate crimes to be dramatically underreported. Also, they said, police departments have widely varying methods of collecting hate crime statistics, further skewing their totals" (Fletcher, 1997, p. A8).

Hate crimes that occur on campus may be even more dramatically underreported because of the college's or university's concern about its image. Even one ugly incident of an attack on a student of color can seriously damage recruitment of minority students and affect sources of revenue such as tuition and private donations. For example, the proportion of campuses reporting hate crimes that meet the official federal definition given at the beginning of this chapter was never higher than 1% in 1992, 1993, and 1994 for colleges and universities in all 50 states and the District of Columbia. The criminal offenses covered in this definition are aggravated assault, forcible sex offenses, and murder. According to Lewis and Greene (1997), "The number of these crimes reported was also very small, ranging from 0 murders in all 3 years to 100 aggravated assaults in 1993" (p. 17). The suspicion is widespread that underreporting of campus crime is a serious problem. Palmer (1993) attempted to estimate the extent of underreporting by comparing statistics reported in conformance to the Campus Security Act with the number of incidents that actually occurred in campus housing at 28 public and 21 private colleges and universities in 1991–1992. The crimes covered violence, vandalism, and harassment (including hate speech) directed at five groups: social and ethnic minorities, gays and lesbians, Jewish students, women students, and resident assistants. According to Palmer (1993):

The respondents identified 5,526 "reported" incidents directed against members of those groups in residence halls during the previous two years, 1989–91. Based on housing officers' estimates of the number of incidents that *actually occurred* [emphasis in original] during that period, I found that the reported incidents represented only about 28 percent of the nearly 20,000 such incidents believed to have taken place. The results of this study (and many others) support the contention that the vast majority of crimes are never reported to college and university officials. (p. B1)

The May 8, 1998 edition of the *Chronicle of Higher Education* included a report of 1996 to 1997 crime data on 487 U.S. colleges and universities that enrolled more than 5,000 students (Fact File, 1998). An accompanying article commented on the paucity of reported hate crimes (Lively, 1998).

Are colleges and universities oases of harmony, without any hate crimes? Or do colleges just hate to report them? The annual campus-security reports submitted by colleges are supposed to tell whether any crimes were motivated by bias. But of the 487 colleges that sent reports for 1996 to the *Chronicle,* just over one-third listed the category of hate crimes. And only a handful of those colleges said they actually had a hate crime occur on campus (p. A57).

Ehrlich, in summarizing the results of a 1991 survey of campus crime, observed: "Who reports their victimization to campus police, student affairs offices, residence hall advisors—or any other school official? The answer is practically no one: Eighty to 94 percent of the student victims under study said that they made no report to any college official or to campus security" (Pincus & Ehrlich, 1994, p. 285). Ehrlich also presented data from the survey on the reasons why students did not report. In descending order of frequency, the students' reasons included: (1) the incident was not serious or important (Ehrlich observed that this response could be seen as a form of denial); (2) the authorities would do nothing; (3) the authorities could do nothing; and (4) they might be retaliated against (p. 285).

Smith and Fossey (1995) reported the results of a 1987 national study conducted by student affairs administrators at Towson State University that mirrored the study reported by Palmer. The investigators collected data on actual versus officially reported campus crimes from student affairs officers, campus security directors, and residence hall directors at 764 institutions. The respondents felt that underreporting was a significant problem in a variety of crimes. For example, they believed that only about one third of campus sexual assaults were reported. Among the three categories of respondents, residence hall directors felt that underreporting was much more of a problem than did campus security directors or student affairs officials (Smith & Fossey, 1995, pp. 13–14). Smith and Fossey attributed these differences as follows: "Residence directors surely are aware of many incidents that victims choose not to report for various reasons, among them being a sense of community among students that discourages 'tattling,' skepticism that the system can respond in a helpful way, unwillingness to devote time and energy to the reporting and prosecuting process, lack of awareness that a cognizable infraction has occurred, and fear of retaliation" (p. 14).

Howard Ehrlich, a leading researcher on ethnoviolence, has conducted and reviewed specific campus surveys of ethnoviolence. In 1987, the National Institute Against Prejudice and Violence headed by Ehrlich conducted an in-depth study of ethnoviolence on the campus of the University of Maryland-Baltimore County (UMBC). The investigators found that about 20% of minority students had been a victim of ethnoviolence on the campus. The attacks were mainly nonphysical, including verbal harassment. There were striking differences by race in awareness of such crimes. "Approximately 80 percent of the Black students were aware of the ethnoviolent events," but of the white students two thirds (65 percent) "were fully unaware of the ethnoviolent events that had occurred, being insulated not only from the events themselves, but presumably even from the coverage of these events in the campus and city news media. . . . These differences in perception may be as significant for the campus culture as the occurrence of violence" (Ehrlich, 1990, pp. 12–13).

In addition to the UMBC study, Ehrlich (1990) reviewed findings from studies done on campuses as diverse as Rutgers; the universities of Virginia, Colorado, and Massachusetts; Stanford University; the Massachusetts Institute of Technology; and St. Cloud State University in Minnesota. Overall, the results were similar—high rates of ethnoviolence against all minority groups, keen awareness of such incidents by minority students, and a contrasting lack of awareness by majority students. Ehrlich concluded, "There can be no doubt . . . that ethnoviolent behavior is a significant part of intergroup relations at American universities" (p. 15).

More recently, Ehrlich (1995) reviewed detailed studies of ethnoviolence victimization on several campuses. He found that the patterns of victimization followed familiar patterns of intergroup relations that varied by the relative status of racial/ethnic groups as well as by the proportion that each group represented on each campus. Among the four campuses studied (Rutgers, the State University of New York at Cortland, the University of Hawaii, and UMBC), the victimization rates varied widely, from 12% for Jewish students at Rutgers to 60% for Hispanic students at Cortland. White students encountered ethnoviolence at rates ranging from 5% to 15%, except at Hawaii where 18% of white students had been victims. Interestingly, the Hawaii campus had virtually no students enrolled from traditionally underrepresented minority groups (blacks, Hispanics, and American Indians).

However, the differential rates of victimization among Asian/Pacific Islanders followed patterns of relative social status and numerical proportionality on the campus. Only 10% of Japanese students, the largest and highest status group on campus, were victims compared to 17% of Chinese, 19% of Native Hawaiians, 36% of Filipinos, and 38% of Pacific Islanders (pp. 10–11).

Finally, Ehrlich provides an insight into the "multiplicative" effect of ethnoviolence referred to earlier through his concept of covictimization (Pincus & Ehrlich, 1994). This concept goes beyond mere heightened awareness of ethnoviolence by minority students compared to whites. He identifies covictims as those who "may have directly *witnessed* an ethnoviolent attack on another person, or they may have *heard* about it from the victim or from others [emphasis in original]. For covictims, attacks on peers are seen as threats not only to the entire group but also to their own personal well-being. Covictims, then, are all of those persons who have seen or heard about an ethnoviolent incident and are emotionally affected as a result" (Pincus & Ehrlich, 1994, p. 283). Ehrlich concludes that covictimization rates range from 35% to 61%, thus greatly expanding the numbers of racial and ethnic minority students directly or indirectly affected by ethnoviolence.

CONCEPTUAL AND THEORETICAL BASES OF ETHNOVIOLENCE

Scholars have studied the theoretical underpinnings of ethnoviolence from many perspectives over the past several decades, but few have examined these concepts in the specific setting of the campus of a college or university. This section touches briefly on general conceptual frameworks and comments on those that seem to be related most closely to the current concerns over campus ethnoviolence.

W.E.B. Dubois and Frederick Douglass sought answers to concerns about racism and ethnic differences as the United States approached the 20th century. It is discouraging, to say the least, that the problems and issues that concerned them persist, albeit in more subtle modes, as we now are about to enter the 21st century. In the 1920s, Robert E. Park, a prominent sociologist, studied ethnic relations from the point of view of assimilation of immigrants. He and other scholars of the era believed in the inevitability of assimilation of differing ethnicities into society. The stages they were expected to pass through included competition, with newer groups at a distinct disadvantage that would disappear over time as they became acculturated and attained a strategically effective measure of economic power. Park and others saw this process as extending also to nonwhite racial groups (Park & Burgess, 1924).

Many basic concepts have been developed and examined in the search for answers to racial and ethnic conflict. These include stereotypes, social distance, multiculturalism (as distinct from assimilation), and differentiating race and ethnicity. Ehrlich (1973), in a classic work on the determinants of prejudice, found that *stereotypes* are not innate, but are incubated by social and psychological processes. These processes can be random, or they can be systematically instilled by societal forces such as mass media for social, political, or economic purposes. The concept of *social distance* has two components: (1) the distance between groups as they are identified by stereotypical views; and (2) personal distance, which is the basic, individual component of social distance. Social distance between groups can be more than the additive value of each group member's personal distance because interactions among individual group members can increase distance by mutual reinforcement of stereotypes.

Multiculturalism is differentiated from assimilation in that it does not assume that the end result of a social process is a single culture that reduces or eliminates diversity. Multiculturalism aims at retaining ethnic identity, but with the goal of developing mutual respect for all other groups' language and culture. Advocates of multiculturalism believe that racial and ethnic social harmony is possible only through development of such respect and that attaining such respect will allow diverse groups to work collaboratively toward common social goals.

Blauner (1994) provides a conceptual basis for clearly differentiating race and ethnicity. He points out that "the peculiarly modern division of the world into a discrete number of hierarchically ranked races is a historic product of Western colonialism" (p. 24). From this perspective, *race* is a concept imposed on persons by society for instrumental purposes. "Thus race is an essentially political construct, one that translates our tendency to see people in terms of their color or other physical attributes into structures that make it likely that people will act for or against them on such a basis" (pp. 24–25). Blauner specifies that *ethnicity* is different and may not necessarily concur with racial definition. He states: "An ethnic group is a group that shares a belief in its common past" (p. 25). Thus, a member of a societally defined racial group such as American Indians may have little or no interest in a common cultural heritage, a heritage that society assumes binds all American Indians together as a race. But a subgroup of American Indians that defines itself by its common heritage, such as the Cherokee Nation, is an ethnic group that is much more concerned with its own cultural structure than it is with commonalities across all tribes, commonalities that society assumes exist by virtue of racial designation.

Confusion between race and ethnicity persists, even among many persons and organizations that seek to remedy problems of hate crimes and ethnoviolence. Such confusion is especially evident in the legislation and reporting requirements cited earlier. Ethnicity is sometimes a more stable basis for solidarity than race. Students of Mexican heritage or Cherokee students may more readily identify the ethnic "tie that binds" them rather than identifying common cause across all "brown" or "red" groups, much less across "all minorities" as a designation.

Persistent confusion about such basic concepts as race and ethnicity is not helpful in attempting to identify theories that could be most applicable to campus ethnoviolence. Feagin and Feagin (1994) provide a useful synopsis of theoretical perspectives in race and ethnic relations. Among the theories that they review, the group of "competition and power-conflict" theories are most relevant to college campuses. They are relevant especially as institutional budget constraints and competition for increasingly scarce scholarships and other gift aid create a "zero-sum" mentality on campus. Given similar financial constraints and scarcity, this mentality would be created on a campus absent racial and ethnic diversity. However, as campuses become more diverse, the presence of readily identifiable groups traditionally viewed to be lower in the social hierarchy create targets and scapegoats. When such readily identifiable groups are perceived as receiving preferential treatment in financial aid and academic support services—and, furthermore, these groups become organized and assertive—the potential for ethnoviolence becomes ominous.

In reviewing conceptual and theoretical frameworks relating to racial and ethnic conflict, the group of constructs that seem highly relevant and that hold significant promise for positive interventions and programs are the "modified or expanded contact hypotheses." These hypotheses are based on the early work of Allport (1954) but modify and expand his theory on the basis of empirical research. Allport proposed that the outcomes of contact between a majority "in-group" and a minority "out-group" would be positive if the contact situation allowed both groups equal status. This proposition assumed that reduction of social distance would result from contact between in- and out-groups that were accorded equal status, especially if there was a common "superordinate" goal.

Empirical testing of Allport's contact hypothesis failed to support this theory, but did lead to a modified or expanded contact hypothesis that has withstood empirical challenges. The subsets or components of the new hypothesis were reviewed by Geranios (1997), who identified five basic factors produced by the extensive empirical testing and theory modification. These are: (1) all members of participating groups have equal status; (2) the contact situation allows each individual of each group to know each member of the other group(s) as an individual; (3) the situation is structured to enable disconfirmation of negative stereotypes; (4) all participants must pursue mutual goals; and (5) the participants cannot be passive, they must cooperate actively (pp. 50–51). These factors, which are widely recognized and used by modern researchers and practitioners in racial and ethnic conflict reduction, were used explicitly in one of the intensive strategies described in detail at the end of this chapter.

RACIAL TENSIONS ON CAMPUS

Heightened tensions and open interracial conflicts have been observed on American college campuses. Between the 1960s and 1980s, there was a significant rise in hostile racial and ethnic stereotyping (Gordon, 1986; Karlins, Coffman, & Walters, 1969), which has continued. Several thousand incidents of racial conflict have been documented by the National Institute Against Prejudice and Violence on an ongoing basis, available for regular updating on the World Wide Web. As Howard Ehrlich, Director of the Institute, observed about this troubling phenomenon, this campus climate of overt hostile perceptions has resulted in "the kind of permission that people feel they have now to act out these racial hostilities openly" (Magner, 1989, p. A28).

Hate crimes that occur on a campus are especially disturbing because this environment should be among the very last places where such mindless violence exists. Is not a college or university campus a place where reasoned dialogue should prevail among persons from different backgrounds and of contrasting attitudes and viewpoints? That is a reasonable presumption, but another perspective maintains that the more liberal and "mixed" the campus, the more potential there is for racial and ethnic violence. For example, Marcus (1994) describes the controversy and bitter hatreds sparked by the invited speech expressing an exclusionary perspective on an urban campus known for its liberality.

> To many, it was ironic that this particular college was torn by intergroup conflict. It had a reputation for being committed in word and deed to diversity and multiculturalism, values which are reflected in the composition of the student body, faculty, staff, administration, and Board of Trustees, as well as in its academic and co-curricular program. Minority students are routinely elected to student government leadership. Yet some believe that it is on precisely such campuses that conflict is likely to occur because of a heterogeneous student body from homogeneous communities; the students' increased and, perhaps, new-found awareness, combined with impatience and frustration by some about the pace of change, no matter how swift or how slow; and the power contests among different groups, particularly on the faculty and staff (p. 226).

According to Hurtado, Dey, and Treviño (1994), who reviewed studies of the relationship between increasing diversity on campus and an overtly hostile racial climate, "The structural diversity of an institution is significantly related to students' perceptions of racial tensions and experiences of discrimination on campus" (p. 7). These authors also identify other factors that bear on institutional racial climate, such as institutional size and the urbanicity of the community in which it is located. They point out that, although increased diversity on campus may elevate the likelihood of "negative interaction that sometimes results from cross-culture contact" (p. 1), it also presents increased "opportunity for cross-race interactions at the institutional and individual level" (p. 25).

Hostile perceptions toward students of color or ethnic minorities are not a new phenomenon on college campuses. In a keynote address to the Council of Colleges of Arts and Sciences at a national meeting, Harvard Dean Henry Rosovsky challenged nostalgic thinking about a past mythological ideal collegiate life. He noted that the infatuation he has observed about the past lack of racial tensions was "for a good reason" (Rosovsky, 1990). As he observed, until this generation, campuses were largely devoid of racial and ethnic minority students. In contrast, there is now more ethnic and racial ethnic diversity on college campuses than ever before (Fact File, 1997).

This new collegiate racial and ethnic student diversity has generated a heightened sense of competition among white, male students, who had been predominant on college campuses throughout most of American history (Ballantine, 1997, pp. 243–249). The current ferment over affirmative action and race-based scholarships are symptomatic of "zero-sum" thinking. Legal challenges to preferential admission to professional schools and universities, and the U.S. Department of Education's stance against race-based scholarships define a potentially explosive campus climate. Levin and McDevitt (1993) observe: "Among students who attend college to pursue lofty goals and embrace the finest ideals of the mind, going to school with different kinds of people may be regarded as a benefit, even as a privilege. In an environment fraught with fierce competition, however, the presence of diversity can represent a *personal threat* [emphasis in original]" (p. 135). They cite a particularly nasty flyer seen on the campus of Brown University, one of the nation's finest institutions: "Once upon a time, Brown was a place where a white man could go to class without having to look at little black faces, or little yellow faces, or little brown faces, except when he went to take his meals. Things have been going downhill since the kitchen help moved into the classroom. Keep white supremacy alive!" (Hamil, 1990, p. 67). Reflecting in part future economic anxieties created during an era of global competition and corporate downsizing, interviews with a number of white, male students on another campus made evident that they tend to view American society as zero sum, in which if someone gains educationally and economically then someone has to lose (Gordon, 1991, pp. 240–241). This is in contrast to the traditional American "plus-sum" view of the future, in which if someone gained, that was believed to add to the productive opportunities for others as well (Williams, 1970, pp. 501–502).

Within this kind of competitive environment and its potential for intergroup conflict, an issue of high concern to students with hostile racial and ethnic attitudes is that of affirmative action. They tend to view affirmative action in terms of what sociologist Nathan Glazer (1988) refers to as the perception of "affirmative discrimination" toward them.

MOVEMENT TOWARD CAMPUS DIVERSITY

The movement toward racial and ethnic diversity on American college campuses is a phenomenon that is barely a generation old. When Wilbur Brookover and David Gottlieb published their widely used college text, *A Sociology of Education* (1964), the civil rights movement was in full process. In their chapter entitled "The Problem of Integration," they addressed the challenge of racial integration within the elementary and secondary schools. Little attention was given to African Americans, Asian Americans, American Indians, or American Hispanics as students in higher education at a time when so few were in public or private colleges.

College campus developments since the 1960s are reflected in Jeanne Ballantine's contemporary text, *The Sociology of Education,* published in 1997. In contrast to Brookover and Gottlieb's 1964 text, Ballantine focuses considerable attention on topics of higher education minority students, desegregation, and integration. The statistical profile of first-year college students in 1997 bears out Ballantine's attention to racial and ethnic diversity in higher education. As college enrollment has increased to the point where half or more of high school graduates go on to college, that enrollment now includes substantial racial and ethnic diversity. The entering 1997 student class included 9.7% African Americans, 5.5% Hispanic Americans, 4.3% Asian Americans, and 2.3% American Indians (Fact File, 1997). No longer is the typical state or private college or university more than 90% white and largely male.

The increase in interracial and ethnic conflict, including violence, has accompanied college campuses experiencing these large numbers of first-generation college students. With significant differences because of the greater acceptance of white Europeans than racial minorities in American society, there are many parallels to the assimilation model of eastern and southern European immigrants into urban centers early in the 20th century. In a classic model developed in the University of Chicago School of Sociology and articulated by Robert Park (1949), the pattern for Irish immigrants and later for Italian Catholics, Polish and Russian Jews, Greeks, and others who entered the United States early in the century was one of first contact, then economic and political competition, followed by overt conflict before there developed intergroup accommodation and assimilation, including higher education assimilation.

In many respects, interracial and cross-ethnic contact in colleges has developed into the competitive and conflict stages of intergroup relations. The current mostly positive intergroup relations between those early-20th-century European immigrant group members has tended to obfuscate the intense conflicts that once characterized relations. As Frederic Thrasher's study (1927) of several hundred ethnically based teen gangs in Chicago in the 1920s documented, intense and sometimes violent conflict occurred between long-resident whites and the newer immigrant white ethnic group members.

The basis for the more intense and longer historically based racial conflict that is now reflected on college campuses was presented in Gunnar Myrdals' detailed, Carnegie Foundation–sponsored, historical sociological analysis, *An American Dilemma* (1944). As Myrdal and his research team observed, the dilemma is the contradiction between the American creed that "all men are created equal" and the practices of first racial slavery and, until this generation, *de jure* and *de facto* racial segregation.

THE CAMPUS ENVIRONMENT TEAM APPROACH

A large number of studies have indicated that no single approach in programming will reduce racial and ethnic conflicts in educational institutions (King, 1986). Rather, successful racial and ethnic conflict resolution and the advancement of a productive environment in educational contexts has been related to the cumulative effects of a clear policy and multiple operational programs. The first prerequisite is an explicit, clear institutional policy on advancing equal educational opportunity. Next, a variety of programs are needed that address the policy in overt and specific ways to change the campus environment into one in which all students can attain the highest levels of achievement of which they are capable.

The rising number of incidents of interracial conflict in the competitive environment of the American campus has precipitated many new policy and program initiatives. The initial response was an expulsionary one to eliminate students as well as faculty who engaged in offensive conduct (Calleros, 1992). The punitive approach, particularly on hostile interracial expressions, raised serious concerns about First Amendment rights and academic freedom. This approach also raised the prospect of using antiharassment codes to penalize African Americans, Hispanics, and others to whom racial and ethnic hostility was most often directed. As a reporter for the *New York Times* noted: "Aside from the inherent problem of defining [verbal] harassment, vilification and intolerance—is the racial hatred preached by [such African-American leaders as] Louis Farrakhan and Angela Davis any worse than that of skin-head neo-Nazis?—the proposed limits on freedom of expression, no matter how well meaning, should raise caution flags" (Barringer, 1989, p. A11).

As hostile racial stereotypes increased (Gordon, 1986; 1991), more institutions developed antiharassment speech codes. The American Civil Liberties Union began to challenge the constitutionality of such codes. The codes at the universities of Michigan and Wisconsin were found to be unconstitutional, which inhibited the use of such codes to advance viable cultural and racial diversity on college campuses (Calleros, 1992).

OUT OF CAMPUS RACIAL CONFLICT: FORMULATING A VIABLE RACIAL AND ETHNIC DIVERSITY POLICY

The challenge for colleges and universities has been to establish antiharassment policies, accompanied by positive intergroup programming, that would meet First Amendment requirements and university academic freedom prerequisites. Many such policies and programs have been developed as a result of interracial conflicts on campuses (Ehrlich, 1992). Major new policies and racial diversity programs were developed after open conflicts at private and public universities across the country including Harvard, Pennsylvania State, Michigan, Wisconsin, Texas, Texas A&M, Stanford, and the University of California-Los Angeles.

Racial conflict occurred at Arizona State University (ASU), the fourth largest university in the United States with more than 49,000 students. The conflict, the resulting campus protests, the negotiations with campus administrators and faculty, and the development of new policies and programs that are still being tested all occurred at ASU as at hundreds of other campuses.

The precipitating conflict—in the context of rising hostile campus racial stereotypes—occurred at ASU on a warm, spring Saturday evening in 1989. Several members of the all-white Sigma Alpha Epsilon fraternity mistakenly identified two passing black students, with a white student friend, as those who had been involved in a verbal altercation between one of the fraternity members and a white female student a few days earlier (Lacey, 1989; Pitzl, 1989). After misidentifying the passing black students, several fraternity members shouted racial epithets, including "nigger" and "porch monkeys," and some spat at them. Shortly afterwards on this partying Saturday evening, more than 500 fraternity members and friends were in the streets. Campus police arrived and arrested the two black students, reportedly handcuffing them and macing one (Lacey, 1989). When it became clear to the police that the black students were the recipients rather than the perpetrators of the violent confrontation, the students were released at another location on the university campus. In a subsequent investigation, 11 white students were cited for disorderly conduct misdemeanor offenses.

This incident precipitated black-led campus protests and the emergence of the Black Student Union (BSU) and an allied Students Against Racism (SAR) composed of black, Hispanic, and Anglo students. Within a week, planned protest activities emerged on campus including a protest march of several hundred students and a sit-in that blocked the main entrance to the student union. Rather than call in campus police or other police reinforcements, the university president and chair of the board of regents, with faculty senate support, met with the student leaders of the BSU and SAR (Pitzl, 1989).

The anger of the black students and their student allies over both the particular riotous event and cumulative rising tensions on campus resulted in initial student demands to adopt the kind of antiharassment codes developed at the universities of Michigan and Wisconsin (Burgess, 1989). The new campus policies were to be developed by a Campus Environment Team (CET) of presidentially appointed faculty and students with the charge to: "(1) work with other persons and organizations on campus to promote a campus environment that values diversity and provides respect for all individuals regardless of their status, and (2) protect free speech and academic freedom" (*Campus Environment Team,* 1991–1992, p. 29).

From the inception of the CET at ASU, restrictive speech provisions similar to those adopted at Michigan and Wisconsin, as well as many other developing university policies, appeared to breach First Amendment rights and the provisions of the American Association of University Professors 1940 Statement on Academic Freedom. The Michigan and Wisconsin provisions were subsequently struck down by federal courts (Calleros, 1995). Based on the perceived need to couple the advancement of racial and cultural diversity policies with the protection of free expression on campus, ASU aimed to offer a policy that would not be successfully challenged. The policy included the following language: "ASU is also strongly committed to academic freedom and free speech. Respect for these rights requires that it tolerate expressions of opinions that differ from its own or that it may find abhorrent" (*Campus Environment Team,* 1991–1992, p. 27).

It was on this basis that ASU established a diversity policy that would be viable on a long-term basis. As it turned out, the America Civil Liberties Union, which had successfully challenged the Michigan and Wisconsin antiharassment codes, cited the ASU policy as being constitutional and did not challenge it (Calleros, 1995). This strategy of combining concerns for cultural diversity with concerns for academic freedom at ASU drew on the sociological and social psychological work of Sherif and Sherif (1953) in their experiments in interracial cooperative efforts to reach common goals, and Lewis Coser (1956) in his theoretical, research-based construct that legitimate conflict, like cooperation, is one of the social processes by which people attempt to solve their problems.

Protection of open and free expression turned out to be important in bringing intergroup conflict within the legitimate norm of campus interaction. There are faculty and student forces on campus who are opposed to many policies and programs designed to address racial diversity issues. As at many other universities and colleges, ASU had a chapter of the National Association of Scholars (NAS), which strongly opposed antiharassment speech codes ("The Wrong Way to Reduce Campus Tension," 1991). By explicitly protecting everyone's right to express his or her views, the university was freed from the diversion to contend with the issues of free speech and academic freedom and to concentrate on advancing proposals and programs designed to advance a racially and ethnically inclusive university campus.

DEVELOPING DIVERSITY PROGRAMMING TO ACHIEVE AN INCLUSIVE CAMPUS COMMUNITY

With a viable cultural diversity policy in place, ASU was in a position to develop effective programming aimed at reducing racial and ethnic campus tensions and promoting an inclusive, productive campus community. As it had long been established that any single approach in reducing intergroup tensions and developing successful academic programs would not be successful alone, a complex variety of programming has taken place. The approaches included both formal academic programming and on-campus extracurricular activities and programs. Although the ASU CET played a role in both areas, it was primarily a stimulus and communication forum for students and faculty.

In the capacity of serving as a communication clearinghouse, the CET took a leading role in working with the university's academic senate in advancing a general studies cultural diversity course requirement, which had been a central concern of the BSU and the SAR during the sit-in protests. The opposition to this requirement proposal, advanced during the 1991–1992 academic year, was led by faculty and students associated with the NAS. The debate in the academic senate at ASU reflected basic issues raised on campuses nationally. The NAS senator and those aligned with him made two major arguments in opposition to a cultural diversity general studies degree requirement.

First was the argument that the cultural diversity requirement proposal was based upon political pressures from the black and other minority communities and their supporters. This was acknowledged and was real. The Phoenix NAACP among other community organizations placed intense pressure upon the university administration. What was pointed out in the debate was that when the university's general studies program was instituted in the 1985–1986 academic year, there was also intense political pressure from Motorola, IBM, and others in the business community to institute a university-wide mathematics and two natural science laboratory requirements. In the case of the math and science requirements, the political pressure helped to produce the requirements, but the academic senate controlled the academic criteria of what courses would be required. This same point was made in relation to the proposed cultural diversity requirement.

The second argument against the proposed cultural diversity requirement was that it abrogated and undermined the basic classical and traditional university pedagogy. It was noted in debate that the university curriculum was a dynamic, not a static one. In the social sciences, political science no longer advanced courses on American and Soviet foreign policy since the collapse of the Soviet Union. In the physical sciences, physics courses no longer concentrated on Newtonian principles but addressed astro- and nuclear physics. In mathematics, the emphasis on descriptive statistics had long been replaced by emphasis on inferential statistics. Similarly, it was argued that the increasing racial and ethnic diversity of the United States, as well as economic and political pluralism on a global basis, resulted in students' need to be more knowledgeable about such differences and their implications.

The diversity requirement was passed and implemented at ASU, as at over one third of all U.S. colleges and universities (Levine & Cureton, 1992). By the 1997–1998 academic year, ASU had instituted degree programs in African-American studies and Chicano studies and had developed courses throughout the university curriculum that addressed economic, historical, literary, and other dimensions of racial and ethnic diversity. Other campus environment changes occurred. Although most black students who lived on campus lived in general dormitories, ASU instituted an Umoja Hall that had an African-American theme and attracted mostly black students but was open to and attracted some others. While only a few dozen students concentrated in such a dormitory, the discussion groups and other programs attracted a wider group of students and became part of the university community as the more traditional fraternities and sororities had long been.

Although the NAS perspective argued that such developments would balkanize and further exacerbate campus racial tensions, the actual number of incidents went down even as the student body numbers went up. Racial and ethnic legitimation appears to be taking place on campus. The ASU student life office's listing of student coalitions and councils in many respects appears the way it did a half century ago with the listing of more than 20 traditional white-based national fraternity and sorority chapters. What is different is the additional listing of more than 20 African-American, Asian, Chicano/Hispanic, and American Indian fraternities, sororities, and associations. Although the Panhellenic and other traditional associations continue to advance programs extensively on campus, these are now accompanied by a variety of cultural diversity programs. When tensions do result in intergroup conflicts, there are now multiple forums to meet and address issues.

INTERGROUP DIALOGUES: VOICES OF DISCOVERY

One of the ASU CET activities was a grant competition supported by internal institutional funds. The competition called for grant proposals aimed at reducing campus social and ethnic tensions, and also for more positive approaches to fostering improved intergroup relations. One of the funded proposals outlined a program of intergroup dialogues to be called "Voices of Discovery."

Voices of Discovery has its theoretical foundation in the modified and expanded contact hypothesis described in an earlier section of this chapter. The program is specifically designed to meet the conditions indicated in the work of Miller and Edwards (1985). Voices of Discovery benefited from the experiences of an earlier, similar program at the University of Michigan. The grant proposal competition in the fall of 1995 resulted in funding for the program through the following academic year. The program was organizationally located in ASU's division of student affairs, beginning in the spring semester, 1996. After the 1996–1997 academic year, the program was accorded permanent status and funding, and was moved into the intergroup relations center in the office of the senior vice president and provost.

The center, and programs like the CET and Voices of Discovery, are programmatic expressions of the university's commitment to improvement of racial and ethnic relations on campus. This commitment is substantial in terms of budget allocations for staff and other resources. The university has also clearly demonstrated that it is committed to these efforts indefinitely, not just in reaction to a particular campus crisis. For example, Leadership 2000 has been an active center program since 1992 and is projected to continue indefinitely. This program includes an intensive 4-day retreat each semester in which students from different ethnicities interact in a "total immersion" mode aimed at thorough exploration of values and attitudes. Other center programs include (1) Faculty Initiatives, in which an academic department engages in structural dialogue about issues relevant to diversity in the classroom; (2) Becoming an Ally, in which students who have participated in Leadership 2000 and/or Voices of Discovery attend a workshop designed to develop proactive diversity initiatives; (3) "Not in Our House," a university-wide campaign to ensure a safe, welcoming community climate for all ethnicities; and (4) Social Justice Award, which acknowledges exemplary contributions by students, faculty, or staff members who have positively influenced social issues on campus.

Of these programs, Voices of Discovery has the greatest potential for influencing attitudes and behavior of undergraduate students of all ethnicities, including the white majority. Geranios (1997) conducted an evaluation of this program and concluded, "Multicultural courses in conjunction with a structured intergroup dialogue produce significant positive cognitive and affective outcomes and . . . a positive change in a student's propensity to interact with diverse groups" (p. iii). Voices of Discovery is a multifaceted, voluntary program requiring participants to (1) enroll in one or more university credit courses on diversity issues; (2) attend an intensive training workshop to acquire skills in active listening, conflict management, and group dynamics; and (3) participate in one of several dialogue groups that meet 2 hours per week for 6 weeks. Each dialogue group comprises two groups of students from different backgrounds (for example, Latino/white, Jewish/Christian, Native American/white, disabled/able-bodied, or gay-lesbian-bisexual/heterosexual).

The dialogues offer a challenging, but safe and controlled, opportunity to explore the attitudes and values of another ethnicity or orientation. The originator of the program, Jesús Treviño (1997), describes the process and outcomes as follows: "Intergroup dialogues can be characterized as face-to-face discussions which are open, honest, challenging, and reflective around issues that exist between the groups. Through interacting with each other, individuals undergo the process of intergroup discovery by getting to know one another both on a personal and group basis." (p. 3)

The theoretically based curriculum for the dialogue includes, in order: (1) group formation and bonding; (2) personal identity versus group identity; (3) ingroup-outgroup dynamics; and (4) closure, affirmation, and action. Trained staff members are essential to guide and control these potentially explosive racial and ethnic encounters. The dialogue session facilitators include graduate students, staff members, and peer (undergraduate) students—all trained in active listening, conflict management, group processes and dynamics, and social contact/identity theory prior to the dialogues. Outcomes defined for the Voices of Discovery program include enhanced active-listening skills, greater understanding of the issues and concerns of another diverse group, knowledge of the concepts of personal versus social identity, and a commitment to improve intergroup relations.

Combining a policy of advancing cultural diversity and academic freedom, the ASU policy and others like it appear to be effective in changing the collegiate environment. There are and will continue to be controversies that will lead to more racial and ethnic conflict. The metropolitan area and the state that provide the setting for ASU continue to experience increases in hate crimes, a trend that is true for the entire nation. On January 9, 1998, the front page of the local headline read: "Hate Crimes on Rise in Arizona." The story began next to a large color picture of a rabbi standing next to a synagogue wall on which two large swastikas had been spray-painted. The story reported the release of the FBI's 1996 figures on hate crimes (Vernon, 1998). The FBI reported 8,759 such crimes nationally, an increase of more than 800 over the previous year. This "rising tide" of hate crimes in the communities and campuses across the nation will need to be addressed on an ongoing basis. The policy to bring conflict within the legitimate framework of the campus has advanced the likelihood of sociologist Robert Park's (1949) classic thesis—progression from interracial contact, competition, and conflict toward accommodation and assimilation of students whose race and ethnicity have historically precluded them from equal higher educational opportunity.

REFERENCES

Allport, G.W. (1954). *The nature of prejudice.* Reading, MA: Addison-Wesley.

Ballantine, J.H. (1997). *The sociology of education.* Upper Saddle River, NJ: Prentice-Hall.

Barringer, F. (1989, April 25). Free speech and insults on campus. *New York Times*, p. A11.

Blauner, B. (1994). Talking past each other: Black and white languages of race. In F.L. Pincus & H.J. Ehrlich (Eds.), *Race and ethnic conflict: Contending views on prejudice, discrimination and ethnoviolence* (pp. 18–28). Boulder, CO: Westview Press.

Brookover, W.B., & Gottlieb, D. (1964). *A sociology of education.* New York: American.

Burd, S. (1996, July 5). Bill would toughen crime-reporting requirements. *The Chronicle of Higher Education,* p. A24.

Burgess, M. (1989, April 24). Officials to discuss plan against racism. *Arizona State University State Press,* pp.1, 6.

Calleros, C.R. (1992). Reconciliation of civil rights and civil liberties after RAV v. St. Paul: Antiharassment policies, free speech, multicultural education, and political correctness at Arizona State University. *University of Utah Law Review,* 1205–1333.

Calleros, C.R. (1995). Paternalism, counterspeech, and campus hate-speech codes: A reply to Delgado and Yun. *Arizona State University Law Journal,* 1249–1280.

Campus environment team: Policy statement supporting diversity and free speech. (1991–1992). Tempe, AZ: Arizona State University.

Coser, L. (1956). *The functions of social conflict.* Glencoe, IL: The Free Press.

Ehrlich, H.J. (1973). *The social psychology of prejudice.* New York: John Wiley & Sons.

Ehrlich, H.J. (1990). *Campus ethnoviolence and the policy options* (Institute Report No. 4). Baltimore, MD: National Institute Against Prejudice and Violence.

Ehrlich, H.J. (1992). *Annual Report of the National Institute Against Prejudice and Violence.* New Brunswick, NJ: National Institute Against Prejudice and Violence.

Ehrlich, H.J. (1995). *Prejudice and ethnoviolence on campus* (HEES Review, Volume 6, No. 2). Baltimore, MD: Higher Education Extension Service.

Fact File. (1997, June 17). This year's freshmen: A statistical profile. *Chronicle of Higher Education,* p. A42.

Fact File. (1998, May 8). Crime data from 487 U.S. colleges and universities. *Chronicle of Higher Education,* pp. 51–57.

Feagin, J.R., & Feagin, C.B. (1994). Theoretical perspectives in race and ethnic conflict. In F.L. Pincus & H.J. Ehrlich (Eds.), *Race and ethnic conflict: Contending views on prejudice, discrimination and ethnoviolence* (pp. 29–48). Boulder, CO: Westview Press.

Federal Bureau of Investigation. (1992). Crime data: The Hate Crimes Statistics Act. *FBI Law Enforcement Bulletin, 61*(5), 24–25.

Federal Bureau of Investigation. (1996, November). The Hate Crimes Statistics Act: 1995 Statistics. *Criminal Justice Information Services Report.* Washington, DC: FBI Natonal Press Office.

Fletcher, M.A. (1997, November 11). Conference explores rise in reports of hate crimes. *Washington Post,* p. A8.

Geranios, C.A. (1997). *Cognitive, affective, and behavioral outcomes of multicultural courses and intergroup dialogues in higher education.* Unpublished Doctoral Dissertation, Arizona State University, Tempe.

Glazer, N. *Affirmative discrimination.* (1988). Cambridge, MA: Harvard University Press.

Gordon, L. (1986). College student stereotypes of blacks and Jews on two campuses: Four studies spanning fifty years. *Sociology and Social Research, 70,* 201–202.

Gordon, L. (1991). Race relations and attitudes at Arizona State University. In P.G. Altbach & K. Lomotey (Eds.), *The racial crisis in American higher education* (pp. 233–248). New York: State University of New York Press.

Hamil, P. (1990, April). Black and white at Brown. *Esquire,* pp. 67–68.

Hogan, D. (1993, April 19). Police reconsider stance on assault case. *Arizona State University State Press,* pp. 1, 8.

Hurtado, S., Dey, E.L., & Treviño, J.G. (1994, April). *Exclusion or self-segregation? Interaction across racial/ethnic groups on college campuses.* Paper presented at the meeting of the American Educational Research Association, New Orleans, LA.

Karlins, M., Coffman T., & Walters, G. (1969). On the fading of social stereotypes: Studies in three generations of college students. *Journal of Personality and Social Psychology, 13,* 1–16.

King, E. (1986). Recent experimental strategies for prejudice reduction in American schools and classrooms. *Journal of Curriculum Studies, 18,* 331–338.

Lacey, M. (1989, May 3). Events of deadly relevance: An armed stand-off at ASU. *New Times,* 5–6.

Levin, J., & McDevitt, J. (1993). *Hate crimes: The rising tide of bigotry and bloodshed.* New York: Plenum Press.

Levine, A., & Cureton, J. (1992). The quiet revolution: Eleven facts about multiculturalism and curriculum. *Change, 24,* 25–27.

Lewis, L., & Greene, B. (1997). *Campus crime and security at postsecondary education institutions.* National Center for Education Statistics Report NCES 97-402. Washington, DC: U.S. Government Printing Office.

Lively, K. (1996, April 26). Education department starts monitoring campus crime reports. *The Chronicle of Higher Education,* p. A49.

Lively, K. (1998, May 8). Most colleges appear unaware of requirement that they track hate crimes. *The Chronicle of Higher Education,* p. A57.

Magner, D.K. (1989, April 26). Blacks and whites on the campuses: Behind the ugly racist incidents, student isolation and insensitivity. *The Chronicle of Higher Education,* pp. A28–32.

Marcus, L.R. (1994). Diversity and its discontents. *Review of Higher Education, 17,* 225–240.

Miller, N., & Edwards, K. (1985). Cooperative interaction in desegregated settings: A laboratory analogue. *Journal of Social Issues, 41,* 63–79.

Myrdal, G. (1944). *An American dilemma.* New York: Harper.

National Victim Center. (1995). Hate crimes legislation. Infolink [On-line], Available: http://www.nvc.org

Office of the Press Secretary. (1997a). Radio address by the president to the nation. The White House [On-line], Available: http:www.hrc.org.html

Office of the Press Secretary. (1997b). Remarks by the president at White House conference on hate crimes. The White House [On-line], Available: http:www.hrc.org.html

Palmer, C. (1993, April 20). Coping with campus crime. *The Chronicle of Higher Education,* p. B1.

Park, R.E., & Burgess, E.W. (1924). *Introduction to the science of society.* Chicago: University of Chicago Press.

Park, R. (1949). *Race and culture.* New York: Free Press of MacMillan.

Pincus, F.L., & Ehrlich, H.J. (1994). *Race and ethnic conflict: Contending views on prejudice, discrimination and ethnoviolence.* Boulder, CO: Westview Press.

Pitzl, M.J. (1989, April 22). ASU students hold sit-in, gain vow of officials to work on anti-bias goals. *Arizona Republic,* pp. 1, 7.

Rosovsky, H. (1990, November). *Academic issues of the 1990s.* Plenary address, Council of Colleges of Arts and Sciences Annual Meeting, New Orleans, LA.

Schmidt, P. (1996, June 14). Lawmakers told that colleges fail to comply with campus-crime law. *The Chronicle of Higher Education,* p. A37.

Sherif, M., & Sherif, C. (1953). *Groups in harmony and tension.* New York: Harper & Row.

Smith, M.C., & Fossey, R.W. (1995). *Crime on campus: Legal issues and campus administration.* Phoenix, AZ: Oryx Press.

Sniffen, M.J. (1996). Race sparks most crimes, FBI says. Available at: http://www.chron.com/content/chronicle/nation/96/11/05/hate.html; accessed May 1998.

Thrasher, F. (1927). *The gang.* Chicago: University of Chicago Press.

Treviño, J.G. (1997). *Overview of the Voices of Discovery program* (Intergroup Relations Center). Tempe, AZ: Arizona State University.

Vernon, A. (1998, January 9). Hate crimes on rise in Arizona. *The Tribune,* p. 1.

Williams, R.M. (1970). *American society: A sociological interpretation.* New York: Knopf.

Wisconsin v. Mitchell, 508 U.S. 476 (1993).

The wrong way to reduce campus tensions: A statement by the National Association of Scholars. (1991, April 24). *The Chronicle of Higher Education,* p. A15.

CHAPTER 9

Women and Violence on Campus

Kay Hartwell Hunnicutt

INTRODUCTION

On September 16, 1997, a 17-year-old female student in her room in a campus residence hall answered a knock at her door. It was 8:00 pm. After seeing two "college-aged" men through the door's peephole, she opened the door. While the men were supposedly asking for directions, one of them grabbed the female student by the throat and forced his way in the room with the other man following inside and closing the door. The student was slashed, raped, beaten, and choked (Anderson, 1997). Police eventually caught two suspects who were students at a nearby high school. This was the second rape on the campus of a female student in 2 days and the sixth reported rape in 2 months. One of the earlier victims was brutally murdered.

At another institution, a first-year female student had been invited to a fraternity party by a member of the fraternity. Later in the evening, she was unable to find her date to take her home. The female student was offered a ride to her home by an alumnus, a graduate attending homecoming events at the university. The alumnus offered her a ride if she would wait until later since he did not want to drive intoxicated. Several fraternity members, the female student, and the alumnus sat in a room on the third floor talking and drinking beer and peach schnapps. Early in the morning, everyone had left the room except the female student and the alumnus. The visiting alumnus then turned up the music and sexually assaulted her (*Mortz v. Johnson,* 1995).

In 1996, a female undergraduate student claimed that a male professor in the English department falsely imprisoned and sexually harassed her in his office while she was protesting a failing grade on an assignment. The student alleged that the professor discussed his prior sexual experiences, made vivid assumptions of how the female student would react to sexual intercourse with him, and directed her to read excerpts from pornographic poetry. In addition, the student claimed she was "impliedly [sic] invited to provide him with sexual favors in exchange for a better grade." A graduate student at this same university also complained of incidents involving the same professor including vulgar and sexually explicit reading material in his classes and conversations by the professor that included vulgar and sexually explicit comments when she went to his office to pick up an assignment (*Pallett v. Palma,* 1996).

No touching took place in either of the scenarios involving the professor, but such situations still produce fear—fear of safety and fear of retaliation and reprisal for not cooperating with the harasser. If the harasser is a faculty member, he or she has power and authority to determine grades, program admission, and often the selection of assistantships.

Rape and sexual assault including "date-rape" or "acquaintance rape," sexual harassment in a faculty office, hostile environment in the classroom, sex-related battery, and peer-to-peer sexual harassment are causing alarm and concern for women on campus.

Most families send young women to college believing that they will be provided a safe environment in which they can learn, explore, and become successful contributors to society. Life in a dormitory room surrounded by a large number of peers, the presence of campus security, and the availability of escort services after night classes or late trips to the computer center or library have fostered, in part, a sense of personal security. Reports citing few crimes against women on some university campuses may also foster a sense of security. Is the campus a safe haven for women? What is the extent of violence against women on campus? For example, do violent acts against women infrequently occur as shown by the published reports of one university in the East: 3 nonforcible sex offenses in 1996, 3 anonymous sexual assault reports in 1994 and 3 in 1996, 1 sexual harassment report in 1995 and 2 in 1996, and 10 total rapes between 1986 and 1996? What is being done to protect women from violence on campus? Why are the statistics on violence against women on campus ambiguous and conflicting? Why do relatively few women victims on campus report violent acts against them?

Because of the growing number of reported crimes regarding violence against women on campus, colleges and universities are becoming increasingly aware of the obligation to protect students and to provide education, policies, and procedures to safeguard individual rights and prevent litigation in the courts.

Federal laws designed, in part, to protect women from violence on campus have generally not been effective in safeguarding their rights in court. With reported increases in forcible-sex offenses (ranging from unwanted fondling to rape) and in nonforcible-sex offenses (consensual sex, statutory rape, and incest), universities have been quick to say that statistics are too imprecise, that colleges cannot be ranked on the basis of safety, and that too many variables can affect an institution's crime rate.

Although universities have developed policies, procedures, and codes of student conduct in dealing with acts of violence against women, perpetrators frequently have not been held accountable for their actions due to unconstitutionally vague and overly broad policies and lack of proper notice of the policies and lack of enforcement. College and university administrators, lawmakers, and the courts continue to face the challenge of coming to terms with the complexity and urgency of campus violence against women.

The purpose of this chapter is to provide an overview of the extent to which women experience violence on campus. Specific types of violence, including both actual and perceived harm, are discussed including rape, "date rape," campus assaults, battery, date aggression, and sexual harassment by faculty and peers. Recent data regarding prevalence and an overview of the legal response are provided.

DIMENSIONS OF RAPE ON CAMPUS

Neft and Levine, in *Where Women Stand: An International Report on the Status of Women in 140 Countries, 1997–1998* (1997), report that rape is thought to be the most underreported of all violent crimes in the United States with estimates ranging from 310,000 to 680,000 rapes and other sexual assaults on women every year.

It is essential to understand that rape has different dimensions to interpret the incidence of rape accurately. For example, if only "aggravated" rapes (rapes that involve more than one attacker, strangers, weapons, or beatings) were reported, then rape would be a relatively rare event. However, considering both aggravated and "simple" rapes (incidents in which a woman is forced to have sex without consent by only one man whom she knows and who refrains from beating her or attacking her with a weapon) reveals rape as a far more common problem, greatly underreported and seriously ignored (Estrich, 1987).

It is not difficult to understand why a significant number of rapes go unreported or are reported too late to bring criminal or civil action against the violators. Almost half of all rape cases are dismissed before trial; there is a perceived and documented lack of male understanding of the effects of rape on a woman; and 82% of female rape victims have been victimized by an acquaintance (53%), current or former boyfriend (16%), current or former spouse (10%), or other relative (3%) (Neft & Levine, 1997).

By the early 1990s rape had become acknowledged as the most common violent crime on U.S. college and university campuses (Neft & Levine, 1997). Several studies have been conducted to determine the extent of rape on college campuses. One of the first studies examining on-campus rape was conducted at Purdue University (Kanin, 1957). The researchers referred to rape as "a progressive pattern of exploitation" and victims as "offended girls." Kanin found that 20% to 25% of the college women surveyed reported being the victim of at least one forceful attempt of intercourse during a date. Little else was done in researching the extent of rape on campus until 1982, when Koss and Oros developed a behavioral self-report survey instrument that included questions about coercive sexual experiences. In 1985, Koss conducted the *Ms.* magazine Campus Project on Sexual Assault, surveying 6,159 students at 35 postsecondary institutions in the United States. The findings were terrifying to female students and faculty. Koss (1988) reported that one fourth of women in colleges and universities have been raped or were attempted rape victims and 90% knew their attackers.

Respondents in a 1987 Towson State University study felt sexual assaults were underreported and believed that only 33.6% of campus sexual assaults were reported to the police (Smith, 1988). Campus rape victims are often away from home for the first time in a new environment. They have convinced parents or relatives that they are in a safe environment and able to take care of themselves (Ehrhart & Sandler, 1985). However, after being victimized by sexual assault, many campus rape victims often feel a loss of trust in themselves or others and transfer to another school or drop out (Holyroyd, 1987).

DATE RAPE

"Date rape" has become a common term for "acquaintance rape" or "friendship rape"; this is unfortunate, because the term is misleading. Not all rapes by someone with whom the victim is acquainted occur on dates. Many occur in the victim's or the perpetrator's domicile during a visit, or elsewhere because of a chance meeting. These are crimes of opportunity, not necessarily a prearranged "date." Koss (1985) found that one fourth of females indicated they had been victims of rape or attempted rape, and that 1 in 15 male students admitted raping or trying to rape a female student during the preceding year of the survey. Women on campus are especially vulnerable to "date rape" because of the frequency with which they date and engage in voluntary social activities with men. Most of the women raped in the United States are between the ages of 18 and 34 (McGarrell & Flanagan, 1985). The greatest incidence of date rape or acquaintance rape is among women aged 15 to 21 years, with 18 being the average age at the time of the rape (Koss, 1985); male assailants are most likely to be less than 25 years old (Koss, 1988).

A review of cases litigated in the courts involving campus rape indicates that alcohol is frequently involved. Towson State University reported that alcohol rather than drugs accounted for 98% of infractions related to personal conduct of students. More than half of the 755 student violations at Virginia Tech stemmed from the use of alcohol. "Virginia's associate dean of students, Sybil R. Todd, who works with victims for date rape, said alcohol has been involved 'in every instance' of date rape on campus. 'I don't know (of any date rape that) involved hallucinogens,' she said" ("Demon rum," 1991).

Date rape or acquaintance rape is reported even less frequently than other forms of sexual assault because few persons identify it as a crime, punishable by law. An unknown number of assaults are not reported due to the nature of the event, the perceived stigma, and the belief that no purpose would be served.

Although acquaintance rape is not confined to the campus setting, many factors make college-aged women the population most vulnerable to rape, particularly those women entering college.

BATTERY AND DATE AGGRESSION

The campus community is embedded in the larger community and within society. A cursory review of crime rates of violence against women on campuses compared with women off campus would lead to the conclusion that female students are safer on campus. Reported crimes not resulting in rape, such as battery and date aggression, are causing concern for the safety and security of female students. Most acts of battery and date aggression go unreported. Those institutions that do report battery, assault, or date aggression report them in a cursory manner. For example, three assaults involving female students not resulting in rape, which were not reported by the Massachusetts Institute of Technology police (*Massachusetts Institute of Technology Crime Report,* 1996), are represented below:

- March 18, 1996: Simple assault—While a female was descending the stairs on a campus building a male grabbed her from behind around the waist, groped her, and fled the area when she screamed. She was unable to identify the suspect.
- August 2, 1996: Simple assault—A visiting student began speaking with an individual while having dinner at the student center. After dinner he walked her back to her dorm and began kissing her. She ran away and told a friend who called campus police. The victim was unable to identify him.
- September 2, 1996: Assault and battery—While a Wellesley College student was attending a party, she was approached by an unknown individual who began asking a few questions. A few minutes later he grabbed her by both arms and shoved her against a wall. Several individuals then demanded that he release her at which time he fled the area. The identity of the male could not be determined.

Battering is the single greatest cause of injury to women in the United States today; it occurs more frequently than car accidents, muggings, and rapes combined (Terpstra, 1996). Physical battering and emotional abuse, which are common among campus women, are frightening, traumatic, and often leave psychological as well as physical scars. One quarter to one third of high school and college students report involvement in dating violence—as recipients, perpetrators, or both (Bergman, 1992; DeMaris, 1987; Makepeace, 1981, 1987; Marshall & Rose, 1987, 1988, 1990; Riggs, O'Leary, & Breslin, 1990).

College women's attitudes in these and similar studies regarding justification for violence included defense against bodily harm and the perception that females are at greater risk of injury due to intimate violence than are men (Roscoe, 1985).

Foo and Margolin (1995), in a multivariate investigation of dating aggression involving 111 male and 179 female undergraduates, reported that, although the majority of both violent and nonviolent students rated dating violence as unacceptable in *most* circumstances, many saw it as acceptable or normal in *some* circumstances. Other studies have found similar results (Henton, Cate, Koval, Lloyd, & Christopher, 1983; Smith & Williams, 1992). Another study involving female college students reported that approximately 70% of the participants listed at least one form of dating violence—such as slapping, punching, and kicking—as acceptable, and more than 80% described dating situations in which physical force between partners was acceptable (Roscoe, 1985).

Makepeace (1986) found that females more than males perceived their dating violence as self-defensive. Dating violence on the part of men appeared to be associated with sensitivities to humiliation, rejection, jealousy, and public embarrassment (Dutton & Browning, 1988; Holtzworth-Munroe, 1992). When studying the personality characteristics of college men who committed dating aggression and violence, Ryan (1995) found that use of threats and verbal abuse were the best predictors of dating aggression, and Hydén (1995) reported that verbal argument was the precipitating event in 18 out of 20 couples interviewed in which the man had battered the woman. Women who are physically abused report a high frequency of psychological abuse in their relationships. In interviews with 234 women with some history of physical abuse (battering), 99% of the female respondents reported being the victims of psychological abuse from a partner (Follingstad, Rutledge, Berg, Hause, & Polek, 1990).

If the percentages reported in these and other studies are any indication of the extent of dating violence, battering, and psychological abuse of female students on campus, additional attention and support need to be made available to victims. It is hoped that appropriate classes, counseling, and education for all students will convince college students that dating violence is unacceptable.

SEXUAL HARASSMENT

Chapter 11 of this book, by Paludi and DeFour, focuses on sexual harassment of students, with an emphasis on harassment by faculty. The following section presents case studies on (1) peer-to-peer harassment and (2) legal and college regulatory aspects of harassment of female students by male faculty.

To a large extent it was Anita Hill's testimony on national television before the U.S. Senate Judiciary Committee in the Clarence Thomas hearings that was responsible for bringing the issue of sexual harassment in the workplace to the attention of the American public. The Navy's "Tailhook" scandal, recently reported harassment of female recruits and enlisted personnel in the Army, and the public allegations against former U.S. Senator Robert Packwood brought further attention to the issue. In addition, popular films, books, and theater representations such as *Disclosure, Oleana,* and *Red Corner* have increased public awareness of the problems of sexual harassment. Peer-to-peer student sexual harassment is a newcomer to the sexual harassment spotlight on college and university campuses, and there is increased litigation by female students against unwanted and offensive attention by their professors. Sexually offensive class assignments and sexually focused speech during class instruction have resulted in numerous court claims and investigations by the Office of Civil Rights (OCR), the enforcement agency of the U.S. Department of Education.

PEER SEXUAL HARASSMENT

Peer sexual harassment on college and university campuses has its roots in social behavior in previous educational settings. The issue of peer-to-peer sexual harassment is a serious problem in public elementary and secondary schools. According to a major national study by the American Association of University Women (AAUW) entitled *Hostile Hallways,* four out of five students attending public schools have been sexually harassed by a present or former student. About half the students experienced the harassment in the middle school/junior high years. Perhaps more surprising, fully one third of the students reported being harassed before the seventh grade. The survey, conducted by Louis Harris and Associates, found that five times as many girls (39%) as boys (8%) were afraid in school because of sexual harassment. Three times as many girls (43%) as boys (14%) felt less confident about themselves after the harassment (AAUW, 1993).

Only a few studies have examined peer-to-peer harassment at colleges and universities. One study found that 90% of undergraduate women report at least one negative experience from male students. The University of Minnesota reported that 52% of female graduate students and 42% of undergraduate students in one study experienced sex-stereotyped references, depictions, and jokes (Benson, 1992). Students most likely to harass female students are members of fraternities, athletic teams, and all-male groups and cliques; males who are defiant or angry at females; and those who have been abused themselves.

In a complaint filed by a female student at Occidental College in 1994, the OCR found that the college failed to provide female students with a prompt and equitable Title IX grievance process involving members of Alpha Tau Omega (ATO) fraternity. One of the complaints alleged that a group of men from ATO marched down a street of student residences chanting lyrics that described violence against women. The men then dropped their pants in front of a sorority house. A second complaint, filed by a campus advocate group called Advocates Working Against Sexual Harassment (AWASH), alleged that an internal newsletter published by ATO violated the college's sexual harassment policy and was part of a hostile learning environment at the college. The college, after being found in violation of Title IX, agreed to take steps to discourage public nudity, organized a forum on "gender, respect and community," revised its sexual harassment policy, trained staff and students on the policy, and advertised the procedures of reporting problems and complaints (OCR, 1994a).

In another OCR complaint, it was alleged that Foothills College failed to respond adequately to a student's complaint that she and other females enrolled in a step-aerobics class were sexually harassed by males who leered at, mimicked, and made derogatory statements to the women as they were exercising. The college did not investigate but agreed to lock the doors to onlookers during class hours, posted signs stating the doors should remain locked, and changed the class time to an earlier slot to avoid spectators. Articles in the campus newspaper were critical of the female complainant for filing a harassment complaint. The college did not defend the complainant and left her vulnerable to retaliation. The OCR found that the college's response was deficient (OCR, 1994b).

In another case of peer harassment, female crew members at Temple University alleged a sexually hostile environment created by coaches and members of a men's athletic team. The female athletes complained that the training room was decorated with lewd photos and that males made lewd comments and obscene gestures to them. An earlier incident in 1994 led to a $5,000 award to a female crew member who alleged a part-time assistant men's coach made lewd gestures to her ("Temple U's Crew Program," 1996).

Cases of sexual harassment on college campuses involving female students have been litigated under Title IX of the Education Amendments of 1972, which has been extended from lack of gender equity in athletic programs to the prohibition against sex discrimination involving educational institutions. Title IX states: "No person in the United States shall, on the basis of sex, be excluded from participation in, be denied the benefits of, or be subjected to discrimination under any education program or activity receiving federal financial assistance" (20 USC, 1681–1686, 1982).

SEXUAL HARASSMENT BY MALE FACULTY

The issue of sexual harassment as presented below focuses on male faculty and female students. The issue of sexual harassment of male students by female professors and same-sex harassment is not covered in this chapter. Most Title IX cases in the courts have applied the principles of Title VII in the workplace to an educational setting. For example, *quid pro quo* principles in an educational setting translates to a lower grade if a student does not comply with a professor's demands for sex or a higher grade if she does comply. A "hostile environment" can result when professors make sexist remarks in the classroom, engage in inappropriate touching, or threaten retaliation when the female refuses sexual advances. The issue of sexual advances by male faculty toward female students is not new, and many believe it is a common pattern. In reviewing Title IX OCR complaints, situational factors, and the extent of faculty discipline, it appears that some universities tolerate sexually harassing behavior to a greater extent than other universities.

In a study by the Institute for Social and Economic Research at Cornell University, 61% of upperclass and graduate women students experienced "unwanted sexual attention from someone of authority within the University." Glaser and Thorpe (1986) received survey responses from 44% (464) of the female members of the Clinical Psychology Division of the American Psychological Association. Of these responses, 31% reported receiving advances from psychology educators either prior to or during working relationships and 17% reported intimate sexual contact (intercourse or genital stimulation) with at least one psychology educator during graduate training.

One group of researchers found that of 235 male faculty members from all departments at one university, 26% reported sexual involvement with women students (Fitzgerald, 1988). And Schneider (1987), in studying the responses of 356 graduate students from a variety of disciplines, reported that 9% reported coercive dating and sex with faculty members. Another 30% were "pressured to be sexual."

In 1980, a study at Arizona State University reported that 13% of the 2,300 female students surveyed indicated they had been sexually harassed by professors, with 46% of those episodes occurring during undergraduate study (Metha & Nigg, 1983).

Many incidents of sexual harassment of female students by male professors go unreported. Women have been motivated not to report because they are made to feel that they are somehow responsible for the sexual advances, or they fear they will be accused of trying to advance their own careers through use of sex rather than ability. Many women fail to report male faculty harassment because they believe that nothing will be done or that the professor's male colleagues will retaliate against them.

There are numerous reported cases of sexual harassment by professors of female students. Cases range from affairs of male faculty with married and unmarried women students, demands for sex with threats of lower grades for noncompliance, restraining female students in faculty offices, to sexually offensive and hostile environments in the classroom.

Recently, a tenured sociology professor was prohibited from working closely with female students although he denied any acts of sexual harassment. One student complained that the professor had repeatedly remarked about her appearance, asked her about her love life, and shared intimate details of his own. He asked her to go to Italy with him. The professor said the student filed the complaint "because she was not cutting it academically and feared she would get a poor grade in his course" ("U. of Texas," 1997, p. A12).

A well-known poet on the faculty of Syracuse University was charged with sexual harassment of a female graduate student when the faculty member threw a drink in the student's face and called her a "Stalinist bitch"; she had also accused him of making a comment about her breasts. These incidents caused a rift in the creative writing program, with some members suggesting the faculty poet and "other faculty stars had abused their power." In other incidents occurring at the faculty member's home and at a club, students reported, "He would drink too much, his mood would darken and he would say abusive things." He made jokes about wanting to sleep with a male student's wife and performing oral sex on women, and he used vulgar language in the classroom. The professor claimed the female student couldn't take criticism of her work. He was suspended for 2 years but left the program to teach at another university ("Star," 1997; "A Prestigious Writing Program," 1997).

Courts, in dealing with students' sexual harassment claims based on a professor's classroom conduct and speech, must balance the freedom of speech protected by the First Amendment of the U.S. Constitution with the principles of academic freedom. A District Court in New Hampshire dealt with these specific issues in *Silva v. University of New Hampshire*. In this case, a writing instructor was disciplined for statements he made in class. Silva made the following statements: "I will put the focus in terms of sex, so you can better understand it. Focus is like sex. You seek a target. You zero in on your subject. You move from side to side. You close in on the subject. You bracket the subject and center on it. Focus connects experience and language. You and the subject become one." As a part of a discussion on definitions, the professor also stated: "Belly dancing is like Jell-O on a plate with a vibrator under the plate." The court found the professor's comments to be protected by principles of academic freedom. However, case law indicates that faculty do not have unlimited freedom in conducting their classes. This is far from being a settled area of law, and the U.S. Supreme Court has never detailed the precise nature of classroom academic freedom. It appears that lower courts will review classroom speech on a case-by-case basis (*Mailloux v. Kiley*, 1971). Because classroom speech of professors, particularly speech of a sexual nature, is the subject of increasing litigation, further discussion is warranted.

In a widely published decision from the Ninth Circuit Court of Appeals, a tenured professor at San Bernardino Valley College had his institutional discipline overturned. Further, the college was enjoined from implementing discipline and required to remove all references to the discipline from his personnel file. The college had disciplined Professor Cohen for violating the college's sexual harassment policy, which prohibited conduct having the effect of unreasonably interfering with students' academic performance or creating an intimidating, hostile, or offensive learning environment. One student, known as "Ms. M," became offended by the professor's repeated focus on topics of a sexual nature, his use of profanity and vulgar language, and his comments, which she believed were directed intentionally at her and other female students in a humiliating and harassing manner. During class discussions on the issues of pornography, obscenity, cannibalism, and consensual sex with children, he asserted controversial viewpoints in a "devil's advocate style." He also stated in class that he wrote for *Hustler* and *Playboy* magazines and read some articles to the class. He also directed students to write essays on pornography. When Cohen assigned a "Define Pornography" paper to Ms. M, she asked for an alternate assignment but Cohen refused the request. Ms. M did not attend any further classes and received a failing grade (Cohen v. San Bernardino Valley College, 1996). This case is important for several reasons. First, the court stated that the college's policy was ambiguous and unconstitutionally vague and the professor's conduct was not within the core region of sexual harassment as defined by the policy; second, the professor had not been given notice that the policy would be applied in classroom situations.

The college's board, the president, and the grievance committee all agreed that Cohen had violated the policy and outlined a plan for discipline. The court did not address Cohen's First Amendment right and stated, "Neither the Supreme Court nor this Circuit has determined what scope of First Amendment protection is to be given a public college professor's classroom speech" (Cohen, p. 972). The court declined to address Cohen's First Amendment speech claims due to the unconstitutionally vague sexual harassment policy as applied to the facts in the case. In a similar case, the Sixth Circuit Court of Appeals found Central Michigan University's policy defining racial and ethnic harassment as unconstitutionally vague and overly broad when the university tried to discipline a coach (*Dambrot v. Central Michigan University,* 1995).

University attorneys defending against student claims of sexual harassment may shortchange the victims of harassment by writing vague and overly broad policies and by recommending ineffective discipline or sanctions in order to protect the university from lawsuits by faculty accused of sexual harassment. Cohen's punishment was to:

1. Provide a syllabus detailing his teaching style, purpose, content and method used in his classes to the department chair;
2. Attend a sexual harassment seminar within ninety days;

3. Undergo a formal evaluation procedure in accordance with the collective bargaining agreement; and
4. Become sensitive to the particular needs and backgrounds of his students, and to modify his teaching strategy when it becomes apparent that his techniques create a climate which impedes the students' ability to learn. (*Cohen*, p. 971)

The federal district court in the eastern district of Virginia held that if a professor was shown to have engaged in *quid pro quo* sexual harassment, the university could be held strictly liable. In this case, a student filed claims against the university under Title IX after a professor spanked her when she failed to achieve an acceptable score on a makeup exam. However, one claim was dismissed due to the student's failure to show that the university intentionally created a hostile environment in violation of Title IX (*Kadiki v. Virginia Commonwealth University*, 1995).

The new guidelines developed by OCR in March 1997 provide detailed guidance to public schools that will assist faculty, administrators, and students on college and university campuses in dealing more effectively with sexual harassment. The importance of developing clearly defined policies, procedures for immediate investigation of claims, and appropriate sanctions for offenders is clearly stated in the OCR guidelines. The "notice" requirement to institutions states that the institution has notice of violations if any reasonable inquiry would have revealed potential violation (OCR, 1997).

STUDENT-PROFESSOR CONSENSUAL RELATIONSHIP

Kaplin and Lee (1995) discussed the difficulty that postsecondary institutions face in drafting sexual harassment policies related to consensual sexual relationship between students and faculty. Proponents of a total ban on all sexual relationships between students and faculty argue that unequal power relationships negate consensuality by the student. Opponents of total banning of such relationships cite institutional infringement on constitutional rights of free association, cite tort claims for invasion of privacy, and argue that students are beyond the legal age of consent.

One southwestern university drafted a policy discouraging sexual relationships between a faculty member and a student in a current class or in situations where the faculty member would supervise the student as a graduate assistant or as chair of a student's program committee.

The American Association of University Professors (AAUP) in its *Policy Documents and Reports* (1990) states: "Faculty members are cautioned against entering into romantic or sexual relationships with their students; so, too is a supervisor cautioned against entering such relationship with an employee. Faculty and staff should be cautious in assuming professional responsibilities for those with whom they have an existing romantic relationship" (p. 114).

A 1971 California appeals court clearly stated its view of the role of professors as models for their students:

> Certain professions. . . impose upon persons attracted to them, responsibilities and limitations on freedom of action which do not exist in regard to other callings. Public officials such as . . . school teachers fall into such a category . . . the integrity of the educational system under which teachers wield considerable power in the grading of students and the granting or withholding of certificates and diplomas is clearly threatened when teachers become involved in relationships with students (*Board of Trustees of Compton Junior College District v. Stubblefield,* 1971).

The Seventh Circuit Court of Appeals in dictum stated that a teacher was not like an ordinary "person on the street" (*Korf v. Ball State University,* 1984). Ball State University had adopted the AAUP's 1987 "Statement on Professional Ethics" and published it in the faculty handbook. The policy was used to terminate a male, tenured professor who had a consensual affair with a male student. The professor claimed that his due process rights were violated because the AAUP statement did not specifically mention sexual conduct and he did not have adequate notice. The court rejected the claim, stating that each and every type of conduct need not be delineated and that the facts and circumstances clearly demonstrate the professor should have understood the standards to which he was being held and the consequences of his conduct (*Korf,* pp. 1277–1278). The *Korf* and *Cohen* cases both revolved, in part, around proper notice to the professor of the policy, whether due process was denied based on the vagueness of the policy, and the appropriateness of the discipline imposed. In *Korf,* the court upheld the university's termination. In *Cohen,* the court overturned the university's disciplinary action.

The *Korf* case, while dealing with male-to-male harassment, is important as to the AAUP statement on professors of either gender.

The AAUP cautions that the threat to academic freedom is implicit in the relationship itself. Others doubt whether a professor can be objective once involved in a consensual relationship with a student. Generally, other students learn of the relationship and complain of favoritism by the professor toward the involved student. It is not unusual for a student involved in a sexual relationship with a professor to manipulate the relationship to his or her advantage in obtaining better grades, letters of recommendation, publishing opportunities, and "networking" among the professor's colleagues for job opportunities. Sometimes the professor becomes the "victim" when the once-consensual relationship turns into a claim of sexual harassment by a student after the professor discontinues the affair.

THE FEDERAL EFFORT TO COMBAT VIOLENCE AGAINST WOMEN

In 1990, the first legislation regarding sexual assault and violence against women on university campuses was passed in the Student Right-to-Know and Campus Security Act. As a result of this act, colleges and universities were required to collect data on campus crime as of September 1, 1991. Title II of the act requires colleges and universities to publish crime statistics every year, to provide current students and employees with information about campus security policies, and to disclose the result of campus disciplinary actions to victims of violent crime (Griffaton, 1993).

There are problems with consistency from institution to institution in reporting specific acts of violence. One university reported predictions of crimes instead of actual crimes on campus. Griffaton (1993) states that while the Crime Awareness and Campus Security Act ensures uniform reporting standards, it fails to remedy the problems surrounding off-campus victimization, reporting of campus crimes to local police authorities, and the release of campus crime reports. However, Congress is currently considering legislation to address these and other issues in the Accuracy in Campus Crime Reporting Act (H.R. 715).

In 1991, Senator Joseph Biden introduced the Violence Against Women Act. The intent of this act was to assist sexual assault victims on college campuses. The bill's sexual assault provisions were incorporated into the Higher Education Act (1992) under Part D. Universities and colleges are eligible for funding if their student code of conduct or other written policy governing student behavior explicitly prohibits all forms of sexual offenses; victims and offenders are notified of the outcome of any sexual assault hearing; and the institution has a sexual assault education program (Bohmer & Parrot, 1993).

Two 1996 court cases regarding the rape of a female student dealt with the constitutionality of the Violence Against Women Act. The cases involved the same facts, but the court published a separate opinion for each issue. Christy Brzonkala was a student athlete at Virginia Polytechnic and State University. She was raped in September 1994 by two members of the football team, in a room in Brzonkala's residence hall. She did not know the name of her assailants until July 1995. She filed a complaint under the university's sexual assault policy. She did not file criminal charges because she did not think she would win. The university did not encourage her to file criminal charges. One of the assailants admitted in the hearing process that Brzonkala had said "no" twice. One assailant was given a 2-

year suspension, and no action was taken against the other assailant who denied any involvement. Brzonkala was not given access to records of the first of two hearings, participated at her own expense, was not given time to get sworn affidavits, and eventually withdrew from the university. In filing her complaint in court, the issues were: Did she state in her complaint that the university's conduct was illegal discrimination based on gender and did Congress have the constitutional authority to pass the Violence Against Women Act? The court found no gender-based claims under Title IX, since it reasoned that the same conduct would have applied on the part of the university had it been an assault on a male. The second decision held that the Violence Against Women Act was unconstitutional since Congress lacked the authority to make such a law under the Commerce Clause. The court held that merely demonstrating that violence against women has an effect on the national economy does not meet the requirement that it have a substantial effect on interstate commerce (*Brzonkala v. Virginia Polytechnic and State University,* 1996). However, in the *Brzonkala* case, the U.S. Court of Appeals for the Fourth Circuit reversed the district court holding and found a rational basis for Congress' finding that violence against women substantially affects interstate commerce (*Brzonkala v. Virginia Polytechnic and State University,* 1997). Interestingly, this same rationale resulted in the U.S. Supreme Court's decision holding the Federal Gun-Free School Zone Act unconstitutional because Congress lacked the authority under the Commerce Clause to enact such legislation. However, an increasing number of states are addressing the prevention of sexual assault on campus through state legislation (Gorton, 1995).

In addition, the federal initiative has had an impact on campus violence against women. The 1991 Campus Sexual Assault Victim's Bill of Rights was added as an amendment to the Higher Education Bill in 1992. The amendment, known as the Ramstad Amendment, requires institutions of higher education to develop and distribute a campus sexual assault policy that describes the institution's sexual assault programs aimed at preventing sex offenses on campus and outlines the procedures that must be followed once a sex offense has occurred (Gorton, 1995).

Litigation involving Title IX has addressed the legal issues of gender discrimination in athletic programs, the employment and salaries of women faculty, and sexual harassment of female faculty and students in academia. The U.S. Supreme Court ruled unanimously that plaintiffs suing under Title IX could claim monetary damages in addition to equitable relief, which increased the number of students and faculty who find Title IX as a viable avenue to challenge alleged discrimination (Kaplin & Lee, 1995).

As case law continues to develop under Title IX of the Education Amendments of 1992, and as OCR continues its commitment to enforce its regulations, victims may find an effective legal remedy in addressing violence against women on university campuses. The law is still evolving.

SUGGESTED STRATEGIES

Colleges and universities can be held accountable for disregarding, failing to investigate, or delaying investigation of allegations of sexual harassment. A copy of OCR's sexual harassment guidelines should be available in each residence hall, each academic department, and other departments and programs throughout the institution.

Institutions must develop and distribute policies regarding consensual relationships. It is doubtful that any policy prohibiting consensual relationships between two adults would pass constitutional review by the courts. However, in the absence of such a policy, a court may use the AAUP policy on ethics as discussed previously in *Korf*.

A sample policy involving "amorous relationships" is provided below:

> In recognition of interests in privacy and free association, university policy does not prohibit fully consensual amorous relationships. Even an apparently consensual amorous relationship, however, may lead to sexual harassment or other breaches of professional obligations, particularly if one of the individuals in the relationship has a professional responsibility toward or is in a position of authority with respect to the other, such as in the context of instruction, advisement, or supervision. Due to the power difference, it may be difficult to avoid the appearance of favoritism or to assure a truly consensual relationship. Amorous relationships may result in conduct that amounts to sexual harassment or that violates the professional duties of even-handed treatment and maintenance of an atmosphere conducive to learning or working.

> In light of these serious risks, every individual in a position of authority should take great care not to abuse that power in personal relationships. Specifically, if involved in an amorous relationship with someone over whom he or she has supervisory authority, the individual must remove himself or herself from any participation in recommendations or decisions affecting evaluation, employment conditions, instruction, or the academic status of the other person in the relationship, and must inform his or her immediate supervisor of the action taken. (Arizona State University, 1996)

Colleges and universities must make it clear that sexually harassing conduct by employees and students will not be tolerated. Violations of sexual harassment policies through conduct that unlawfully interferes with an individual's work or educational performance or unlawfully creates an intimidating, hostile, or offensive working, learning, or residential environment may result in disciplinary action up to and including termination for employees and in sanctions including suspension or expulsion for students.

All policies prohibiting harassing conduct are subject to constitutionally protected speech rights and principles of academic freedom and should be carefully reviewed by the institution's legal department.

REFERENCES

American Association of University Professors. (1990). *AAUP Policy Documents and Report*. Washington, DC: Author.

American Association of University Women. (1993). *Hostile Hallways: The AAUW Survey on Sexual Harassment in America's Schools*. Washington, DC: Author.

Anderson, A. (1997, September 18). 2nd rape in 48 hours reported; situations similar. *The Arizona State University State Press*, pp. 1–2.

Arizona State University. (1996, July 1). Amorous relationships policy. In *Academic Affairs Policy and Procedures Manual*. Tempe, AZ: Author.

Benson, K. (1992). Sexual harassment on a university campus: The influence of authority relations, sexual interest, and gender stratifications. *Social Problems, 29,* 236–251.

Bergman, L. (1992). Dating violence among high school students. *Social Work, 37*(1), 21–27.

Board of Trustees of Compton Junior College District v. Stubblefield, 16 Cal. App. 3d 820 (1971).

Bohmer, C., and Parrot, A. (1993). *Sexual assault on campus*. New York: Lexington Books.

Brzonkala v. Virginia Polytechnic and State University, 935 F.Supp. 772, 779 (W.D. Va. 1996).

Brzonkala v. Virginia Polytechnic and State University, 132 F.3d 949, 968 (4th Cir. 1997).

Cohen v. San Bernardino Valley College, 92 F.3d 968 (9th Cir. 1996).

Cornell University Office of Equal Opportunity. (1988). *On campus with women*. Ithaca, NY: Author.

Dambrot v. Central Michigan University, 55 F.3d 1177 (6th Cir. 1995).

DeMaris, A. (1987). The efficacy of a spouse abuse model in accounting for courtship violence. *Journal of Family Issues, 8*(3), 291–305.

Demon rum, not drugs major evil on campus. (1991, March 31). *The Oregonian:* p. A17.

Dutton, D.G., and Browning, J.J. (1988). Concern for power, fear of intimacy, and aversive stimuli for wife abuse. In G. Hotaling, D. Finkeihor, J.T. Kirkpatrick, & M.A. Straus (Eds.), *Family abuse and its consequences: New directions in research* (pp. 163–175). Newbury Park, CA: Sage.

Ehrhart, J., and Sandler, B. (1985). *Campus gang rape: Party games?* Washington, DC: Project on the Status and Education of Women and the Association of American Colleges.

Estrich, S. (1987). *Real rape*. Cambridge, MA: Harvard University Press.

Fitzgerald, L.F. (1988). Sexual denial in scholarly garb. *Psychology Women Quarterly, 12,* 329–332.

Follingstad, D., Rutledge, L., Berg, B., Hause, E., & Polek, D. (1990). The role of emotional abuse in physically abusive relationships. *Journal of Family Violence, 5,* 107–120.

Foo, L., & Margolin, G. (1995). A multivariate investigation of dating aggression. *Journal of Family Violence, 10,* 351–377.

Glaser, R.D., & Thorpe, J.S. (1986). Unethical intimacy: A survey of sexual contact and advances between psychology educators and female graduate students. *American Psychologist, 43,* 49.

Gorton, C.A. (1995). *A policy analysis of sexual assault policies and procedures at America's twenty largest four year public universities.* Unpublished doctoral dissertation, Arizona State University, Tempe, AZ.

Griffaton, M. (1993). Forewarned is forearmed: The Crime Awareness and Campus Security Act of 1990 and the future of institutional liability for student victimization. *Case Western Reserve Law Review, 43,* 525.

Henton, J.M., Cate, R., Koval, J., Lloyd, S., & Christopher, S. (1983). Romance and violence in dating relationships. *Journal of Family Issues, 4*(3), 467–482.

Holtzworth-Munroe, A. (1992). Attributions and marital violent men: The role of cognition in marital violence. In J.H. Harvey, T.L. Orbuch, & A.L. Weber (Eds.). (1992). *Attributions, accounts, and close relationships* (pp. 165–175). New York: Springer-Verlag.

Holyroyd, J. (1987). He knows my name and he knows where I live. *Center Against Sexual Abuse Newsletter, 3,* 1.

Hydén, M. (1995). Verbal aggression as prehistory of women battering. *Journal of Family Violence, 10,* 55–71.

Kadiki v. Virginia Commonwealth University, 892 F. Supp. 746 (E.D. Va. 1995).

Kanin, E.J. (1957). Male aggression in dating—courtship relations. *American Journal of Sociology, 63,* 197–204.

Kaplin, W.A., & Lee, B.A. (1995). *The law of higher education.* San Francisco: Jossey-Bass.

Korf v. Ball State University, 726 F.3d 1222 (7th Cir. 1984).

Koss, M.P. (1985). The hidden rape victim: Personality, attitudinal, and situational characteristics. *Psychology of Women Quarterly, 9,* 193–212.

Koss, M.P. (1988). Hidden rape: Sexual aggression and victimization in a national sample in higher education. In A.W. Burgess (Ed.), *Rape and sexual assaults II* (pp. 3–25). New York: Garland.

Mailloux v. Kiley, 448 F.2d 1242, 1243 (1st Cir. 1971).

Makepeace, J.M. (1981). Courtship violence among college students. *Family Relations, 30,* 97–102.

Makepeace, J.M. (1986). Gender differences in courtship violence victimization. *Family Relations, 35,* 383–388.

Makepeace, J.M. (1987). Social factors and victim-offender differences in courtship violence. *Family Relations, 36,* 87–91.

Marshall, L.L., & Rose, P. (1987). Gender, stress and violence in the adult relationships of a sample of college students. *Journal of Social Personal Relationships 4,* 299–316.

Marshall, L.L., & Rose P. (1988). Family of origin violence and courtship abuse. *Journal of Counseling and Development, 66,* 414–418.

Marshall, L.L., & Rose, P. (1990). Premarital violence: The impact of family of origin on violence, stress, and reciprocity. *Violence and Victims, 5*(1), 51–64.

Massachusetts Institute of Technology Crime Report. (1996). Cambridge, MA: Massachusetts Institute of Technology.

McGarrell, E.F., & Flanagan, J.J. (Eds.). (1985). *Source book for criminal justice statistics—1984.* Washington, DC: U.S. Government Printing Office.

Metha, A., & Nigg, T. (1983). Sexual harassment on campus: An institutional response. *Journal of the National Association for Women Deans, Administrators and Counselors, 46*(2), 9–15.

Mortz v. Johnson, 651 N.E.2d 1163 (Indiana, 1995).

Neft, N., and Levine, A.D. (1997). *Where women stand—An international report on the status of women in 140 countries, 1997–1998.* New York: Random House.

Office of Civil Rights. (1994a, May 2). Occidental College (CA). OCR letter of finding, Complaint No. 9-93-2100.

Office of Civil Rights. (1994b, June 10). Foothill College (CA). OCR letter of finding, Complaint No. 9-94-2021.

Office of Civil Rights. (1997, March). Sexual harassment guidance: Harassment of students by school employees, other students, or third parties. Federal Register, *62*(49), 12042.

Pallett v. Palma, 914 F. Supp. 1018 (S.D.N.Y. 1996).

A prestigious writing program and its big-name authors face charges of sexual harassment. (1997, February 21). *The Chronicle of Higher Education,* pp. A8–A10.

Riggs, D.S., O'Leary, K.D., and Breslin, F.C. (1990). Multiple correlates of physical aggression in dating couples. *Journal of Interpersonal Violence, 5*(1), 61–73.

Roscoe, B. (1985). Courtship violence: Acceptable forms and situations. *College Student Journal, 19*(4), 389–393.

Ryan, K.M. (1995). Do courtship-violent men have characteristics associated with a "battering personality"? *Journal of Family Violence, 10,* 99–120.

Schneider, B.E. (1987). Graduate women, sexual harassment, and university policy, *Journal of Higher Education, 46,* 51–52.

Silva v. University of New Hampshire, 888 F. Supp. 293 (D.N.H. 1994).

Smith, J.P., & Williams, J.G. (1992). From abusive household to dating violence. *Journal of Family Violence, 7*(2), 153–165.

Smith, M. (1988). *Coping with crime on campus.* New York: American Council on Higher Education, MacMillian Series.

Star of Syracuse's writing program to leave for Stanford. (1997, March 28). *The Chronicle of Higher Education,* p. A12.

Temple U.'s crew program embroiled in sexual harassment lawsuit. (1996, June 7). *The Chronicle of Higher Education,* p. A33.

Terpstra, J.C. (1996). Social responsibility and personal accountability: Gender violence, discrimination, and sexual harassment. In National University Center for Law, Education and Public Policy, *Legal Challenges Facing Higher Education in the 90s: Conference Proceedings,* p. 1.

U.S. Congress. (1990). Student Right-to-Know and Campus Security Act, Public Law 101-542.

U.S. Congress. (1992). Campus Sexual Assault Victim's Bill of Rights Act of 1991. 20 U.S.C.A. § 1145h (Supp. 1993).

U of Texas finds sociologist guilty of sexual harassment. (1997, May 16). *The Chronicle of Higher Education,* p. A12.

CHAPTER 10

Heterosexism and Campus Violence: Assessment and Intervention Strategies

Nancy J. Evans and Sue Rankin

INTRODUCTION

Hate crimes are words or actions that are intended to harm or intimidate a person or group due to race, gender, sexual orientation, religion, or other group identification. They include violent assaults, murder, rape, and property crimes motivated by prejudice, as well as threats of violence or other acts of intimidation (Herek, 1989, 1994; Levin & McDevitt, 1993). Hate crimes legislation varies immensely across the country (Jenness & Grattet, 1993; National Institute Against Violence and Prejudice, 1993). Therefore, what is deemed a criminal act varies by jurisdiction. Regardless of the legal system's response, hate crimes carry a message, not only to the victim, but also to the entire stigmatized minority community—that victimization is the punishment for stepping outside culturally accepted boundaries. Although there are no accurate data on the number of hate crimes committed each year, most reports suggest that violence of this nature is increasing exponentially (Berk, Boyd, & Hammer, 1992; Center for Democratic Renewal, 1992; Fenn & McNeil, 1987; Fernandez, 1991; Hernandez, 1990) and that some of the most frequent, visible, violent, and culturally legitimated bias-motivated conduct is directed toward lesbian, gay, bisexual, and transgendered (LGBT) persons (Berrill, 1990, 1992; Comstock, 1989, 1991; Herek, 1989, 1994; Herek & Berrill, 1992; Jenness, 1995). For example:

- Former marine John Knight slashed the throat of Honolulu resident William Rowe just "because he was gay." Knight boasted to friends, "I got a faggot. I killed him. . . . Do you want to see the body?" ("The year in review," 1994).
- Allen Schindler, a sailor, was beaten to death by two shipmates because he was gay ("The year in review," 1994).

- A lone assailant stalked and shot two lesbians hiking in Pennsylvania, killing one woman and critically wounding the other (Brenner, 1995).

These acts of violence are fueled by heterosexism—the assumption of the inherent superiority of heterosexuality, an obliviousness to the lives and experiences of LGBT people, and the presumption that all people are, or should be, heterosexual. Rich (1980) calls the ideology of heterosexism "compulsory heterosexuality," a systematic set of institutional and cultural arrangements that reward and privilege people for being or appearing to be heterosexual while establishing potential punishments or lack of privilege for being or appearing to be lesbian, gay, bisexual, or transgendered.

Like racism, sexism, and other ideologies of oppression, heterosexism is manifested both in societal customs and institutions and in individual attitudes and behaviors (Herek, 1990). Heterosexism is preserved through the routine operation of major social institutions such as employment, where discrimination on the basis of sexual orientation remains legal in many states; marriage, where lesbian and gay couples generally are denied the community recognition, legal protection, and economic benefits accorded to married heterosexual partners; the law, where sexual intimacy between same-sex partners remains illegal in one half of the states, and the constitutionality of such laws is upheld by the U.S. Supreme Court; and religion, where the United Methodist, Presbyterian, Baptist, Roman Catholic, and other churches condemn homosexuality as incompatible with Christian teaching. The maintenance and expression of heterosexism is possible primarily because it is in keeping with current social norms.

Higher education institutions mirror the societal climate and contribute to the maintenance of institutionalized heterosexism. Recent evidence indicates an increase in the number of hate crimes directed toward LGBT students, faculty, and administrators (D'Augelli, 1989c, 1990; D'Emilio, 1990; Herek, 1994; Herek & Berrill, 1990; Rankin, 1994, 1998; Tierney & Rhoads, 1993). This chapter discusses the anti-LGBT prejudice evident on college campuses and proposes intervention strategies to address the violence.

Given the heterosexist values underlying higher education, the work involved in proactively addressing violence against LGBT individuals and building a community inclusive and welcoming of LGBT persons is controversial and demanding. Advocates do not have an easy task. Systematically examining and publicizing the extent of the problem and thoughtfully developing a comprehensive intervention strategy are necessary steps to build support for change. The data provided in this chapter, along with the intervention strategies and suggestions for effecting change, were gathered from many sources. Building coalitions across a number of colleges and universities is a way of gathering information and ideas, as well as obtaining encouragement and support, when working to end oppression and create a university climate in which all students feel safe, welcome and nurtured.

HETEROSEXISM, CAMPUS CLIMATE, AND ANTI-LGBT VIOLENCE

One of the primary missions of higher education institutions is unearthing and disseminating knowledge. Academic communities expend a great deal of effort fostering an environment where this mission is nurtured, with the understanding that institutional climate has a profound effect on the academic community's ability to excel in research and scholarship (Boyer, 1990; Peterson & Spencer, 1990; Rankin, 1994; Tierney & Dilley, 1998). Recent investigations suggest that the climate on college campuses not only affects the creation of knowledge, but also has a significant impact on members of the academic community who, in turn, contribute to the creation of the campus environment (Kuh & Whitt, 1988; Peterson & Spencer, 1990; Rankin, 1994, 1998; Tierney, 1990). Therefore, preserving a climate that offers equal learning opportunities for all students and academic freedom for all faculty—an environment free from discrimination—should be one of the primary responsibilities of educational institutions. Yet, the climate on many college campuses is not equally supportive of all of its members. As noted earlier, the literature suggests that LGBT members of the academic community are subjected to physical and psychological harassment, discrimination, and violence reflective of the anti-LGBT prejudice prevalent in American society. A manifestation of heterosexism, anti-LGBT prejudice obstructs the pursuit of knowledge in academe and creates a hostile climate that allows or countenances victimization. A review of the campus climate for LGBT students, faculty, and staff is presented in the following section.

CAMPUS CLIMATE REVIEW*

For several years the National Gay and Lesbian Task Force (NGLTF) documented reported incidents of harassment and violence against lesbian, gay, and bisexual students around the country. In a 1988 review of campus victimization, 1,411 anti-LGBT incidents, including threats, vandalism, harassment, and assault were reported to NGLTF. When asked if anti-LGBT violence had increased on

---

*Portions of this section were adapted with permission from S. Rankin (1998). Campus climate for lesbian, gay, bisexual, and transgendered students, faculty, and staff: Assessment and strategies for change. In R. Sanlo (Ed.), *Working with lesbian, gay, and bisexual students: A guide for administrators and faculty* (pp. 203–212). Westport, CT: Greenwood Publishing.

their campus since the previous year, 32% responded affirmatively (Berrill, 1988).

In response to the heightened awareness of LGBT victimization and to issues of LGBT inequity prevalent on college campuses, several university administrators appointed task forces or ad-hoc committees to investigate the institutional climate for LGBT members of their academic community. In other instances, concerned LGBT students, faculty, and staff initiated the investigation. The campuses surveyed included public and private institutions that varied in size and geographic location. Each institution's purpose for assessing the campus climate was unique, prompted by a particular set of circumstances occurring on that campus. The methodologies used to examine the campus climate also were varied. The methods utilized were generally influenced by the purpose and goals of the committee as well as by the type of questions the institution was examining. Of the 30 college and university reports reviewed, 13 conducted surveys, 6 conducted focus groups or interviews, and 5 opted for a combination of both quantitative and qualitative methodology (6 reports did not indicate their method of assessment). Just as there were a variety of stimuli for writing the reports and various methods employed to complete the assessment, the populations sampled also differed. For example, the University of Arizona queried 600 faculty and staff regarding their perceptions of the campus climate for lesbian, gay, and bisexual people. In contrast, the University of Massachusetts conducted three surveys purposefully sampling lesbian, gay, and bisexual students, resident assistants, and student service personnel. Open forums and public hearings where all members of the academic community were encouraged to share their voice were held at Vanderbilt University, Rutgers University, and the University of Wisconsin at Madison. Pennsylvania State University and the University of California-Davis conducted extensive interviews with LGBT faculty and staff.

Although it is difficult to compare the investigations due to differences in research methodologies, instruments, and samples, it is clear that anti-LGBT prejudice is prevalent in higher education institutions. For example, in studies where surveys were used as the primary tool, LGBT students indicated that they were the victims of anti-LGBT prejudice ranging from verbal abuse (2% to 86%) to physical violence (6% to 59%) to sexual harassment (1% to 21%) (Table 10–1).

In those investigations that used qualitative data, analogous findings were reported indicating the invisibility, isolation, and fear of LGBT members of the academic community. Their lives are filled with "secret fears." For professors, counselors, staff assistants, or students who are LGBT, there is the constant fear that, should they "be found out" they would be ostracized, their careers would be destroyed, or they would lose their positions. Although the reports indicate differences among the experiences of these individuals, their comments suggest that regardless of how "out" or how "closeted" they are, all have fears that prevent

Heterosexism and Campus Violence 173

Table 10–1 Percentages of Persons Reporting Anti-LGBT Victimization on Campus Surveys

	NGLTF, 1988 (n=2074)	U Arizona, 1992 (n=600)	CU-Chico, 1993 (n=682)	UC-Santa Cruz, 1990 (n=733)	U Colorado Boulder, 1991 (n=1004)	Emory, 1987 (n=51)	U Illinois Chicago, 1995 (n=1161)	U Illinois Urbana, 1987 (n=92)
Verbal harassment	86	12	23	2	23	67	42	58
Hear anti-gay remarks	n/a	n/a	42	4	n/a	n/a	n/a	79
Threats of physical violence	44	n/a	n/a	1	6	22	44	n/a
Property damaged/grafitti	19	n/a	18	3	12	22	22	n/a
Objects thrown	27	n/a	n/a	n/a	n/a	12	n/a	n/a
Followed or chased	35	n/a	n/a	n/a	n/a	16	n/a	n/a
Spat upon	14	n/a	n/a	1	n/a	2	n/a	n/a
Assaulted/wounded with weapon	9	n/a	n/a	n/a	n/a	0	n/a	n/a
Physical confrontation/ assault	19	n/a	32	n/a	2	6	5	n/a
Sexually harassed/assaulted	n/a	n/a	n/a	1	n/a	0	n/a	n/a
Anticipate future victimization	83	n/a	n/a	n/a	n/a	n/a	n/a	n/a
Fear for safety	62	n/a	n/a	16	n/a	n/a	n/a	55
Know others who have been victimized	84	n/a	n/a	n/a	n/a	n/a	n/a	n/a
Nonreporting of at least one incident	n/a	n/a	n/a	n/a	95	n/a	n/a	90
Harassed by roommate	n/a	n/a	n/a	n/a	n/a	n/a	n/a	n/a
Pressure to be silent/ threatened with exposure	n/a	n/a	n/a	3	n/a	n/a	4	91
Negative effects on job advancement	n/a	35	n/a	n/a	30	n/a	15	88

continues

Table 10-1 continued

	U Mass Amherst, 1985 (n=445)	Michigan State, 1992 (n=113s)	Oberlin, 1990 (n=267f)	U Oregon, 1990 (n=105s); 1994 (n=514f)	Penn St, 1987, 1994 (n=671s) (n=1078f)	Rutgers, 1989 (n=213)	U Virginia, 1989 (n=1244s)	U Wisconsin Milwaukee, 1994 (n=366)	Yale, 1986 (n=215)
Verbal harassment	45	51s/35f	14s/19f	n/a	72	55	48	67	65
Hear anti-gay remarks	92	83s/79f	n/a	96s/44f	98	90	18	90	n/a
Threats of physical violence	21	n/a	6	54s/32f	25	16	16	59	25
Property damaged/grafitti	33/88	70s/71f	70s/51f	86s/53f	16	22	n/a	56	10
Objects thrown	n/a	n/a	n/a	n/a	13	12	n/a	n/a	19
Followed or chased	n/a	n/a	n/a	n/a	22	18	n/a	n/a	25
Spat upon	n/a	1	n/a	n/a	6	1	n/a	n/a	3
Assaulted/wounded with weapon	5	n/a	n/a	n/a	1	2	n/a	n/a	1
Physical confrontation/ assault	n/a	8s/0f	n/a	24s/3f	4	5	7	6	5
Sexually harassed/assaulted	21	12s/7f	3s/0f	18s/11f	15	8	7	16	10
Anticipate future victimization	n/a	n/a	n/a	n/a	93	86	n/a	n/a	92
Fear for safety	n/a	n/a	n/a	61s/57f	58	35	n/a	58	57
Know others who have been victimized	n/a	n/a	n/a	n/a	n/a	57	41	55	76
Nonreporting of at least one incident	67	n/a	most s	69s/0f	93	88	n/a	50	90
Harassed by roommate	50	n/a	n/a	n/a	n/a	42	n/a	n/a	22
Pressure to be silent/ threatened with exposure	29	64s/41f	70s/19f	69s/57f	n/a	n/a	n/a	56/20	n/a
Negative effects on job advancement	n/a	n/a	0s 22f	n/a	n/a	n/a	n/a	8	n/a

Note: s = student response; f = faculty response; n/a = not available.

Source: Copyright © Sue Rankin.

them from acting freely. Three themes emerged from the interviews, focus group comments, and open-forum statements presented in several of the reports: (1) invisibility/ostracism, (2) isolation/self-concealment, and (3) consequences to the university.

Invisibility/Ostracism

Institutionalized heterosexism on college campuses creates an oppressive situation for LGBT persons. The university environment negates their existence, thereby promoting further invisibility. The fear of rejection has a tremendous impact on the way that these individuals lead their lives. Typical statements included:

- "I have tenure, but if the faculty in my department found out that I am a lesbian, I would be ostracized." (Faculty member, Pennsylvania State University, 1994)
- "What has to be the most painful to me is the invisibility I have had to face as a gay student. I have often felt unwelcome in the place that is my home. What I can say is that except for the university gay community, I have not participated in campus life. It may be an unwritten rule, but it is clear that gays are not welcome in fraternities or on athletic teams." (Undergraduate student, University of Minnesota, 1993)
- "The predominant feeling I have is one of exclusion. Much of the coursework and research in which I am currently involved addresses issues of children and families. We are lucky to get a footnote. Professors may comment that families including GLB members do exist, but the discussion revolved completely around the lives of heterosexuals." (Graduate student, University of Minnesota, 1993)
- "At work, I am not a human being. At 5:00 pm, that's when I can be who I am. I remember thinking about finishing school and starting work as being a freeing experience. Now I know better; it makes me bitter at the University and the world." (lesbian, gay, or bisexual staff member, University of Colorado-Boulder, 1991)

Isolation/Self-Concealment

In order to prevent what they anticipate will be rejection by their colleagues or peers, many LGBT members of the academic community choose to conceal their sexual orientation. As one faculty member noted, "Being gay is a part of myself that I've learned to hide in order to survive" (Tierney, 1992, p. 52). To protect themselves from discrimination, LGBT persons do a lot of lying. The university climate communicates the message that being honest about one's sexual orientation may have direct negative effects on salary, tenure, promotion, and emotional well-being, so they choose to remain "in the closet," where it is safe. Here is how they feel:

- "I know a lesbian who had to leave her position . . . because of gossip about her sexual orientation. In her new position she had to stay in the closet and be very secretive about her private life. She eventually quit the university." (Non-lesbian, gay, or bisexual staff member, University of Colorado-Boulder, 1991)
- "A friend of mine confessed to a campus priest that he was "homosexual" and was told to change or he'd 'go to hell.'" (Undergraduate student, University of Massachusetts, 1985)
- "An untenured faculty member has been afraid to attend meetings of the Faculty/Staff Lesbian, Gay, Bisexual Caucus for fear that senior colleagues might hear about it." (Tufts University, 1993)
- "I completed your survey at home so no one would see me." (lesbian, gay, or bisexual staff member, University of Colorado-Boulder, 1991)

Consequences to the University

The university suffers from its own heterosexism. Talented LGBT students, faculty, and staff feel "forced" to leave the university, and students (both LGBT and heterosexual) are deprived of role models and academic growth. Following are some specific examples:

- "I'd be scared to be an advisor to the lesbian, gay, and bisexual student association because that is being too obvious. I am close to tenure, and I don't want them to find a reason not to give it to me." (Faculty member, Pennsylvania State University, 1994)
- "A student enrolled in a history of sexuality course told the instructor that her roommates tried to talk her out of it because 'only lesbians take that course.'" (Tufts University, 1993)
- "I regret that I can't be more 'out' in order to be a role model for gay students." (Faculty member, University of Minnesota, 1993)
- "A heterosexual athlete told his advisor that he wanted to drop a dance course because he was afraid that his Tufts teammates might think he was gay." (Tufts University, 1993)
- "I would feel better about myself if I could be more supportive to students and colleagues. Living a dual life is painful." (lesbian, gay, or bisexual staff member, University of Colorado-Boulder, 1991)

- "It's a very good school, or at least it's trying to be. I'd like to stay if this place were better for gays—but it's not." (Faculty, Pennsylvania State University, 1992)

Heterosexism is a form of prejudice that renders LGBT people invisible. Indeed, most LGBT members of the academic community are invisible. Due to fear of harassment and discrimination, LGBT individuals deliberately conceal their sexual orientation. The pervasive heterosexism of the university not only inhibits the acknowledgment and expression of LGBT perspectives, it also affects curricular initiatives and research efforts. Examples of this include the lack of LGBT content in university course offerings and the absence of LGBT studies programs. Further, the contributions and concerns of LGBT people are often unrecognized and unaddressed, to the detriment of the education not only of LGBT students, but of heterosexual students as well.

In summary, the results of the climate studies reveal two important themes. First, institutions of higher education do not provide an empowering atmosphere for LGBT faculty, staff, and students—an atmosphere where their voices are heard, appreciated, and valued. Second, and perhaps more significant, the results suggest that the climate on college campuses acts to silence the voices of its LGBT members with subtle and overt discrimination.

IMPACT OF HETEROSEXISM AND ANTI-LGBT VIOLENCE

Victims of anti-LGBT crimes face the same negative psychosocial consequences as the victims of other hate crimes. Victimization shatters three basic assumptions: the illusion of invulnerability, the view of oneself in a positive light, and the perception of the world as a meaningful place (Janoff-Bulman & Frieze, 1983).

According to Slater (1993), the impact of victimization on LGBT students is related to the amount of support that the student has had throughout his or her life. Those who have had little support have more trouble coping with negative situations and experiences than those who have previously received understanding and assistance in dealing with issues related to their sexual orientation.

Common problems experienced by victims of LGBT-related violence include a heightened sense of vulnerability and fear for their safety; chronic stress; depression, feelings of helplessness, anxiety, and anger; somatic disturbances; low self-esteem; and internalized homophobia (negative feelings experienced by LGBT persons about their sexual orientation) (D'Augelli, 1992; Herek, 1994, 1995; Hershberger & D'Augelli, 1995; Norris & Kaniasty, 1991; Savin-Williams &

Cohen, 1996; Slater, 1993). In addition, criminal victimization is often followed by posttraumatic stress disorder (Herek, 1994, 1995).

Individuals who have been the targets of violence often experience further victimization in the form of accusations that they deserved what happened to them (Berrill & Herek, 1992). They may also experience harassment and discrimination if their sexual orientation becomes known as a result of the crime (D'Augelli, 1992). For example, "In a 1988 case involving the beating death of an Asian-American gay man, a Broward County [Florida] circuit judge jokingly asked the prosecuting attorney, 'That's a crime now, to beat up a homosexual?' The prosecutor answered, 'Yes sir. And it's also a crime to kill them.' The judge replied, 'Times have really changed'"(Hentoff, 1990).

Lack of support from others is a common occurrence that leads victims to isolate themselves and avoid reporting or talking about what they have experienced (Savin-Williams & Cohen, 1996). In the findings of the campus climate review discussed earlier, 50% to 90% of those who responded indicated that they did not report at least one incident.

Development of a healthy LGBT identity is often retarded by victimization (D'Augelli, 1992). Individuals who have experienced violence directed at them are less open about their identities (D'Augelli, 1989c; Savin-Williams & Cohen, 1996). They report changing their behavior to avoid locations known to attract LGBT people, including even LGBT student organization meetings and cultural and educational events with LGBT themes (D'Augelli, 1992). They also avoid associating with openly LGBT people. Academic performance is affected in negative ways. Students report withdrawing from school and doing poorly because of the stress related to their experiences (Slater, 1993).

More serious mental health problems also result from experiences of violence or fear of violence (D'Augelli, 1993; Savin-Williams & Cohen, 1996). Substance abuse is a significant problem for LGBT persons (Glaus, 1988; Remafedi, 1987), and suicide attempts are three times more likely among LGBT youths than among heterosexual youths (Gibson, 1989).

The previous campus review suggests that both the subtle and overt forms of harassment and discrimination experienced by LGBT persons on college campuses are supported by institutionalized heterosexism. The norms that assume and reward presumed heterosexuality condone or even cause the victimization of LGBT people. D'Augelli (1989c) wondered:

> Why hasn't this problem made its way through the usual streams of anointment as a campus "problem?" The answer is locked in "the closet" of lesbian and gay life on campus. It is a "closet" inhabited not only by self-identified lesbian, gay, and bisexual students, faculty, and staff, although they constitute most of the inhabitants. The "closet" is shared by heterosexual people on campus who know the needs of lesbians and gay men but do not speak out on their behalf. (p. 129)

The remainder of this chapter discusses strategies for implementing change on campus to provide for a more supportive climate and to reduce hate crimes directed at LGBT persons.

STRATEGIES FOR CHANGE: A COMPREHENSIVE PROGRAM OF INTERVENTION

To successfully address a problem as endemic as violence against LGBT students, a comprehensive program of intervention is needed. In a revision of the "cube" model, originally developed by Morrill, Oetting, and Hurst (1974), Hurst and Jacobson (1985) identified intervention targets, methods, and purposes to guide program development. To encourage individual and interpersonal development, they believe that interventions must be targeted both at individual students and at the campus ecology using a variety of methods. The Hurst and Jacobson model is used below to structure the discussion of necessary interventions to address violence directed at LGBT students. While a number of these interventions do not focus on violence directly, they are designed to improve the campus climate for LGBT individuals and to educate the campus about the experiences and concerns of LGBT people. Such interventions do much to provide LGBT individuals with a voice and raise awareness about the oppression LGBT people endure in this society. They target the cultural norms that foster and condone violence.

Interventions Aimed at Individuals

Interventions must address the needs of LGBT students as well as heterosexual students on campus. LGBT students who are victims of violence must be provided with support and assistance in dealing with their emotional reactions. They must also be encouraged to report crimes to appropriate authorities. LGBT students must be shown that the issues they face are taken seriously and that the institution cares about them as people.

Heterosexual students must be given clear messages that harassment and violence directed toward LGBT individuals is unacceptable and that action will be taken against perpetrators. Anyone found guilty of violent actions should be appropriately sanctioned. Educational efforts should also be directed to heterosexual students to raise their awareness about LGBT lifestyles and the experience of being an LGBT person.

Interventions Aimed at the Campus Ecology

Hurst and Jacobson (1985) identified five aspects of the campus ecology at which interventions can be directed: physical, social, cultural, academic, and organizational/structural.

Physical interventions would include modification of the physical setting to make the campus safe for LGBT students. Options include identifying safe locations for students to congregate, such as a center for LGBT students, or a meeting location within the student union building or cultural center. Providing appropriate lighting, patrolling the campus at night, quickly removing graffiti, and taking down harassing flyers are other ways of ensuring that the campus feels safe for LGBT people.

Social interventions are those targeted to improve the interpersonal climate on campus. Creating a hot line where students can anonymously receive support, information, and advice from individuals who are familiar with the concerns of LGBT students is one possibility. A network of "gay-friendly" faculty, students, and staff can also be established and publicized to LGBT students. Peer support should be available in the form of organizations and groups aimed at students who are at different levels of development. For instance, one group might address the concerns of individuals who are questioning their identity, another might provide social activities for individuals who have identified as LGBT, and a third may be more political in nature. Support groups for victims of violence would more directly address the needs of individuals who have been attacked. Social activities that normalize the inclusion of LGBT people are also needed. These might include dances, parties, or dinners for LGBT students.

Cultural interventions help to raise awareness and demonstrate inclusiveness. Including LGBT artists and musicians and films and plays with LGBT themes and characters in the cultural programming of an institution sends a message that LGBT individuals are a welcome part of the campus. LGBT cultural programming also helps to educate the rest of the campus population about the issues and experiences of LGBT people.

Academics are central to the mission of the institution. Addressing LGBT issues in classes provides legitimacy to these topics and helps to break down heterosexist values and norms. Certainly LGBT topics should be an integral part of the curriculum in disciplines such as psychology, sociology, education, literature, history, and the arts and humanities. Using inclusive language should be encouraged in all classrooms. Encouraging research on LGBT-related topics, including the causes and prevention of violence against LGBT persons, is another way that academic personnel can intervene to fight oppression.

Organizational and structural interventions are imperative if systemic and stable change is to occur (Manning & Coleman-Boatwright, 1991). Highly visible commissions and/or committees with authority to take action are needed to direct LGBT advocacy efforts. Staff must be hired with specific responsibility for addressing the needs of LGBT people on campus. They must have authority to develop and implement educational and social programming, provide support for LGBT students, and conduct outreach activities. Ideally, these administrators would staff an LGBT office, the mission of which is to develop and carry out educational, support, and advocacy functions. Specific roles of such an office might be to handle reports of victimization and to provide counseling and support for LGBT individuals who have experienced violent attacks. Having widely distributed and publicized policies in place that prohibit discrimination against LGBT people and that specifically address harassment and violence directed against LGBT individuals is also crucial. These policies must ensure the confidentiality of victims. Documentation of hate crimes against LGBT people must also be required.

Methods of Intervention

Hurst and Jacobson (1985) proposed four methods of intervention: (1) administrative and resource management; (2) media; (3) consultation and training; and (4) counseling, teaching, and advising.

Administrative leadership and allocation of sufficient resources are necessary for the implementation of all of the interventions suggested above. Top-level administrators must support development, implementation, and enforcement of policy and programs designed to address violence against LGBT people. They must establish, listen to, and support the efforts of LGBT task forces and committees. Resources must be provided to establish advocacy offices, to hire LGBT staff, and to fund educational efforts.

Media can play a key role in raising awareness and promoting the rights of LGBT people. Articles in the campus newspaper, bulletins, and flyers advertising LGBT-related events; posters stating campus policies against violence and harassment; publication and distribution of pamphlets and brochures stating policies and sanctions; programs on campus radio and television stations; and inclusion in handbooks and promotional materials of information about services for LGBT students all help to send messages about the institution's position with regard to violence against LGBT students and about the climate for LGBT people.

Consultation and training for staff and faculty are vital if a positive climate is to be accomplished. Studies have demonstrated that many candidates for residence life positions (D'Augelli, 1989a), faculty (DeSurra & Church, 1994; LaSalle, 1992), university staff (LaSalle, 1992), campus police (Yeskel, 1985), and student service staff (Yeskel, 1985) exhibit homophobic and heterosexist behavior. Training front-line staff—including campus police officers, residence life staff, medical personnel, and counselors—to sensitively handle and to take seriously reports of LGBT violence is key to building trust within the LGBT community. Only when people feel safe and respected will they report incidents of victimization and seek the support they need and deserve. Consultation and training provided by individuals with credibility, such as fellow police officers or other doctors, is typically more effective than that provided by LGBT advocates. Faculty development efforts provided by academic scholars will certainly attract more participants than workshops given by the campus diversity officer.

Direct intervention through counseling, teaching, and advising is a necessary component of any comprehensive strategy. Counseling can take the form of psychological counseling at the campus counseling center, crisis counseling provided by a hot line or an LGBT advocate, or peer counseling made available by trained LGBT undergraduates. All counselors must be knowledgeable about general LGBT issues as well as specific issues faced by victims of violence. Teaching can occur either in the classroom or in educational workshops. Providing information about the extent and impact of violence against LGBT individuals both in formal and informal settings is vital to increasing awareness. Including information in orientation programs for new students and faculty helps to convey the university's expectations regarding treatment of LGBT individuals. High-profile guest lecturers often have a greater impact than local staff and faculty. Advising can take many forms. Supportive faculty and staff advisors serve as important role models for LGBT student organization members. Informal advising provided by members of a gay-friendly support network or by staff of a LGBT advocacy office can assist students who are seeking advice about victimization or other issues they are facing.

CREATING CHANGE

Proposing needed interventions does not make them happen. Creating change, especially when attacking oppressive structures, is difficult. Based on qualitative and quantitative analyses of planned change in higher education, Creamer and Creamer (1990) developed a theoretical model of change, the "probability of the adoption of change (PAC)" model. In this model they identify nine key variables within an organization that influence the potential for change: (1) leadership, (2) championship, (3) top-level support, (4) circumstances, (5) value compatibility, (6) idea comprehensibility, (7) practicality, (8) advantage probability, and (9) strategies.

Three leadership variables—leadership, championship, and top level support—were found to play a central role in effecting change (Creamer & Creamer, 1990). Successful projects had a clearly identified leader who was strongly invested in the project. Project champions were advocates who developed support for the project in both the planning and the implementation stages. Top-level support from the chief executive officer was instrumental in creating a climate for change. These three variables, in turn, influenced the other six variables. The key factor, then, in initiating change appears to be ensuring that sufficient leadership exists to sponsor, advocate for, and carry out the steps necessary to get a program under way.

Advocacy must come from within the LGBT community and their allies. Unfortunately, fear of discrimination, harassment, or violence leads many LGBT individuals to adopt a very low profile on campus (D'Augelli, 1989b). Senior-level staff and faculty with tenure are logical persons to provide leadership (Rankin, 1998). LGBT student organizations are also often instrumental in creating awareness and advocating for change on campus (D'Augelli, 1989b, Rankin, 1998). Rankin (1998) advocates a "top-down, bottom-up" approach.

Gaining the support of top-level administrators, including the president, usually takes a sustained and well-coordinated effort on the part of advocates. D'Augelli (1989b) reported on a successful effort to get sexual orientation included in a university nondiscrimination policy. Building on gains made by other minority groups was part of this strategy. Many campuses are responding to pressure to address multicultural issues more effectively. National education reports calling for inclusive, welcoming environments on campus (for example, Association of American College and Universities, 1995; Boyer, 1990) can also be used as leverage to gain the attention of senior university officials. Demonstrating that the concerns of LGBT students parallel those of other student groups can be effective when working for equitable treatment.

Collecting information in a systematic way about harassment and violence experienced by LGBT students also helps to build the case that the institution must attend to this problem (D'Augelli, 1989b). Federal requirements to report incidents of victimization can be used to ensure that hate crimes against LGBT persons are publicized. Such reports encourage universities to demonstrate that they are doing something to address the issue of crime and violence.

Leadership to carry out interventions should be provided by an individual who is knowledgeable about the issues and invested in making change happen. A task force or commission on LGBT concerns can often play this role. A key administrator, such as the chief student affairs officer (assuming that this person is qualified), can also guide development and implementation of strategies. In some cases, a campus diversity officer might provide leadership.

Once the leadership variables are in place, individuals playing these key roles need to work to ensure that circumstances, such as felt need and timing, favor implementation of change (Creamer & Creamer, 1990). Underlying values compatible with those of the project need to be identified. Ideas must be clearly stated for all constituents. The practical details of resources and staffing must be addressed. Finally, the benefits of the intervention for the university community must be identified, and strategies must be developed to gain support for the project. Failure to address these variables systematically can lead to unsuccessful outcomes.

REFERENCES

Association of American Colleges and Universities. (1995). *The drama of diversity and democracy: Higher education and American commitments*. Washington, DC: Author.

Berk, R., Boyd, E.A., & Hammer, K.M. (1992). Thinking more clearly about hate motivated crimes. In G. Herek & K. Berrill (Eds.), *Hate crimes: Confronting violence against lesbians and gay men* (pp. 19–35). Newbury Park, CA: Sage.

Berrill, K. (1988). Organizing for equality. *Newsletter of NGLTF Campus Project*, pp. 334–349.

Berrill, K. (1990). Anti-gay violence and victimization in the United States. *Journal of Interpersonal Violence, 5*, 274–294.

Berrill, K. (1992). Anti-gay violence and victimization in the United States: An overview. In G. Herek & K. Berrill (Eds.), *Hate crimes: Confronting violence against lesbians and gay men* (pp. 19–45). Newbury Park, CA: Sage.

Berrill, K.T., & Herek, G.M. (1992). Primary and secondary victimization in anti-gay hate crimes: Official response and public policy. In G.M. Herek & K.T. Berrill (Eds.), *Hate crimes: Confronting violence against lesbians and gay men* (pp. 289–305). Newbury Park, CA: Sage.

Boyer, E. (1990). *Campus life: In search of community*. Princeton, NJ: The Carnegie Foundation for the Advancement of Teaching.

Brenner, C. (1995). *Fight bullets: One woman's story of surviving anti-gay violence*. Ithaca, NY: Firebrand Books.

Center for Democratic Renewal. (1992). *When hate groups come to town*. Montgomery, AL: The Black Belt Press.

Comstock, G.D. (1989). Victims of anti-gay/lesbian violence. *Journal of Interpersonal Violence, 4*, 101–106.

Comstock, G.D. (1991). *Violence against lesbians and gay men*. New York: Columbia University Press.

Creamer, D.G., & Creamer, E.G. (1990). Use of a planned change model to modify student affairs programs. In D.G. Creamer & Associates, *College student development: Theory and practice for the 1990s* (pp. 181–192). Alexandria, VA: American College Personnel Association.

D'Augelli, A.R. (1989a). Homophobia in a university community: Views of prospective resident assistants. *Journal of College Student Development, 30*, 546–552.

D'Augelli, A.R. (1989b). Lesbians and gay men on campus: Visibility, empowerment, and educational leadership. *Peabody Journal of Education, 66*, 124–142.

D'Augelli, A.R. (1989c). Lesbians' and gay men's experiences of discrimination and harassment in a university community. *American Journal of Community Psychology, 17*, 317–321.

D'Augelli, A.R. (1992). Lesbian and gay male undergraduates' experiences of harassment and fear on campus. *Journal of Interpersonal Violence, 7*, 383–395.

D'Augelli, A.R. (1993). Preventing mental health problems among lesbian and gay college students. *Journal of Primary Prevention, 13*, 245–261.

D'Emilio, J. (1990). The campus environment for gay and lesbian life. *Academe, 76*(1), 16–19.

DeSurra, C.J., & Church, K.A. (1994, February). *Unlocking the classroom closet: Privileging the marginalized voices of gay/lesbian college students*. Paper presented at the Annual Meeting of the Speech Communication Association, New Orleans, LA.

Fenn, P., & McNeil, T. (1987). *The response of the criminal justice system to bias crime: An exploratory review*. Cambridge, MA: Abt Associates.

Fernandez, J. (1991). Bringing hate crimes into focus. *Harvard Civil Rights-Civil Liberties Law Review, 26,* 261–292.

Gibson, P. (1989). Gay male and lesbian youth suicide. In ADAMHA, *Report of the Secretary's Task Force on Youth Suicide* (Vol. 3, pp. 110–142; DHHS Publication No. ADM 89-1623). Washington, DC: U.S. Government Printing Office.

Glaus, K.O. (1988). Alcoholism, chemical dependency, and the lesbian client. *Women and Therapy, 8,* 131–144.

Hentoff, N. (1990, September 25). The violently attacked community in America. *The Weekly Newspaper of New York.*

Herek, G. (1989). Hate crimes against lesbians and gay men: Issues for research and policy. *American Psychologist, 44*(6), 948–955.

Herek, G. (1990). The context of anti-gay violence. *Journal of Interpersonal Violence, 5,* 316–333.

Herek, G. (1994). Heterosexism, hate crimes, and the law. In M. Costanzo & S. Oskamp (Eds.), *Violence and the law* (pp. 121–142). Newbury Park, CA: Sage.

Herek, G.M. (1995). Psychological heterosexism in the United States. In A.R. D'Augelli & C.J. Patterson (Eds.), *Lesbian, gay, and bisexual identities over the lifespan* (pp. 321–346). New York: Oxford University Press.

Herek, G., & Berrill, K. (1990). Documenting the victimization of lesbians and gay men. *Journal of Interpersonal Violence, 5,* 301–315.

Herek, G., & Berrill, K. (1992). *Hate crimes: Confronting violence against lesbians and gay men.* Newbury Park, CA: Sage.

Hernandez, T.A. (1990). Bias crimes: Unconscious racism in the prosecution of 'racially-motivated' violence. *Yale Law Journal, 99,* 832–864.

Hershberger, S.L., & D'Augelli, A.R. (1995). The impact of victimization on the mental health and suicidality of lesbian, gay, and bisexual youth. *Developmental Psychology, 31,* 65–74.

Hurst, J.C., & Jacobson, J.K. (1985). Theories underlying students' needs for programs. In M.J. Barr, L.A. Keating, & Associates (Eds.), *Developing effective student service programs: Systematic approaches for practitioners* (pp. 113–136). San Francisco: Jossey-Bass.

Janoff-Bulman, R., & Frieze, I. (1983). A theoretical perspective for understanding reactions to victimization. *Journal of Social Issues, 39*(2), 1–17.

Jenness, V. (1995). Social movement growth, domain expansion, and the framing processes: The gay/lesbian movement and violence against gays and lesbians as a social problem. *Social Problems, 42*(1), 145–170.

Jenness, V., & Grattet, R. (1993). *The criminalization of hate: The social context of hate crimes in the United States.* Paper presented at the annual meeting of the American Sociological Association, Miami, FL.

Kuh, G.D., & Whitt, E.J. (1988). *The invisible tapestry: Culture in American college and universities* (ASHE-ERIC Higher Education Report No. 1). Washington, DC: Association for the Study of Higher Education.

LaSalle, L.A. (1992, April). *Exploring campus intolerance: A textual analysis of comments concerning lesbian, gay, and bisexual people.* Paper presented at the meeting of the American Educational Research Association, San Francisco, CA.

Levin, J., & McDevitt, J. (1993). *Hate crimes: The rising tide of bigotry and bloodshed.* New York: Plenum.

Manning, K., & Coleman-Boatwright, P. (1991). Student affairs initiatives toward a multicultural university. *Journal of College Student Development, 32,* 367–374.

Morrill, W.H., Oetting, E.R., & Hurst, J.C. (1974). Dimensions of intervention for student development. *Personnel and Guidance Journal, 52,* 354–359.

National Institute Against Violence and Prejudice. (1993). *Striking back at bigotry: Remedies under federal and state law for violence motivated by racial, religious, or ethnic prejudice.* Baltimore, MD: Author.

Norris, F.H., & Kaniasty, K. (1991). The psychological experience of crime: A test of the mediating role of beliefs in explaining the distress of victims. *Journal of Social and Clinical Psychology, 10,* 239–261.

Peterson, M., & Spencer, M. (1990). Understanding academic culture and climate. In W. Tierney (Ed.), *Assessing academic climates and cultures* (pp. 3–18). San Francisco, CA: Jossey-Bass.

Rankin, S. (1994). *The perceptions of heterosexual faculty and administrators toward gay men and lesbians.* Unpublished doctoral dissertation, The Pennsylvania State University, University Park, PA.

Rankin, S. (1998). Campus climate for lesbian, gay, bisexual, and transgendered students, faculty, and staff: Assessment and strategies for change. In R. Sanlo (Ed.), *Working with lesbian, gay, and bisexual students: A guide for administrators and faculty* (pp. 203–212). Westport, CT: Greenwood Publishing.

Remafedi, G. (1987). Adolescent homosexuality: Psychosocial and medical implications. *Pediatrics, 79,* 331–337.

Rich, A. (1980). Compulsory heterosexuality and the lesbian existence. *Signs, 5*(4), 631–650.

Savin-Williams, R.C., & Cohen, K.M. (1996). Psychological outcomes of verbal and physical abuse among lesbian, gay, and bisexual youths. In R.C. Savin-Williams & K.M. Cohen (Eds.), *The lives of lesbians, gays, and bisexuals: Children to adults* (pp. 181–200). Fort Worth, TX: Harcourt Brace.

Slater, B.R. (1993). Violence against lesbian and gay male college students. *Journal of College Student Psychotherapy, 8,* 177–202.

Tierney, W.G. (Ed.). (1990). *Assessing academic climates and cultures.* San Francisco, CA: Jossey-Bass.

Tierney, W.G. (1992). *The report of the Committee for Lesbian and Gay Concerns.* University Park, PA: The Pennsylvania State University.

Tierney, W.G., & Dilley, P. (1998). Constructing knowledge: Educational research and gay and lesbian studies. In W. Pinar (Ed.), *Queer theory in education* (pp. 49–71). Mahwah, NJ: Lawrence Erlbaum.

Tierney, W.G., & Rhoads, R. (1993). Enhancing academic communities for lesbian, gay, and bisexual faculty. In J. Gainen & R. Boice (Eds.), *Building a diverse faculty* (New Directions for Teaching and Learning, no. 53, pp. 43–50). San Francisco: Jossey-Bass.

The year in review. (1994, January 25). *The Advocate,* 48–54.

Yeskel, F. (1985). *The consequences of being gay: A report on the quality of life for lesbian, gay and bisexual students at the University of Massachusetts at Amherst.* Unpublished report. Amherst, MA: University of Massachusetts at Amherst.

CHAPTER 11

Sexual Harassment of Students: The Hidden Campus Violence

Michele A. Paludi and Darlene C. DeFour

INTRODUCTION

Colleges and universities are violent institutions for students, especially for women students (Paludi, 1997b). Acts of violence and sexual harassment are alarmingly frequent on campuses. Chapter 9 focuses on violence against women on campus, with an emphasis on physical violence, including rape and other coercive acts. This chapter discusses sexual harassment of students, which may or may not include physical violence. The chapter begins by briefly setting the overall context of sexual violence against students and then discusses sexual harassment.

Research has suggested that approximately one third of all American college students report using or being victims of courtship violence, with pushing, grabbing, shoving, slapping, and throwing of objects being the most common forms of violent behavior in dating relationships (Follingstad et al., 1992; Ryan, Frieze, & Sinclair, in press; White & Koss, 1991). In addition, between 8% and 15% of college women have disclosed that they were raped according to the legal definition of rape, which includes acts involving nonconsensual sexual penetration obtained by physical force; by threat of bodily harm; or when the victim is incapable of giving consent by virtue of mental illness, mental retardation, or intoxication. The majority of women report that they knew their attacker and that the rape occurred while on a "date" (Koss, 1993; Rozee, in press).

Just as disturbing as these rates of sexual assault is that these acts of violence and sexual harassment are perpetrated on college campuses—suggesting that colleges and universities are violent institutions for students, especially women (Paludi, 1997b). Male violence against women on college campuses crosses lines of sexual orientation, age, ethnicity, and economic status (Goodman et al., 1993). Many acts of sexual violence on college campuses involve the use of alcohol and other drugs (Paludi, 1997b). College women victims of sexual violence show identifiable degrees of impairment on standard psychological tests. Sexual victimization has been recognized as a significant etiology in eating disorders, borderline syndrome, and multiple personality disorder (Goodman et al., 1993). Women students who have been victimized while at college typically change their educational program or career goals as a consequence of being victimized; many leave the campus, never to complete their degree. In addition, women have reported emotional and physical reactions to being victimized, including depression, insomnia, headaches, helplessness, and decreased motivation (Koss, 1993; Rozee, in press). Violence on college campuses occurs in a social context in which women experience physical hardship, loss of scholarships, administrative neglect, and isolation (Paludi, 1997b).

Recently, Paludi, DeFour, Attah, & Batts (in press) reported that college students, when asked to identify activities that would fall under the heading "campus violence," generated the following terms: hazing, rape, assault, battering, emotional abuse, and date rape. The term *sexual harassment* was not mentioned. This finding may be surprising, since current estimates of sexual harassment of college students by professors are high, ranging from sexual come-ons to sexual coercion (such as threatening to lower a student's grades for noncompliance with a faculty member's requests for sexual activity) (Paludi, 1996). Students' omission of sexual harassment from a list of behaviors that constitute campus violence reflects the silence that still surrounds sexual harassment on college campuses. Sexual harassment remains a "hidden issue," as the Project on the Status and Education of Women referred to it in 1978.

Paludi (1995, 1996, 1997b) has offered the following explanation for why sexual harassment is the hidden campus violence. Sexual harassment confronts victims with perceptions that are often invalidated by those around them. Victims' experiences get relabeled as anything *but* sexual harassment. Most victims of sexual harassment doubt their experiences; they are encouraged to conform to their professors' (and campus administrators') relabeling of the experiences. Professors are seen as essential because they generate power. They give grades, write letters of recommendation and speak to colleagues about students' performance in classes. Perhaps professors' greatest power lies in the capacity to enhance or diminish students' self-esteem (Carr, 1991; MacPike, 1996; Paludi, 1990; 1996; Payne, 1991; Riggs & Murrell, 1995; Watts, 1996). This power can motivate students to learn or convince them to end their college career (Zalk, Paludi, & Dederich, 1991). Consequently, students remain silent about their experiences with sexual harassment. Thus, the actual incidence of sexual harassment on college campuses remains hidden.

College students are typically not informed about the incidence of sexual harassment in general or on their campus, specifically. A "null environment" is set up for them on campus, in which they are not provided with necessary information, especially concerning the sanctions for sexual harassment (Paludi & Barickman, in press). Thus, they may be more likely to reinterpret their experiences as not sexual harassment and to remain silent about their victimization.

Most students learn about faculty harassers informally from other students. For example, Adams, Kottke, and Padgitt (1983) reported that 13% of the women students they surveyed stated that they had avoided taking a class or working with certain professors because of the risk of being subjected to the professors' sexual advances. Bailey and Richards (1985) found that of 246 women graduate students, 21% had not enrolled in a course to avoid being sexually harassed by the professor.

This chapter discusses sexual harassment on college campuses. Specifically, the chapter discusses the legal and behavioral definitions of sexual harassment, the impact of sexual harassment on students and professors, and explanatory models for why sexual harassment exists on college campuses. The chapter concludes with suggestions for making the issue of sexual harassment more visible on college campuses by using effective policy statements, investigatory procedures, and campuswide educational programs. In addition, the chapter summarizes other forms of sexual harassment on college and university campuses (such as men as victims rather than perpetrators of sexual harassment).

LEGAL DEFINITION

Sexual harassment of students is prohibited within colleges and universities as a form of sexual discrimination under Title IX of the 1972 Educational Amendments. Title IX prohibits discrimination on the basis of sex and covers all educational institutions that receive federal financial assistance and all federally funded educational programs in noneducational institutions. It also covers institutions whose students receive federal financial aid.

The Office of Civil Rights (OCR) at the U.S. Department of Education is the federal agency that ensures that schools and campuses comply with Title IX. According to OCR: "Title IX protects students from unlawful sexual harassment in all of a school's programs or activities, whether they take place in the facilities of the school, on a school bus, at a class or training program sponsored by the school at another location, or elsewhere. Title IX protects both male and female students from sexual harassment, regardless of who the harasser is."

OCR recognizes that sexual harassment may take two forms: *quid pro quo* sexual harassment and *hostile environment* sexual harassment. The major elements of *quid pro quo* sexual harassment are as follows: (1) the sexual advances are unwanted; (2) the harassment is sexual; (3) the submission is explicitly or implicitly tied to a term or condition of school status or is used as a basis for making decisions about the student's college status (e.g., a professor threatens to fail a student unless the student agrees to comply with the professor's sexual requests).

Hostile environment sexual harassment refers to unwelcome behaviors that are pervasive and/or severe and that affect a student's ability to participate in or benefit from an educational program or create an abusive educational environment for the student. Examples of hostile environment sexual harassment include making sexual innuendoes, comments, and remarks; brushing up against the body; leering or ogling; making insulting, degrading, or sexist comments; and using pornographic teaching materials in the classroom. According to OCR, a hostile environment can be created by a school employee, another student, or even someone visiting the school, such as a student or employee from another school.

Sexual harassment is prohibited within colleges and universities for employees under Title VII of the Civil Rights Act. The Equal Employment Opportunity Commission (EEOC) is the federal agency designated by Title VII to interpret the law and handle sexual harassment complaints. The EEOC guidelines define sexual harassment as unwelcome sexual advances, requests for sexual favors, and other verbal or physical conduct of a sexual nature when any one of the following criteria is met:

- Submission to such conduct is made either explicitly or implicitly a term or condition of the individual's employment.
- Submission to or rejection of such conduct by an individual is used as the basis for employment decisions affecting the individual.
- Such conduct has the purpose or effect of unreasonably interfering with an individual's work or creating an intimidating, hostile, or offensive work environment.

Quid pro quo sexual harassment is represented by the first two statements; hostile environment sexual harassment is represented by the third statement. This definition has been extended to college students as well (Watts, 1996). For example, in 1992, the U.S. Supreme Court cited an employment sexual harassment case and made the analogy between a student's relationship to a professor and an employee's relationship to a supervisor. The Court said: "When a supervisor sexually harasses a subordinate because of the subordinate's sex, that supervisor 'discriminates' on the basis of sex. . . . We believe the same rule should apply when a teacher sexually harasses and abuses a student" (112 S.Ct. 1028,1033).

INCIDENCE OF FACULTY-STUDENT SEXUAL HARASSMENT

Researchers have reported that approximately 50% of undergraduate women and approximately 10% of college men experience sexual harassment from at least one of their instructors during their 4 years of college. When definitions of sexual harassment include sexist remarks and other forms of "gender harassment," the percentage of undergraduate women who experience sexual harassment nears 70% (Fitzgerald & Omerod, 1993; Paludi, 1996).

Fitzgerald and colleagues (1988) investigated approximately 2,000 women at two major state universities. Half of the women respondents reported experiencing some form of sexually harassing behavior. The majority of this group reported sexist comments directed toward them by faculty; the next largest category of harassment behavior was seductive behavior, including faculty who invited them for drinks and a backrub, brushed up against them, or showed up uninvited to their hotel rooms during out-of-town conventions.

Adams (1997) reported that the incidence of academic sexual harassment of women of color is even greater than that reported by white women. DeFour (1996), Barak (1997), and Adams (1997) suggested that women of color are more vulnerable to receiving sexual attention from professors; they are subject to stereotypes about sex, are viewed as sexually mysterious and inviting, and are less sure of themselves in their careers (DeFour, 1996). Adams (1997) suggested that the sexual harassment of women of color is even more hidden than that of white women as a consequence of institutionalized racism.

Similar conclusions were reached by Tang, Yik, & Cheung (1996) from their study of sexual harassment of Chinese college students. Their results indicated that compared to men, twice as many women said they had been sexually harassed, approximately 25% of women students experienced various levels of sexual harassment, and 1% were coerced into sexual activities by professors.

Barickman, Paludi, and Rabinowitz (1992) noted that, in addition to women of color, the following women are more likely than other women to experience sexual harassment by professors:

- graduate students, whose future careers are often determined by their association with a particular faculty member
- students in small colleges or small academic departments, where the number of faculty available to students is quite small
- students in male-populated fields (such as engineering)
- students who are economically disadvantaged and work part-time or full-time while attending classes
- lesbians who may be harassed as part of homophobia

- physically or emotionally disabled students
- victims of past sexual abuse
- inexperienced, unassertive, or socially isolated students, who may appear more vulnerable and appealing to those who would intimidate or entice them into an exploitive relationship

According to Sandler and Paludi (1993), campuses most likely to have a high incidence of faculty-student sexual harassment (as well as other forms of sexual harassment) have the following characteristics:

- Have no effective and enforced policy prohibiting sexual harassment.
- Have no effective and enforced investigatory procedures for dealing with sexual harassment.
- Have no training programs on sexual harassment awareness.
- Do not support victims of sexual harassment with therapeutic assistance and assistance with courses.
- Do not give sanctions to individuals who engage in sexual harassment.

HOW STUDENTS DEFINE THEIR EXPERIENCES

Fitzgerald, Gold, and Brock (1990) classified individuals' responses into two categories: internally focused strategies and externally focused strategies. Internal strategies represent attempts to manage the personal emotions and cognitions associated with the behaviors they have experienced. The following are internally focused strategies:

- *Detachment:* Individual minimizes situation or treats it as a joke.
- *Denial:* Individual denies behaviors and attempts to forget about it.
- *Relabeling:* Individual reappraises situation as less threatening and offers excuses for harasser's behaviors.
- *Illusory control:* Individual attempts to take responsibility for harassment.
- *Endurance:* Individual puts up with behavior because of the belief that no help is available or because of fear of retaliation.

Externally focused strategies focus on the harassing situation itself, including reporting the behavior to the individual charged with investigating complaints of sexual harassment. Fitzgerald, Gold, and Brock (1990) classified externally focused strategies into the following categories:

- *Avoidance:* Individual attempts to avoid situation by staying away from the harasser.

- *Assertion/confrontation:* Individual refuses sexual or social offers or verbally confronts the harasser.
- *Seeking institutional/organizational relief:* Individual reports the incident and files a complaint.
- *Social support:* Individual seeks support of others to validate perceptions of the behaviors.
- *Appeasement:* Individual attempts to evade the harasser without confrontation; attempts to placate the harasser.

Ormerod and Gold (1988), using this classification system, noted that internal strategies represented by far the most common response overall. Most people do not tell a harasser to stop. Their initial attempts to manage the initiator are rarely direct. Typically harassers are more powerful—physically and organizationally—than the victims, and sometimes the harasser's intentions are unclear. The first or first few harassing events are often ignored by victims—especially when they are experiencing hostile-environment sexual harassment and when there are no effective and enforced policy statement and investigatory procedures that make it "safe" to report sexual harassment (McKinney, 1994; Paludi & Barickman, in press; Payne, 1991; Riggs & Murrell, 1995). Victims may interpret or reinterpret the situation so that the incident is not defined as sexual harassment. Victims of sexual harassment are concerned about retaliation should they confront the harasser. The reality for most students is that they cannot simply leave the college where they are being sexually harassed. This is underscored for students who are receiving financial aid and scholarships (DeFour, 1996).

Malovich and Stake (1990) found that women students who were high in performance self-esteem and who held nontraditional gender role attitudes were more likely to report incidents of sexual harassment than women who were high in self-esteem and who held traditional gender role attitudes, or women who were low in self-esteem. Brooks and Perot (1991) found that reporting behavior was predicted by the severity of the offense and by feminist attitudes on the part of the student. Reilly, Lott, and Gallogly (1986) found that 61% of women college students ignored the sexual harassment or did nothing, and 16% told the faculty member to cease the sexual harassment.

PROFESSOR ABUSE OF AND DENIAL OF THE IVORY POWER

Professors who sexually harass students typically do not label their behavior as sexual harassment, despite the fact that they report they frequently engage in initiating personal relationships with students. They deny the inherent power differential between themselves and their students as well as the psychological power conferred by this differential that is at least as salient as the power derived from their role as evaluators (Fitzgerald & Ormerod, 1993; Levy & Paludi, 1997; Paludi, 1996).

Kenig and Ryan (1986) reported that male professors are less likely than female professors to define sexual harassment as including jokes, teasing remarks of a sexual nature, and unwanted suggestive looks or gestures. In addition, female professors are more likely than male professors to disapprove of romantic relationships between students and faculty. Male professors are also more likely than female professors to agree with the following statements: "An attractive woman has to expect sexual advances and learn how to handle them"; "It is only natural for a man to make sexual advances to a woman he finds attractive"; and "People who receive annoying sexual attention usually have provoked it."

Male professors are also more likely than women to agree with the following statements, taken from Paludi's (1995) survey of "attitudes toward victim blame and victim responsibility":

- Women often claim sexual harassment to protect their reputations.
- Many women claim sexual harassment if they have consented to sexual relations but have changed their minds afterwards.
- Sexually experienced women are not really damaged by sexual harassment.
- It would do some women good to be sexually harassed.
- Women put themselves in situations in which they are likely to be sexually harassed because they have an unconscious wish to be harassed.
- In most cases when a woman is sexually harassed, she deserved it.

Kenig and Ryan (1986) also reported that male professors were more likely than female professors to believe that individuals can handle unwanted sexual attention on their own without involving the college or university. Thus, male professors view sexual harassment as a personal, not an organizational issue; they minimize the responsibility of the college administration.

Bernice Lott (1993) and her colleagues have also found empirical support for a widely accepted assumption among researchers—that sexual harassment is part of a larger and more general misogyny. This hostility toward women includes extreme stereotypes of women such as the idea that sexual harassment is a form of seduction and that women secretly need/want to be forced into sex (Zalk, 1996).

Dziech and Weiner (1984) classified academic sexual harassers in two categories: public and private. Public harassers are professors who engage in flagrant sexist and/or seductive behavior toward women students. Private harassers are restrained, intimidating professors who coerce students with their formal authority. In addition to these two categories, Dziech and Weiner suggest other roles that may be assumed by a professor who sexually harasses students: counselor/helper; confidant; intellectual seducer; and opportunist. The intellectual seducer requires self-disclosure on the part of students (such as through journal writing), which he then uses to gain personal information about the student. The opportunist uses the physical setting (such as a laboratory) to gain intimacy with students.

Men who sexually harass are not distinguishable from their colleagues who do not harass with respect to the following variables: age, marital status, occupation, or job status (Fitzgerald & Weitzman, 1990). Men who harass do so repeatedly to many women, especially when they have not received any training and counseling about the impact of their behavior on other individuals.

Paludi (1995) noted that theoretical accounts to explain sexual harassment of women students by male professors should include men's attitudes toward other men, competition, and power. Many of the men act out of extreme competitiveness and concern with ego or out of fear of losing their positions of power. They do not want to appear weak or less masculine in the eyes of other men so they sexually harass women. Women are the objects of the game to impress other men. When men are encouraged to be obsessively competitive and concerned with dominance, it is likely that they will eventually use violent means to achieve dominance (Paludi, 1995). Male professors who sexually harass students are also likely to be verbally abusive and intimidating in their body language.

For many men, aggression is one of the major ways of proving their masculinity, especially among those men who feel some sense of powerlessness in their lives (Doyle & Paludi, 1998). The male-as-dominant or male-as-aggressor is a theme so central to many men's self-concept that it literally carries over into their interpersonal communications, especially with women. Sexualizing a professional relationship may be the one area where a man can still prove his masculinity when few other areas can be found for him to prove himself in control, or the dominant one in a relationship. Thus, sexual harassment is not a deviant act, but rather an over-conforming to the masculine role in this culture (Fitzgerald & Weitzman, 1990; Paludi, 1996).

CONSENSUAL RELATIONSHIPS

Neither Title IX nor Title VII prohibit consensual sexual relationships between faculty and students. However, college administrators have been developing policies on this matter because they recognize that consensual relationships cause serious problems for the following reasons (Sandler & Paludi, 1993):

- The situation involves one person exerting power over another.
- The seduction of a much younger individual is usually involved.
- Conflicts of interest arise (for example, how can a teacher fairly grade a student with whom he or she is having a sexual relationship?).
- The potential for exploitation and abuse is high.

- The potential for retaliatory harassment is high when the sexual relationship ceases.
- Other individuals may be affected and claim favoritism.

For example, the University of Minnesota, University of Connecticut, New York University, and Massachusetts Institute of Technology have "discouragement policies," in which consensual relations are not strictly prohibited but discouraged. The University of Iowa, Harvard University, Yale University, and Temple University have policies that prohibit sexual relationships between faculty and students over whom the professor has some authority (advising, supervising, grading, teaching). However, a faculty member does not have to be the student's professor in order to be powerful and to have the potential to abuse that power over the student. A professor's relationship with a student (as a mentor, for example) can often involve emotional complexities and power differentials similar to the relationship of a counselor/therapist and a client (Stites, 1996; Zalk, Paludi & Dederich, 1991).

Keller (1988) suggests that when a college does not address consensual relationships in its sexual harassment policy, professors mask sexual harassment as mentoring or being a role model and consequently make it difficult for students to label the behavior as sexual harassment. DeChiara (1988) argues that students will be spared the stress of coping with unwanted sexual attention from professors when a college has a consensual relationship policy. This issue is addressed in more detail later in the chapter.

OTHER FORMS OF SEXUAL HARASSMENT ON COLLEGE CAMPUSES

Research findings suggest that male faculty members are more likely to engage in sexual harassment of women students than any other faculty-student combination. This section summarizes research on other types of sexual harassment that may occur on college and university campuses. For example, it is possible for female professors to harass male students. The incidence of this form of sexual harassment, however, is quite small (Fitzgerald & Omerod, 1993). The incidence of women sexually harassing other women is also small. Many of men's experiences with sexual harassment are with other men (Berdahl, Magley, & Waldo, 1996; U.S. Merit Systems Protection Board, 1981); men may be reluctant to disclose this information due to homophobic concerns (Levy & Paludi, 1997), thereby keeping this form of sexual harassment hidden.

Several issues have been raised in research studies on the sexual harassment of men. For example, Berdahl, Magley, and Waldo (1996) found that men report being significantly less offended or threatened by behaviors that women find sexually harassing. These authors also reported that, for women and men, situations that are experienced as sexually harassing are those that signify a perceived loss of control and security in the workplace. The actual sexually harassing behaviors, however, differ for women and men. Behaviors that men label sexually harassing are ones that represent challenges to current constructions of masculinity (such as ridiculing and name-calling a man who wants to take parental leave following the birth of a child).

Pryor and Whalen (1997) assert that male-to-male sexual harassment may be more common in male-populated occupations. The U.S. Department of Defense survey of active-duty military personnel reported that approximately 64% of the women and 17% of the men indicated they experienced sexual harassment from a coworker (Martindale, 1992), with sexual teasing, jokes, or remarks being the most commonly experienced sexually harassing behavior.

Saal, Johnson, and Weber (1989) found that men frequently interpret women's behavior as sexual even when it is not. It is likely that men may mislabel women's behavior as sexual and consequently, if it is unwanted, as sexual harassment.

Student-to-student or peer sexual harassment continues to receive considerable attention in the research literature. Ivy and Hamlet (1996), for example, reported that peers were a main source of sexual harassment for women students they sampled. The main verbal form of peer sexual harassment experienced by women students involved "lewd" comments or sexual comments in college classroom settings. Tang, Yik, and Cheung (1996) also reported that peer sexual harassment occurred twice as frequently as faculty-student sexual harassment in their sample of 491 women and 358 men Chinese college students.

"Contrapower" sexual harassment involves subordinates as harassers. Women professors have reported experiencing sexual harassment from male students (Grauerholz, 1989; Quina, 1996). Most women professors choose to remain silent about this type of sexual harassment for fear of professional repercussions, underscoring their tenuous existence on campus (Quina, 1996).

IMPACT OF SEXUAL HARASSMENT ON INDIVIDUALS

The impact of sexual harassment and the failure of an institution to take action once it has been notified about sexual harassment has been examined from three main perspectives: work/study outcomes, psychological outcomes, and physical outcomes. Researchers have documented decreased morale, increased absenteeism, decreased school satisfaction, lowered grades, and damage to interpersonal relationships on campus (Dansky & Kilpatrick, 1997; Koss, 1990; Quina, 1996; Rabinowitz, 1990). The consequences of sexual harassment for individuals' emotional well-being include depression, helplessness, extreme sadness, strong fear reactions, loss of control, worry, disruption of their lives, and decreased motivation (Dansky & Kilpatrick, 1997; Rabinowitz, 1990).

The following physical symptoms have been reported as a result of academic sexual harassment: headaches, sleep disturbances, eating disorders, gastrointestinal disorders, nausea, and crying spells (Dansky & Kilpatrick, 1997; Koss, 1990; Quina, 1996; Rabinowitz, 1990).

These responses are influenced by disappointment and self-blame in the way others react and the stress of sexual harassment-induced life changes, such as loss of student loans, loss of teaching or research fellowships and disrupted educational or work history. For example, Schneider (1987) found that 29% of women reported a loss of academic or professional opportunities and 14% reported lowered grades or loss of financial support because of sexual harassment.

METHODOLOGICAL ISSUES IN RESEARCH ON SEXUAL HARASSMENT

It is important for researchers who are investigating sexual harassment to be aware of the implications of the questions they are asking and the underlying assumptions they are making (Lengnick-Hall, 1995; Paludi, DeFour, Attah, & Batts, in press). For example, one research formulation is that sexually harassed women suffer from helplessness. Evidence outside of the laboratory suggests, however, that sexually harassed students and employees seek help from friends, family, trusted faculty, therapists, and—when the institution has an effective and enforced policy statement and investigatory procedures—they seek help from the campus or workplace. Fitzgerald & Omerod (1993) identified a variety of coping styles—internal and external. Thus, although it may be true that many victims remain silent, they are not helpless. This research-based assumption of helplessness contributes to revictimizing women who do come forward. Disclosure to friends, family, and faculty is significant; this information should be used in investigation as data about the impact of sexual harassment on the individual.

Other research suggests that it is the responses of others (especially college administrators) to individuals' requests for resolution of the sexual harassment that are inadequate. Rather than continuing to research the problems of sexually harassed students, it might be more profitable to systematically address effective interventions, training programs, and investigatory procedures. This suggestion will be expanded later in the chapter.

In addition, much of the research on academic and workplace sexual harassment has given little attention to construct validity issues (Lengnick-Hall, 1995). Researchers have not typically asked questions that sexually harassed students and employees themselves might ask. Researchers have not typically investigated sexually harassed individuals' own conceptions of their experiences. The research studies rely heavily on descriptive situations as research stimuli. Students, for example, are asked if they believe the description of an incident is sexual harassment and are asked how they would respond to the situation presented. To date, there is no empirical research to suggest that how individuals respond to descriptive situations on paper in any way reflects how they would respond in an actual situation (Lengnick-Hall, 1995).

In these research studies, ambiguous situations are typically presented to college participants with no criteria for determining the occurrence of sexual harassment. Consequently, individuals use their own definitions of sexual harassment in making their judgments (Frazier, Cochran, & Olson, 1995). Would we obtain similar findings if we provided individuals with criteria for determining sexual harassment and less ambiguous situations? Barak (1997) points out that this methodological problem is even more apparent when trying to compare the incidence of sexual harassment across cultures; researchers have not used the same definition or the same measuring instrument in collecting data.

Furthermore, researchers have pointed out that the paper descriptions provided to research participants do not capture the depth of emotions and interactions that transpire in a real sexual harassment incident (Lengnick-Hall, 1995; Paludi, DeFour, Attah, & Batts, in press). The percentages reported by researchers based on students' responses to these paper descriptions tell us nothing about the severity or the pervasiveness of the sexual harassment; these are important issues, especially for seeking resolution of sexual harassment through the courts (Paludi, 1997b).

In addition, researchers are not paying attention to the reactivity of the measures used. In the process of measuring students' opinions about sexual harassment, students may change due to the increased sensitization to issues surrounding sexual harassment. Furthermore, studies have not examined whether students chose to file a complaint as a consequence of what they are learning by participating in a research study.

In order to eliminate the dichotomy between research and practice, Paludi, DeFour, Attah, and Batts (in press) offer several suggestions:

- Make data collection include a small training program on sexual harassment, including distributing the campus's policy statement on sexual harassment.
- Invite the individual charged with investigating sexual harassment complaints to the research session.
- Provide information to research participants about counseling available for victims of sexual harassment.

- Notify the counseling center and the investigator of sexual harassment complaints that data will be collected on this issue and may result in students wanting to discuss their own experiences and/or topics raised by the research program.
- Share results of studies with campus administrators and sexual harassment investigators.

DEALING WITH SEXUAL HARASSMENT ON COLLEGE CAMPUSES

To deal effectively with sexual harassment of students on college campuses (as well as the sexual harassment of faculty and nonfaculty employees), the following components have been recommended: an effective policy statement, an effective investigatory procedure, and education/training programs for all members of the campus (see Biaggio & Brownell, 1996; Levy & Paludi, 1997; OCR, 1997; Sandler & Paludi, 1993).

In order to promote the effective and equitable resolution of problems involving sexual harassment, campuses should have an explicit policy statement. Such a policy allows the campus to uphold and enforce its policy against sexual harassment within its own community. The components of an effective policy statement for campuses include: statement of purpose, legal definition of sexual harassment, behaviors that constitute sexual harassment, statement of impact of sexual harassment on students and the campus, statement of campus's responsibility in investigating complaints, statement of individual's responsibility for reporting sexual harassment, statement of sanctions, statement concerning sanctions for retaliation, statement concerning false complaints, identification of individual(s) responsible for hearing complaints (Paludi, 1996; Paludi & Barickman, in press). The policy statement should be reissued each year by the college president and displayed prominently throughout the campus. In addition, the policy statement must be published in employee, faculty, and student handbooks.

Procedures for investigating complaints of sexual harassment must take into account the psychological issues involved in the victimization process, including individuals' feelings of powerlessness, isolation, changes in social network patterns, and wish to gain control over their personal and professional lives. These procedures must be confidential. Students must believe that once a complaint is filed, it will not become the subject of campus gossip. The procedures should maintain secrecy, and students should be made aware of this. Students should also be assured that complaints will be taken seriously and that an investigation will be undertaken promptly by an individual trained to handle such situations. Students will always feel more encouraged to become involved in a process when they understand fully what the process entails. Fear of the unknown can discourage reporting (Levy & Paludi, 1997).

The most important feature of an effective policy statement on sexual harassment for college campuses is the training programs designed to implement this policy (Biaggio & Brownell, 1996; Levy & Paludi, 1997; Paludi & Barickman, 1991; Salisbury & Jaffe, 1996; Wagner, 1996). Effective training programs send a clear message to all individuals that the sexual harassment policy must be taken seriously and that sexual harassment will not be tolerated by the campus administrators. Successful training programs are mandatory and held annually and have the active support and participation of the college president. Additional suggestions for training programs may be found in Paludi (1997a), Levy and Paludi (1997), and Paludi and Barickman (in press).

In addition to developing campuswide sexual harassment awareness programs, institutions are encouraged to: (1) include information about sexual harassment in new student orientation classes; (2) encourage sororities and fraternities to present programs on sexual harassment; (3) include information on sexual harassment in materials for transfer students, adjunct faculty, graduate teaching assistants, and part-time faculty; and (4) encourage faculty to incorporate discussions of sexual harassment in their courses.

Recently, Paludi, Doling, and Gellis (in press) have outlined ways psychology faculty can integrate information about sexual harassment in their courses. There are several benefits to teaching about sexual harassment in the undergraduate (and graduate) curriculum, including (1) assisting students in accurately labeling whether their experiences are sexual harassment; (2) providing information about the campus's policy statement and investigatory procedures; (3) indicating how the campus is committed to maintaining a harassment-free and retaliatory-free learning environment; and (4) providing necessary information that can carry over to students' experiences in the workplace. Pedagogical techniques, including audiovisual material, role-playing exercises, and case studies are presented in Paludi (1996), Paludi and Barickman (1991; in press) and in Paludi, Doling, & Gellis (in press).

An adequate program for dealing with sexual harassment on college campuses requires more than a general policy against the behavior. It requires the efforts and support of the campus administration, faculty, employees, and students and the continual training of all members of the campus community, as well as a procedure that *encourages,* not merely allows, complaints. Once this program is in place, the entire campus will benefit from an environment of cooperation, respect, and dignity. Such a program will increase the visibility of issues surrounding sexual harassment on college campuses, and this form of victimization will remain hidden no longer.

REFERENCES

Adams, J. (1997). Sexual harassment and black women: A historical perspective. In W. O'Donohue (Ed.), *Sexual harassment: Theory, research, and treatment* (pp. 213–224). Boston: Allyn & Bacon.

Adams, J., Kottke, J., & Padgitt, J. (1983). Sexual harassment of university students. *Journal of College Student Personnel, 24,* 484–490.

Bailey, N., & Richards, M. (1985, August). *Tarnishing the ivory tower: Sexual harassment in graduate training programs.* Paper presented at the Annual Meeting of the American Psychological Association, Los Angeles, CA.

Barak, A. (1997). Cross-cultural perspectives on sexual harassment. In W. O'Donohue (Ed.), *Sexual harassment: Theory, research, and treatment* (pp. 263–300). Boston: Allyn & Bacon.

Barickman, R.B., Paludi, M.A., & Rabinowitz, V.C. (1992). Sexual harassment of students: Victims of the college experience. In E. Viano (Ed.), *Victimization: An international perspective* (pp. 153–165). New York: Springer.

Berdahl, J.L., Magley, V., & Waldo, C. (1996). The sexual harassment of men? *Psychology of Women Quarterly, 20,* 527–547.

Biaggio, M., & Brownell, A. (1996). Addressing sexual harassment: Strategies for prevention and change. In M.A. Paludi (Ed.), *Sexual harassment on college campuses: Abusing the ivory power* (pp. 215–234). Albany, NY: State University of New York Press.

Brooks, L., & Perot, A. (1991). Reporting sexual harassment: Exploring a predictive model. *Psychology of Women Quarterly, 15,* 31–47.

Carr, R. (1991). Addicted to power: Sexual harassment and the unethical behaviour of university faculty. *Canadian Journal of Counselling, 25,* 447–461.

Dansky, B., & Kilpatrick, D. (1997). Effects of sexual harassment. In W. O'Donohue (Ed.), *Sexual harassment: Theory, research, and practice* (pp. 152–174). Boston: Allyn & Bacon.

DeChiara, P. (1988). The need for universities to have rules on consensual relationships between faculty members and students. *Columbia Journal of Law and Social Problems, 21,* 137–162.

DeFour, D.C. (1996). The interface of racism and sexism on college campuses. In M.A. Paludi (Ed.), *Sexual harassment on college campuses: Abusing the ivory power* (pp. 49–55). Albany, NY: State University of New York Press.

Doyle, J., & Paludi, M.A. (1998). *Sex and gender: The human experience* (4th ed.). New York: McGraw-Hill.

Dziech, B., & Weiner, L. (1984). *The lecherous professor.* Boston: Beacon Press.

Fitzgerald, L.F., Gold, Y., & Brock, K. (1990). Responses to victimization: Validation of an objective policy. *Journal of College Student Personnel, 27,* 34–39.

Fitzgerald, L.F., & Omerod, A. (1993). Sexual harassment in academia and the workplace. In F.L. Denmark & M.A. Paludi (Eds.), *Psychology of women: Handbook of issues and theories* (pp. 553–581). Westport, CT: Greenwood.

Fitzgerald, L.F., Shullman, S., Bailey, N., Richards, M., Swecker, J., Gold, Y., Omerod, A., & Weitzman, L. (1988). The incidence and dimensions of sexual harassment in academia and the workplace. *Journal of Vocational Behavior, 32,* 152–175.

Fitzgerald, L.F., & Weitzman, L. (1990). Men who harass: Speculation and data. In M.A. Paludi (Ed.), *Ivory power: Sexual harassment on campus* (pp. 125–140). Albany, NY: State University of New York Press.

Follingstad, D., Rutledge, L., McNeill-Harkins, K., & Polek, D. (1992). Factors related to physical violence in dating relationships. In E.C. Viano (Ed.), *Intimate violence: Interdisciplinary perspectives* (pp. 121–135). Washington, DC: Hemisphere.

Frazier, P., Cochran, C., & Olson, A. (1995). Social science research on lay definitions of sexual harassment. *Journal of Social Issues, 51,* 21–37.

Goodman, L., Koss, M., Fitzgerald, L., Russo, N., & Keita, G. (1993). Male violence against women: Current research and future directions. *American Psychologist, 48,* 1054–1058.

Grauerholz, E. (1989). Sexual harassment of women professors by students: Exploring the dynamics of power, authority, and gender in a university setting. *Sex Roles, 21,* 789–801.

Ivy, D., & Hamlet, S. (1996). College students and sexual dynamics: Two studies of peer sexual harassment. *Communication Education, 45,* 149–166.

Keller, E. (1988). Consensual amorous relationships between faculty and students: The constitutional right to privacy. *Journal of College and University Law, 15,* 21–42.

Kenig, S., & Ryan, J. (1986). Sex differences in levels of tolerance and attribution of blame for sexual harassment on a university campus. *Sex Roles, 15,* 535–549.

Koss, M.P. (1990). Changed lives: The psychological impact of sexual harassment. In M.A. Paludi (Ed.), *Ivory power: Sexual harassment on campus* (pp. 73–92). Albany, NY: State University of New York Press.

Koss, M.P. (1993). Rape: Scope, impact, interventions, and public policy responses. *American Psychologist, 48,* 1062–1069.

Lengnick-Hall, M. (1995). Sexual harassment research: A methodological critique. *Personnel Psychology, 48,* 841–864.

Levy, A., & Paludi, M.A. (1997). *Workplace sexual harassment.* Englewood Cliffs, NJ: Prentice Hall.

Lott, B. (1993). Sexual harassment: Consequences and realities. *NEA Higher Education Journal, 8,* 89–103.

MacPike, L. (1996). Sexual harassment and academic power. In B. Lott & M.E. Reilly (Eds.), *Combating sexual harassment in higher education* (pp. 188–196). Washington, DC: National Education Association.

Malovich, N.J., & Stake, J.E. (1990). Sexual harassment of women on campus: Individual differences in attitude and belief. *Psychology of Women Quarterly, 14,* 63–81.

Martindale, M. (1992). *Sexual harassment in the military: 1988.* Arlington, VA: Defense Manpower Data Center.

McKinney, K. (1994). Sexual harassment and college faculty members. *Deviant Behavior, 15,* 171–191.

Office of Civil Rights. (1997). *Policy guidance.* Washington, DC: U.S. Department of Education.

Ormerod, A., & Gold, Y. (1988, March). *Coping with sexual harassment: Internal and external strategies for coping with stress.* Paper presented at the annual conference of the Association for Women in Psychology, Bethesda, MD.

Paludi, M.A. (Ed.). (1990). *Ivory power: Sexual harassment on campus.* Albany, NY: State University of New York Press.

Paludi, M.A. (1995, November). *Sexual harassment in higher education: Abusing the ivory power.* Keynote address for the Canadian Association Against Sexual Harassment in Higher Education, Saskatoon, Saskatchewan, Canada.

Paludi, M.A. (Ed.). (1996). *Sexual harassment on college campuses: Abusing the ivory power.* Albany, NY: State University of New York Press.

Paludi, M.A. (1997a). Sexual harassment in the schools. In W. O'Donohue (Ed.), *Sexual harassment: Theory, research and treatment* (pp. 225–249). Boston: Allyn & Bacon.

Paludi, M.A. (1997b, September). *The campus as a violent institution for women: From the research lab to campus and community intervention.* Paper presented at the Higher Education Conference, San Antonio, TX.

Paludi, M.A. & Barickman, R.B. (1991). *Academic and workplace sexual harassment: A manual of resources.* Albany, NY: State University of New York Press.

Paludi, M.A., & Barickman, R.B. (in press). *Sexual harassment, work, and education: A resource manual for prevention.* Albany, NY: State University of New York Press.

Paludi, M.A., DeFour, D.C., Attah, K., & Batts, J. (in press). Sexual harassment in education and the workplace: A view from the field of psychology. In M.A. Paludi (Ed.), *Psychology of victimization: A handbook.* Westport, CT: Greenwood Press.

Paludi, M.A., Doling, L., & Gellis, L. (in press). Teaching about sexual harassment in the undergraduate psychology curriculum. In M.A. Paludi (Ed.), *Psychology of victimization: A handbook.* Westport, CT: Greenwood Press.

Payne, K. (1991, April). *The power game: Sexual harassment on the college campus.* Paper presented at the Annual Meeting of the Southern States Communication Association, Tampa, FL.

Project on the Status and Education of Women (1978). *Sexual Harassment: A hidden issue.* Washington, DC: Association of American Colleges.

Pryor, J., & Whalen, N. (1997). A typology of sexual harassment: Characteristics of harassers and the social circumstances under which sexual harassment occurs. In W. O'Donohue (Ed.), *Sexual harassment: Theory, research and treatment* (pp. 129–151). Boston: Allyn & Bacon.

Quina, K. (1996). Sexual harassment and rape: A continuum of exploitation. In M.A. Paludi (Ed.), *Sexual harassment on college campuses: Abusing the ivory power* (pp. 183–197). Albany, NY: State University of New York Press.

Rabinowitz, V.C. (1990). Coping with sexual harassment. In M.A. Paludi (Ed.), *Ivory power: Sexual harassment on campus* (pp. 199–213). Albany, NY: State University of New York Press.

Reilly, M.E., Lott, B., & Gallogly, S. (1986). Sexual harassment of university students. *Sex Roles, 15,* 333–358.

Riggs, R., & Murrell, P. (1995). Sexual harassment in the community college: The abuse of power. *New Directions in Community Colleges, 23,* 57–66.

Rozee, P. (in press). Stranger rape. In M.A. Paludi (Ed.), *Psychology of victimization: A handbook.* Westport, CT: Greenwood Press.

Ryan, K., Frieze, I., & Sinclair, H. (in press). Physical violence in dating relationships. In M.A. Paludi (Ed.), *Psychology of victimization: A handbook.* Westport, CT: Greenwood Press.

Saal, F., Johnson, C., & Weber, N. (1989). Friendly or sexy? It depends on whom you ask. *Psychology of Women Quarterly, 13,* 263–276.

Salisbury, J., & Jaffe, F. (1996). Individual training of sexual harassers. In M.A. Paludi (Ed.), *Sexual harassment on college campuses: Abusing the ivory power* (pp. 141–152). Albany, NY: State University of New York Press.

Sandler, B., & Paludi, M.A. (1993). *Educator's guide to controlling sexual harassment.* Washington, DC: Thompson.

Schneider, B. (1987). Graduate women, sexual harassment, and university policy. *Journal of Higher Education, 58,* 46–65.

Stites, M.C. (1996). What's wrong with faculty-student consensual sexual relationships? In M.A. Paludi (Ed.), *Sexual harassment on college campuses: Abusing the ivory power* (pp. 115–139). Albany, NY: State University of New York Press.

Tang, C., Yik, M., & Cheung, F. (1996). Sexual harassment of Chinese students. *Archives of Sexual Behavior, 25,* 201–215.

U.S. Merit Systems Protection Board. (1981). *Sexual harassment of federal workers: Is it a problem?* Washington, DC: U.S. Government Printing Office.

Wagner, K.C. (1996). Training counselors in issues relating to sexual harassment on campus. In M.A. Paludi (Ed.), *Sexual harassment on college campuses: Abusing the ivory power* (pp. 279–292). Albany, NY: State University of New York Press.

Watts, B. (1996). Legal issues. In M. A. Paludi (Ed.), *Sexual harassment on college campuses: Abusing the ivory power* (pp. 9–24). Albany, NY: State University of New York Press.

White, J., & Koss, M. (1991). Courtship violence: Incidence in a national sample of higher education students. *Violence and Victims, 6,* 247–256.

Zalk, S. (1996). Men in the academy: A psychological profile of harassers. In M.A. Paludi (Ed.), *Sexual harassment on college campuses: Abusing the ivory power* (pp. 81–113). Albany, NY: State University of New York Press.

Zalk, S.R., Paludi, M.A., & Dederich, J. (1991). Women students' assessment of consensual relationships with their professors: Ivory power reconsidered. In M.A. Paludi & R.B. Barickman. *Academic and workplace sexual harassment: A resource manual* (pp. 99–114). Albany, NY: State University of New York Press.

CHAPTER 12

Substance Abuse and Violence

David S. Anderson and Carol Napierkowski

INTRODUCTION

Violence and the use of alcohol and other drugs on college campuses have been consistently correlated. However, the connections between them have very rarely been examined. This chapter defines violence, explores data on the extent of violence on college campuses, discusses the use of alcohol on college campuses, and examines connections between violence and substance abuse on campus. The chapter also examines primary and secondary prevention efforts on college campuses.

Definitions and types of violence have been documented elsewhere in this book. For the purposes of this chapter, the authors use the definition of violence developed by Roark (1995), who defines violence as "behavior by intent, action, and/or outcome which harms another person." The kinds of behavior included in this definition include murder; assault; robbery; vandalism; domestic violence; sexual harassment; dating violence; and harassment against minorities, gays, and lesbians. Since the 1980s, there has been the perception that violence has increased on college campuses. Although exact figures are difficult to ascertain due to different interpretations or reporting procedures, national surveys conducted in 1986 and 1987 reported numerous incidents of sexual assault, vandalism, and physical assault (Cockey, Sherrill, & Cave, 1989). These incidents of violence have been of increasing concern to administrators who have been searching to identify the underlying causes and potential interventions to address them. What has become apparent is that drug and alcohol usage and violent behavior are linked. For example, Cockey, Sherrill, and Cave (1989) found that in 1986, 5.4% of sexual assaults involved the use of alcohol and 9.3% involved the use of other drugs, 53.9% of physical assaults involved alcohol and 14% involved other drug use, 48.2% of reported acts of vandalism involved alcohol and 6.9% involved other drug use. For 1987, these authors reported that 32% of sexual assaults involved alcohol and 12% involved drug use; for physical assault, 40% involved alcohol and 13% involved drug use; for vandalism, alcohol was involved in 52% of the cases and drugs were involved in 13%. Clearly, violent behavior is frequently associated with the use of drugs and alcohol, and the strength of the relationship seems to vary considerably over time.

THE NATURE OF THE SOCIETAL PROBLEM

Consideration of the relationship between substance abuse and violent behavior should include a focus on the high school setting, as well as the overall societal setting. It is not appropriate to place the entire responsibility upon the college environment for the linkages between substance use and violent behavior. Institutions of higher education certainly have a responsibility to address the issue from as many perspectives as possible. However, substance abuse and violent behavior are not specific to the college environment.

In a study of substance use and violent behavior in the high school setting conducted in 11 states, the Johnson Institute (1993) found that one third of students reported problems as a result of their use of drugs or alcohol. This report compared nonusers, problem users, and those with "dependence risk." When compared with nonusers, problem users were twice as likely to get into physical fights and four times more likely to commit vandalism. A careful understanding of the behaviors of these school-age students is very helpful, particularly as colleges prepare to "inherit" them as their entering student population.

Collins and Schlenger (1988) found a similar association between alcohol use and violence within the overall society. Consumption of more than five drinks at a setting increases the likelihood of an individual being involved in a violent incident, whether as a victim or perpetrator. Collins and Messerschmidt (1993) reported that presence of alcohol has been identified in more than one half of all incidents of domestic violence.

What we can conclude from this brief review is that, indeed, there is a linkage between violence and the use of substances. What helps to explain this? Although many of these studies do not prove causality, the constant linkages between the use of controlled substances and violent acts, coupled with what we know about how these substances act on the human body, clearly suggest that there is a causal linkage.

DRUGS AND THE HUMAN BODY

It is important to understand some basic information about how drugs affect an individual. Seven principles determine the effects of drugs or alcohol on an individual. These include (1) properties of the drug, (2) the means of preparation, (3) the method of use, (4) dosage, (5) characteristics of the individual, (6) the context of use, and (7) distance from the plant. Understanding these basic principles will help determine the effects resulting from the ingestion of a specific substance by a particular person.

The primary factor is the properties of the drug. People are generally affected in a similar way by particular drugs, thus permitting us to define a drug group or category. For example, hallucinogens generally cause distortions in one's sense of reality, and stimulants raise energy output. Related to the nature of the substance itself are three related, yet distinct factors—means of preparation, method of use, and dosage. The means of preparation identifies how diluted the substance is. A substance can be prepared in various ways to control concentration as well as the method of use. Drugs can be deliberately diluted to increase the quantity sold. For example, crack is a more concentrated form of cocaine.

The method of use is the way that the substance enters the human body; whether it is injected, snorted, drunk, eaten, or inhaled has some influence upon how the substance affects the user. Generally, when substances are injected or smoked, they have greater personal reinforcement and thus more addictive potential. Dosage, or the amount of the substance consumed, also influences the effects; higher dosages potentially bring a different quality to the experience. For example, with alcohol, lower dosage may result in an individual demonstrating greater sociability, caring, and openness; higher dosage may result in aggression and irritability. With marijuana, low use may result in giddiness and playfulness, and higher use may result in paranoia.

A fifth factor defining the effects of a drug are characteristics inherent in the individual. Based on various predisposing factors (such as the presence of alcohol dehydrogenase or acetaldehyde dehydrogenase), some individuals are able to consume larger amounts of alcohol than others. With marijuana, some individuals enjoy this substance more than others, in part due to their unique individual characteristics. In a related way, the context of use helps determine what type of experience a person has following the use of a substance. For example, there are notable differences between using the substance in a social context and using it when alone.

A seventh principle is distance from the plant, meaning the extent to which the drug has been changed through various processes. This principle is based on the fact that the closer a substance is to the original plant, the less damaging to the body and the less addictive it is. For example, morphine is the natural substance extracted from the poppy plant; generally, benign effects result from its use. When morphine is refined, heroin results; this does more damage to the body, is a more potent pain killer, and is more addictive.

CLASSIFICATION OF DRUGS

Countless drugs exist in our world culture. Although some of these are natural, others are synthetic or semisynthetic. Regardless of their origin, the effects of the substances can be classified into several discrete categories. Some authors have designated these as either "uppers," "downers," or "all-arounders." Other, more conventional approaches distinguish the drug classifications as sedative-hypnotics, depressants, hallucinogens, stimulants, marijuana, and narcotics. This section contains a brief description of each of these effects and their potential linkage with violent behavior.

There are distinct categories of drugs based on their overall effects on the human body. *Narcotics* (also called the opiates) such as morphine, methadone, opium, and heroin cause euphoria, drowsiness, and respiratory depression. These sedatives depress the body's central nervous system and are typically used to relieve pain and to promote sleep. The linkage with violence may occur when the brain's functioning is depressed, causing the individual to make a judgment that he or she might otherwise not make. Another link with violence is found when the body is withdrawing from the use of the substance; with heroin, for example, the user may experience insomnia, anxiety, irritability, and restlessness.

Depressants such as barbiturates, tranquilizers, and alcohol represent another category. Often, these are divided into alcohol and sedative-hypnotic drugs. The use of these substances can result in disorientation, slurred speech, and drunken behavior. Just as their title suggests, these substances can fully depress the human body (even to the extent of ultimate depression—death). The use of these substances can result in dependence, with abrupt withdrawal after heavy use resulting in substantial health risk due to convulsions. The relationship between violence and depressants is similar to that found with narcotics. Both substances alter judgment and involve severe withdrawal.

Stimulants include substances such as cocaine, amphetamines, and their "look-alikes." They result in an overstimulation of the nervous system, energized muscles, increased alertness, excitation, euphoria, increased heart rate, increased blood pressure, and decreased appetite. A common use of stimulants is to reduce feelings of fatigue or drowsiness. Although the user may remain awake, concern remains about the quality of the individual's decision-making ability. Extreme irritability, and often violence, is cited with the use of stimulants, particularly when the user is highly fatigued or when the user, seeking more effect, becomes quite hostile and irritated due to unavailability of the substance.

Hallucinogens include substances such as LSD (lysergic acid diethylamide), PCP (phencyclidine), MDA (methyleneoioxyamphetamine), and psilocybin. These substances result in illusions and hallucinations, rambling speech, and poor perception of time and distance. One of the other principles—context of use—is particularly important with the use of hallucinogens to minimize unpleasant experiences (or "bad trips"). Because of the erosion of ego boundaries with the use of some hallucinogens, individuals may engage in behavior that they may not otherwise have done.

Inhalants typically are substances that are intended for one purpose but misused to "get high." Included in this category are aerosol products, paint thinner, airplane glue, and lighter fluid. They cause slurred speech, poor coordination, impaired judgment, and drunken behavior. Users often demonstrate impulsiveness, excitement, and irritability. Violent behavior could result through unintended consequences of behavior or impaired ability to function normally.

The final category of substances is *cannabis,* which includes marijuana, hashish, and hashish oil. The general results include relaxed inhibitions, euphoria, and loss of perceptions. Although violent behavior can result from the use of cannabis products due to poor judgments or perceptions, this generally is not expected.

Violent behavior can result from the use of substances in most of the categories discussed above. Although some substances are likely to produce violent reactions, others may or may not depending upon circumstantial factors and the specific behavior considered to be violent. Although just how this occurs is not clearly understood, some initial insights will be helpful.

THE EXTENT OF THE PROBLEM

In a discussion of the issue of substance abuse among college students, one of the first considerations is what is meant by "substance abuse." When used here, the term *substance* includes illicit drugs and alcohol; the term refers to any item that, when ingested, has a high probability of helping individuals cause harm to themselves or others. The substance causes physical changes in a person, and typically has physical effects on the brain and influences the person's judgment. Alcohol, marijuana, heroin, cocaine, amphetamines, and similar products can easily cause harm to an individual and, often, others. Substance use may become addictive as indicated by changes in tolerance, withdrawal symptoms, loss of control, and a wide variety of negative consequences. Harm may also be that inflicted upon others, such as the harm caused by a drunk or impaired (through drugs or alcohol) driver or by a smoker (with secondhand smoke), and upon loved ones affected by a substance user's negative behavior exhibited while or after being under the influence of a substance.

Thus, substance *abuse* is the involvement with a substance where negative consequences to oneself or others occur. The use of most of these substances can result in abuse, in part due to the very nature of the product. Tobacco may be the product with the greatest probability of causing harm when used as designed. Other products (such as paint thinner) are perfectly acceptable when used as expected (however, when paint thinner is used as an inhalant, substance abuse occurs).

Clarity about these terms is important since substance abuse is believed to be a major contributing factor in violent behavior by college students. It is difficult to understand precisely what abuse is. One individual may view a specific behavior as "normal," while another person may view the same behavior as "abusive." Similarly, individuals may look at their behavior and see no problems, while those around them are quite troubled by it and define it as abusive. In our society, clear articulation of what is "normal," is quite elusive, particularly surrounding drugs, and most particularly surrounding alcohol. This lack of clarity contributes to many of the problems surrounding the relationship between substance use (and abuse) and violence.

Thus, the consequences associated with the use of drugs and alcohol provide a basis for understanding the association between substance use and violent behavior. Two data sources will be explored as a basis for understanding the relationship between drug and alcohol usage and violent behavior: (1) surveys of student behaviors and attitudes and (2) surveys of college administrators reporting campus data and their perceptions of effects of usage on their campus.

According to Wechsler et al. (1995b), many students report that they have been victimized by a violent act or its consequences during the previous year. These violent acts include pushing, hitting, or assault (13%); property damage (12%); insult or humiliation (27%); or unwanted sexual advances (21%). It is clear that violent behavior is already a part of many students' lives. The existence of this behavior is a consideration when students make decisions about ways in which they interact in social situations or when they observe a behavior they may not like.

Another factor in the problem of drug and alcohol use is the benign neglect that exists concerning substance abuse. Consider the relationship between substances and violence. If, indeed, this relationship has some causal aspect (and it is very reasonable to expect this), then a prudent campus leader would want to implement strategies to reduce or eliminate the causative agent (the substances). One would hope to see campus personnel taking seriously any behavior demonstrating the misuse of drugs or alcohol. Thus, limiting alcohol or drug use may reduce some of the consequent behaviors of violence. Benign neglect occurs when campus leaders observe some problematic alcohol or drug use and treat it as "no big deal."

The use or abuse of drugs or alcohol on campus is often viewed as "expected." Colleges traditionally have been viewed as a place where students experiment with new behaviors, "let loose," and act in defiance of traditional norms of responsibility. A typical response by many students is that they will "grow out of" their use of drugs or alcohol. Students, parents, alumni, and administrators often view the use of drugs or alcohol as a "rite of passage."

How many students actually use drugs or alcohol? A measure that helps to monitor regular substance use is the extent of consumption during the previous 30 days. Johnston et al. (1995) gathered data on students' use of drugs or alcohol during the previous 30 days, thus providing a helpful pulse of students' routine use of drugs or alcohol (Table 12–1). To begin with a positive trend, they found a dramatic decrease in the use of alcohol and illicit drugs between 1985 and 1994. In 1994, 15% of college students reported using marijuana and 68% reported using alcohol. Although these percentages are still not acceptable, they represent a downward trend in the use of these substances.

In another national study, Presley, Meilman, and Lyerla (1996) examined college students' use of drugs and alcohol on a regular (daily or monthly) and irregular (yearly) basis, clearly demonstrating the high use of alcohol among college students (Table 12–2).

Both surveys indicate substantial use of drugs and alcohol by college students. Wechsler and his colleagues (1995a) conducted a similar study to examine more closely the issue of heavy drinking by college students. They found that heavy drinking (called "binge drinking" in the article) begins, for many students, during their high school years. The authors define "binge drinking" for men as consuming five or more alcoholic drinks in one setting at least once in the previous 2

Table 12–1 Percentage of College Students Using Drugs and Alcohol During Previous 30 Days

Substance	1985	1986	1987	1988	1989	1990	1991	1992	1993	1994	1995
Alcohol	80.3	79.7	78.4	77.0	76.2	74.5	74.7	71.4	70.1	67.8	67.5
Marijuana	23.6	22.3	20.3	16.8	16.3	14.0	14.1	14.6	14.2	15.1	18.6
Cocaine	6.9	7.0	4.6	4.2	2.8	1.2	1.0	1.0	0.7	0.6	0.7
Hallucinogens	1.3	2.2	2.0	1.7	2.3	1.4	1.2	2.3	2.5	2.1	3.3

Note: College students in this study are 1 to 4 years beyond high school.

Source: National Survey Results on Drug Use from *The Monitoring the Future Study 1975–1995*, National Institute on Drug Abuse, NIH, Pub. No. 98-4140, 1997.

Table 12–2 Percentage of College Students Using Drugs and Alcohol

Substance	Daily	Monthly	Yearly
Alcohol	0.9	62.5	83.6
Marijuana	1.4	13.4	27.7
Cocaine	0.1	0.9	4.1
Steroids	0.1	0.3	0.7

Source: Data from *Alcohol and Drugs on American College Campuses, Vol. IV: 1992–94*, C.A. Presley, P.W. Meilman, and J.R. Cashin, The Core Institute, Southern Illinois University at Carbondale, October 1996.

weeks (for women, it is four alcoholic drinks). The authors demonstrate that this heavy drinking is widespread among college students, with one in five college students engaging in this behavior three or more times during a 2-week period. They also analyzed the issue of secondhand effects of alcohol use, demonstrating that many students feel that they are inconvenienced (through noise, disruption, violence, damage, and other problematic behavior) by others' heavy use of alcohol. As drinking quantity and frequency gets higher (from abstainer to frequent, heavy drinker), so does the extent of the adverse consequences.

THE NATURE OF THE CAMPUS PROBLEM

This section reviews some of the relationships between substance use and violent behavior. The *College Alcohol Survey* (Anderson & Gadaleto, 1997) found that alcohol is believed to be involved in one half to three quarters of various campus incidents such as damage to residence halls and other campus property, violation of campus policies, violent behavior, physical injury, and acquaintance rape (Table 12–3). The survey indicates that alcohol's involvement with each of these incidents persists in spite of the implementation of a wide variety of strategies and other initiatives designed to address these problems.

In a study conducted by Towson State University, of the students who indicated that they had committed crimes such as armed robbery, theft, vandalism, or sexual assault, 46% stated they had been using alcohol at the time. Further, of the students who indicated that they had been victims of crime, 46% reported that they had used alcohol or other drugs before the crime was committed ("Campus Crime Linked," 1990).

Looking specifically at the linkage between acquaintance rape and alcohol consumption, Abbey (1991) observed that these behaviors frequently occur simulta-

Table 12–3 Percentage of Involvement of Alcohol in Campus Situations, *College Alcohol Survey*

Situation	Alcohol Involvement
Damage in residence halls	65%
Damage to other property on campus	58%
Violation of campus policies	56%
Violent behavior	64%
Physical injury	42%
Acquaintance rape	75%

Courtesy of George Mason University, Fairfax, Virginia.

neously. Abbey noted linkage between alcohol consumption by the perpetrator and alcohol consumption by the victim. In addition, Abbey cited the implications of these findings for prevention programs.

The use of alcohol in combination with other illegal substances such as Rohypnol (also known as "roofies" or "roopies") is an example of a contemporary challenge faced on college campuses. This drug is cheap (less than $5 per dose) (U.S. Drug Enforcement Administration, 1998) and, when used with alcohol, it produces a rapid and very dramatic high and reduced sexual inhibitions. The drug is associated with date rape (Metro Toronto Research Group, 1996). Police departments indicate women have reported waking up with no clothes on and having been sexually assaulted after ingesting the drug (Staaten, 1998).

LINKING SUBSTANCE USE AND VIOLENCE: SOME THEORIES

Numerous theories exist regarding the linkage between substances and violence. This section focuses primarily upon alcohol because of its prevalence as the "drug of choice" on college campuses as well as the greater availability of published research on the substance. In one synthesis of the research regarding the linkage between alcohol and aggressive behavior, White, Hansell, and Brick (1993) cite three theoretical frameworks that exist regarding this relationship. One theory suggests that alcohol use causes aggressive behavior; this would occur because alcohol has some chemical effects on the brain, thereby affecting judgment, perception, attention, sociability, and related consequences that subsequently contribute to aggression. A second theory is the converse: aggressive behavior leads to heavy alcohol use. In this model, the assumption is that some individuals are naturally aggressive, and that they are more likely than others to place themselves in situations where heavy alcohol use is encouraged. These individuals may drink heavily to relax, and/or they may drink so that they have an excuse to behave in an aggressive manner. The third theoretical approach is that there is *not* a causal link between alcohol use and aggression; rather, these two behaviors are linked because they share a common cause. Proponents of this approach suggest that, for some individuals, a common factor or set of factors may cause both alcohol use and aggression.

Several approaches to the study of the relationship between violence and substance abuse are available. Pernanen (1993) highlights five theories that often underlie the explanatory research attempting to identify ways in which consuming alcohol may result in violent behavior.

1. *Impulse-driven approaches* assume that alcohol consumption affects the body without the overlay of the brain. Alcohol may interact with certain

physical abnormalities (such as the presence of hypoglycemia or low levels of the neurotransmitter serotonin) to result in violence.
2. With *psychodynamic impulse-driven processes,* attention is given to subconscious and internal symbolic processes.
3. *External-stimulus-driven approaches* incorporate strategies that influence behavior such as frustrating, threatening, and provoking cues.
4. *Cognitive-driven approaches* are based on the belief that external cues are perceived and processed by the individual who has been drinking; expectancies of people and situations, and social and cultural definitions of drinking events contribute to these approaches.
5. *Cognitive-impairment approaches* incorporate the effects that alcohol has on how a drinking individual processes situations and people, including misperceptions, depletion of one's cognitive resources, and attention span.

COLLEGE STUDENT DEVELOPMENT CONTEXT

In 1989, Smith reported that although patterns of violence and drug and alcohol use on college campuses may mirror society at large, college campuses have always been seen as a retreat from society. There have been two complementing assumptions about college life. One is that of the university as a place of restraint and pursuit of scholarly activities. The other is a place where students complete their identity through experimentation. Chickering and Reisser (1993) reported that the psychological tasks one must accomplish during this period include developing competence, mature relationships, purpose, and integrity. It has been assumed that, in completing both scholarly pursuits and psychological tasks, college students engage in a wide variety of activities and relate to a wide variety of groups.

Students, particularly those of traditional age, may see experimentation with alcohol and other drugs as part of these experimental activities. A "party culture" exists within the university where experimentation with drugs and alcohol is perceived as part of the norm. As part of this norm, evidence of violent behavior also exists. Wechsler et al. (1995b) found that white males with college-educated parents and who are majoring in business, are residents of a fraternity, are engaging in risky behaviors, are involved in athletics, indulged in binge drinking in high school, and view parties as very important are associated with binge drinking in college. Marlatt, Baer, and Larimer (1995) found that those who revealed a history of conduct problems also engaged in excessive drinking. They also reported that men who lived in fraternity houses reported more alcohol-related problems than other respondents.

The results of these studies suggest that, for a certain segment of the college population, alcohol-related activities are associated with important developmental activities of college life. However, this assumption may be a perceived rather than an actual norm for all college students. Berkowitz and Perkins (1996) found that students tend to persistently perceive the peer norms of drug and alcohol use as more liberal than their own usage. They also found that students in certain situations may behave according to inaccurately perceived peer pressure. Thus, although there are "at-risk" groups on campus for substance abuse, the conclusion that most students perceive experimentation as a necessary part of their development may not be accurate. Any attempt to address the use of alcohol on campus must first examine the various groups, often called subcultures and herein called cocultures, that exist and the developmental needs that are addressed by those cocultures.

A DIFFERENT PERSPECTIVE

It has long been assumed that alcohol use and violence are linked. Furthermore, there is general agreement that violence and alcohol use by college students mirrors usage patterns of U.S. society. However, to address effectively the issue of alcohol and violence on college campuses, one must take a careful look at campuses today and how they serve the developmental needs of college students. For example, what are the basic attitudes regarding the use of alcohol and violence on campus? Are alcohol and violence considered unavoidable evils, given a population of young adults and the college culture? Is there a clear cause and effect between usage and violence? Or, are substance abuse and violence merely symptoms of a larger problem on college campuses? The answers to such questions will largely determine the intervention strategies that a particular college will employ.

Any examination of the issue must include an identification of the developmental and personal needs that are served by alcohol and other drug use and by engaging in violent behavior. Evidence exists that both alcohol and violence may serve some of the same needs. In fact, some overlap exists between the primary and secondary prevention strategies for alcohol abuse and violence. Sherrill (1989), Rickgarn (1989), and Napierkowski and Gadaleto (1996) reported that approaches for violence prevention include zero tolerance to violence, education, careful recordkeeping, staff training, anger management, mediation, and other conflict resolution skills. Marlatt, Baer, and Larimer (1995) noted that approaches to alcohol prevention include education, training, developmental programs, and interventions appropriate to the severity of use. It would seem reasonable to reexamine patterns of alcohol usage and violent behavior to look for common core issues and to begin programs that comprehensively address these issues.

A conference hosted in 1995 by the University of Notre Dame (Coleman, Kigar, & Shoup, 1995) examined some underlying core issues for college students. This conference focused on drug and alcohol abuse and built upon many of the same themes, concerns, frustrations, and sensitivities felt by those guiding campus substance abuse programs. The challenge was to take a new look at an old problem. Conference attendees participated in "think-tank" groups to identify approaches that would work. During their time together, participants worked through a five-step visioning process to determine what they wanted campuses to look like in the years ahead. They envisioned campuses without problems associated with drugs or alcohol. These discussions identified approaches that would be appropriate to reach this end—that is, strategies that would be helpful in reducing drug and alcohol abuse.

Seven life health principles were identified at the 1995 conference (Coleman, Kigar, & Shoup, 1995). These principles include values, self-care, relationship health, ecological connections, optimism, service learning, and community. According to Coleman and Anderson (in press), these principles point the way to a healthier life for the total campus community and serve as the foundation for a wider discussion of this topic.

The principle-based approach is an improved paradigm for addressing issues of substance abuse. This is not to suggest that current efforts are not appropriate or helpful; in fact, they may be necessary but not sufficient to address the problems associated with drugs, alcohol, and violence. Current efforts would benefit from being supplemented by these seven life health principles.

CAMPUS-BASED EFFORTS

With this background and alternative framework, it is helpful now to look at the specific strategies currently being used to help address drug and alcohol abuse on the college campus. If violent behavior is linked to drug and alcohol abuse, then it makes sense to integrate these issues theoretically and programmatically. In this way, some violence prevention activities may be drug and alcohol prevention efforts; and some drug and alcohol prevention efforts may have the beneficial consequence of reducing violent behavior.

Historically, colleges and universities have done much to address drug and alcohol abuse, with the majority of the emphasis on alcohol abuse. Formal attention on violence prevention has occurred to a much lesser degree. The lessons to be learned from campus-based efforts, then, are primarily found with alcohol abuse prevention activities. These activities initially focused on affecting individuals through strategies designed to influence knowledge, attitudes, and behavior. More recently, however, efforts have addressed the overall campus environment. A blend of these approaches, incorporating prevention as well as intervention strategies, is appropriate for campus efforts.

Ultimately, however, any strategies that address substance abuse must take into consideration the personal and societal factors underlying the use of drugs or alcohol. For example, individuals tend to use substances based on their perceptions of others' use. In fact, many of these norms are not perceived correctly. Berkowitz and Perkins (1996) demonstrate that students typically think that other students are using drugs or alcohol at levels higher than they are actually using these substances. A framework of understanding use of drugs and alcohol might include social, personal, emotional, cognitive, and physical reasons.

Efforts to address alcohol and other drug issues are captured in three nationwide, college-based surveys or projects. The *College Alcohol Survey* has been conducted every 3 years since 1979 (Anderson & Gadaleto, 1997). Its focus was solely on alcohol issues until 1988, when it expanded to include other drugs; in 1997 the focus changed to include campus violence prevention initiatives. The *Drug and Alcohol Survey of Community, Technical and Junior Colleges,* conducted in 1991, emphasized campus-based strategies used at 2-year campuses (Anderson & Pressley, 1991). The third, to be discussed later in this section, is *Promising Practices: Campus Alcohol Strategies* (Anderson & Milgram, 1997).

Overall, it is clear that college and university campuses are undertaking a tremendous amount of effort to address drug and alcohol abuse. There is no question that today's campuses are undertaking more efforts to combat abuse than the campuses of 20 years ago, but recently there have been reductions in campus-based efforts to address drug and alcohol issues (Anderson & Gadaleto, 1997). Although the question of whether alcohol is permitted on the campus has not changed during this period (73% of 4-year institutions allow alcohol on the campus), what has changed are the circumstances under which this may occur. For example, 87% of campuses require that an alternative beverage also be served at functions where alcohol is served; this increased continuously from 54% in 1979 to 95% in 1991, after which it dropped to the 1997 levels (Anderson & Gadaleto, 1997). In 1979, 24% of the nation's campuses required that food be served at public functions where alcohol is available. Currently, 78% of the nation's colleges require this. Other policies concerning advertising an event; whether alcohol can be the primary focus of an event; and the permissibility of advertising on campus bulletin boards, campus radio stations, or student newspapers also have become more restrictive during this period.

In addition to the permissibility of alcohol on the campus, Anderson and Pressley (1991) reported other parallel findings for the 2-year campus. They reported that alcohol was permitted on 21% of the college campuses, with prior registration at 44% of the campuses. Nonalcoholic beverages are required by 68% of the respondents, with food required by 56% of the respondents.

Regarding prevention and education efforts, virtually every 4-year institution has a drug and alcohol education and prevention effort on campus; this contrasts with 69% in 1979 (Anderson & Gadaleto, 1997). By comparison, 92% of 2-year institutions report such an effort. Task forces or other group efforts to address alcohol education are found on the majority of 4-year campuses (compared with 37% in 1979) and 44% of the 2-year campuses. Student programming groups (72%, up from 38% in 1982) assist in the implementation of educational activities.

A wide variety of other prevention and education initiatives are found on 4-year campuses, including locations for resource materials (98%), a designated alcohol/substance abuse coordinator or specialist (73%), an undergraduate course (66%), alcohol-awareness weeks or days (81%), and drug-awareness weeks or days (36%). In a similar vein, many 2-year campuses have a resource center (87%), a designated coordinator (66%), undergraduate courses (45%), alcohol-awareness weeks or days (73%), and drug-awareness weeks or days (64%). Employee assistance programs exist on 80% of 4-year campuses and 57% of 2-year campuses (Anderson & Gadaleto, 1997; Anderson & Pressley, 1991).

Treatment and referral services on the campus include locations where students who perceive that they have a drinking problem receive assistance. At 4-year institutions, support groups for those whose lives are negatively affected by alcohol are found on 53% of campuses, up from 21% in 1979. A group counseling experience specifically for students who are problem drinkers is found on 53% of the campuses, up from 33% in 1979. Peer counselors whose primary focus is alcohol and other drug abuse are found on 64% of the campuses, an increase from 31% in 1988 (Anderson & Gadaleto, 1997). For 2-year institutions, support groups are found on 38% of the campuses, and support groups for alcoholics or drug abusers are found on 44% of the campuses (Anderson & Pressley, 1991).

Targeting substance abuse prevention initiatives to those cocultures (or subpopulations) often seen as needing greater substance abuse support services (due to their high risk) and as "victims" of violence (as evidenced by other chapters in this volume), the results of one question on the *College Alcohol Survey* (Anderson & Gadaleto, 1997) are quite informative. Respondents were asked: "To what extent does your campus' alcohol and/or other substance abuse education and prevention effort pay special attention to the unique needs of each of the following groups: (Scale: 1 = none; 2 = a little; 3 = some; 4 = a lot)." Freshman students, fraternity and sorority members, and athletes received scores of 3.39, 3.10, and 2.89 respectively; other populations received noticeably lower scores. Women rated 2.43, people of color rated 2.00, gays/lesbians/bisexuals received 1.78, and persons with disabilities received 1.67. These findings document the fact that these last four cocultures have, traditionally, not had attention focused on their unique substance abuse needs.

During the past 3 years, colleges and universities have demonstrated increasing attention in their substance abuse initiatives to issues directly related to violence prevention (Anderson & Gadaleto, 1997). For example, when college administrators are asked the extent to which various life skills are incorporated in their overall campus programming on drug and alcohol issues (on a 1-to-5 scale, with 1 =

"not at all important" and 5 = "very important"), conflict management increased in importance from 3.14 to 3.40 and violence reduction increased from 2.97 to 3.37.

When asked to summarize the extent of education and prevention efforts dealing with violence on their campus compared with several years before, college administrators note increases on 68% of campuses and decreases on only 3% (Anderson & Gadaleto, 1997).

Clearly, the efforts on the 4-year college campuses have increased in a variety of areas, from policies and education to treatment. There are more efforts on 4-year college campuses than there were two decades ago. A wide variety of initiatives are currently being undertaken by colleges and universities across the country. One gains some insight into the extent of these efforts, however, with the following observation: the average annual expenditure of funds for campus alcohol and other drug efforts (excluding personnel costs) for the 2-year college is much less than that of the 4-year institution.

A recent initiative identified the best practices to address alcohol problems on campus. This project, *Promising Practices: Campus Alcohol Strategies,* solicited college and university presidents and chief student affairs officers to identify their most helpful and appropriate efforts to address alcohol abuse. The process examined efforts in 10 distinct areas: awareness and information, environmental/targeted approaches, peer-based efforts, policies and implementation, enforcement, assessment and evaluation, support and intervention services, curriculum, training, and staffing and resources. Campuses were asked to describe their efforts in any or all of these areas. Campuses that had numerous approaches were identified as "comprehensive" campus efforts, since they exemplified the overall aim of having a broad-based effort that attempts to meet the needs of numerous populations through a variety of approaches. The resulting document was a *Sourcebook* that was distributed widely to all college and university presidents, complete with findings and recommendations for action (Anderson & Milgram, 1997).

Included in the *Sourcebook's* findings and recommendations are additional insights that help one understand the "state of the art" on college campuses. The report recommended that campus efforts have clearly defined outcomes, evaluation of efforts, adequate funding and resources, and sufficient support.

Which of these efforts, or which efforts in what combination, have been effective on the college campus? Unfortunately, evaluation data are very limited. Clear solutions and answers to these problems currently are not available. However, we do have some clues. Intervention efforts designed for the campus have increased in their scope and extent over the past several decades. More is being done. However, while student use of some substances has declined, it may not have fallen to a level low enough to call these efforts successful. Some behaviors—in particular, the heavy drinking—have not changed appreciably over time.

Does this mean that our efforts of the past 20 years have been unsuccessful? Does it mean that it is meaningless to continue these efforts? The response is a qualified "no." These efforts have been necessary and appropriate. It was necessary to develop and implement these strategies on campus to get to the current point. Regardless of what new directions we take in the future, it will be necessary to continue many of these approaches. What is important now is to determine ways in which the concept of drug and alcohol abuse prevention can be broadened. It is necessary to identify strategies that can be implemented, in a meaningful and successful way, to address the variety of perspectives surrounding drug and alcohol abuse. The challenge today is to determine new strategies for moving forward. This can be viewed as a type of "crossroads"; we know that we should not continue to do only what we have been doing and that we should be moving toward newer and expanded strategies that incorporate some of the current efforts and identify, creatively, other more meaningful efforts.

A MODEL FOR CAMPUS PREVENTION EFFORTS: THE MULTIDIMENSIONAL FRAMEWORK INITIATIVE

As we move to integrate the vision of a healthy campus, it is important to provide some specific focus regarding the ways in which the campus drug and alcohol abuse and violence prevention strategies can be conceptualized and implemented. A campus-based approach is more than a single event, a single resource, or a single strategy.

Simple and singular approaches rarely are sufficient to address meaningful long-term (or short-term) change. A multidimensional framework that we introduce here for collegiate drug and alcohol initiatives suggests that campus-based efforts should incorporate a very wide variety of approaches and foundations. A campus-based effort is a complex tapestry that integrates six distinct, yet overlapping dimensions: (1) institution-wide, (2) comprehensive, (3) broad-based, (4) clear outcomes, (5) multitargeted, and (6) cultural awakening.

The *institution-wide* approach emphasizes that campus efforts address drug and alcohol issues without restriction to a single office, task force, or individual. The efforts should reflect the institutional goals; involve a planning committee; and include leadership by the alcohol and drug coordinator, the president, academic units, athletics, counseling services, campus security, student government, the health service, and others. Infusing the themes of drug and alcohol abuse prevention throughout the campus makes good sense.

The *comprehensive* element suggests the need for a wide variety of campus efforts, including large-scale and small-scale programming, policies and procedures, training, curricular infusion, referral and support, and research and evaluation. This is very important for several reasons. First, different individuals learn by different approaches, and to reach a specific individual there must be an approach that matches his or her learning style. Second, it is helpful to have different implementation elements reinforce one another; if a student hears a message repeated in a totally different context, there is greater likelihood that it will be retained and acted upon.

The *broad-based* emphasis recognizes that different members of the audience have different levels of use of alcohol and other drugs, and thus need different kinds of messages. For example, efforts should include prevention, health promotion, and a system to respond to problems. Such diverse efforts are critical since a prevention message may not be heard by someone with a problem, and a referral message would be inappropriate for someone who does not use substances.

The fourth element is the emphasis on *clear outcomes*. Typically, efforts focus on knowledge or attitude change, rather than the ultimate outcome of behavior. Thus, it is appropriate to have some clear understanding of how efforts that address knowledge are likely to affect behavior.

The effort should be *multitargeted*. Student subpopulations should be identified and efforts be targeted to different subpopulations including male and female students, traditional-age and nontraditional-age students, people of color, commuters and residential students, members of fraternities and sororities, athletes, and adult children of alcoholics. The multitargeted approach would also address groups beyond students—including faculty and staff, graduates, community members, community leaders, and visitors.

The *cultural awakening* theme focuses on the factors underlying students' misuse of drugs and alcohol. It is based on the fact that most drug and alcohol efforts are dealing directly with alcohol and other drug issues by emphasizing topics such as accurate information, healthy attitudes, informed choices, and decision making and related skills. The broader focus includes underlying personal, social, and environmental issues.

In addition to these six dimensions, a rigorous evaluation component is essential. After tremendous efforts are expended on behalf of changing behavior, some attention must be given to whether the initiatives are effective. Evaluation activities can be conducted using quantitative measures, qualitative techniques, or, preferably, both.

The key point for this process is that the campus-based effort must encompass many aspects of the campus. Attempts that are limited in scope, approach, audience, or involved participants will result in limited change. Thus, it is incumbent upon the higher education community to redirect many of the processes and initiatives that are inherent on the college campus. A sense of renewal is critical if we are to meet the inherent goal of the *institution* of higher education, revised and revitalized to meet current societal and cultural needs. No single approach is sufficient or appropriate for all campuses, nor for any one campus each year. There is an inherent need for ongoing attention to this issue, and reassessment of approaches, foundations, and resources allocated to the effort.

CAMPUS IMPLEMENTATION STEPS

In view of the previous theory and discussions, we offer 10 recommendations for campus planners and programmers who are developing and implementing strategies to prevent drug and alcohol abuse and violence on campus. These suggestions are based on the assumption that substance abuse and violence are often linked. Further, regardless of whether they are formally linked through a cause-and-effect relationship, they may share underlying factors that affect their expression on college campuses.

1. Campus leadership personnel should implement a comprehensive, multifaceted initiative with a long-term perspective. Campus efforts to address substance abuse and violence must acknowledge that no "quick fix" exists. Efforts should engage key stakeholders throughout the campus and its community, and incorporate a variety of approaches to reflect various learning and behavioral management styles. Approaches should include education, environmental interventions, policy, assessment, and procedures.
2. A task force inclusive of key constituencies should be charged with providing ongoing leadership and vision. This task force should maintain needs-based strategies and efforts, ensuring that duplication among services and strategies (such as substance abuse prevention efforts and violence prevention efforts) does not exist. The group should work with campus-gathered data and external (state, regional, and national) data to monitor and revise strategies as appropriate. This group should help provide ongoing leadership on issues such as a more careful examination of theories underlying substance abuse and violence, their linkages, and alcohol-specific issues (such as differences between problem drinking and experimental drinking).
3. Measures of students, faculty, and staff should be gathered on an ongoing basis. Information about students' attitudes, perceptions, and experiences should be systematically collected and reviewed to guide program efforts and resource allocation. Faculty and staff needs should be assessed to identify approaches for training and services provided. Program outcomes and effectiveness should be monitored to assist in reviews of programs.
4. Faculty, staff, and students should receive training and skill-building as necessary. Training opportunities should be provided for faculty and staff on campus as well as at association meetings and at national and state conventions. Topics for faculty, staff, and students may include problem iden-

tification, intervention and referral strategies, resources, curriculum linkages, class projects and assignments, and related needs. Students should receive training in conflict resolution, conflict mediation, intervention skills, assertiveness, anger management, and related issues.
5. Communications strategies should be widely used to communicate campus environmental messages. A wide variety of communication approaches (such as public service announcements, posters, electronic information services, media coverage) should be incorporated to disseminate awareness about the desired behaviors and attitudes for the campus. These should include resources, policy information, referral sources, and implications of engaging in behaviors of concern.
6. Environmental interventions should be implemented as deemed necessary. Specific efforts to affect the physical environment, such as employment of security personnel, use of metal detectors, and screening procedures should be carefully considered and implemented as appropriate.
7. The president and board of trustees should provide active leadership and support. Their role is critical for the success of this effort, as they provide key policy approaches for the campus. Their attention to and concern with substance abuse and violence can be demonstrated by their commitment, funding, and direction.
8. Attention to strategies for the surrounding community should be encouraged. The campus can serve as a resource for its surrounding schools and communities, based on its insights and processes implemented. Further, working with local school students can help prepare them for appropriate behaviors as they matriculate to institutions of higher education.
9. Engage all groups in the process of addressing substance abuse and violence issues. Groups that are often recipients of violence and at high risk for substance abuse should be actively involved in the planning and implementation of the campus and community initiatives. In addition, other groups such as alumni, parents, and community leaders should be involved to generate their support, insights, and leadership.
10. Generate creative solutions to address causes as well as symptoms. Although it is important to continue strategies that appear to be working, it is imperative that new, viable approaches be considered. These creative approaches should address not only symptoms but also the underlying causes of violence and substance abuse.

SOME FINAL CONSIDERATIONS

The issue of substance abuse and violence prevention is complex. Intuitively, we know that there are linkages between the two. Some initial research highlights some deeper understanding of the linkages, including attention to common underlying factors.

Although college campuses have actively addressed substance abuse for decades, violence prevention efforts are much more recent. The importance of having these efforts linked is demonstrated by the work undertaken on campuses where there is more attention devoted to violence prevention. The insights gained from substance abuse prevention efforts can inform violence prevention efforts. A case can be made for combining these (and perhaps other issues as well) under an overall umbrella or matrix of services that assist with promoting the quality of life on the campus.

Ultimately, the question is one of what we envision for our campuses. We can use the issues associated with substance abuse and violence to stimulate discussion and dialogue on campuses. Each campus has its own history, and tradition, and needs, as determined by its student population, surrounding community, faculty and staff, and related constituencies.

As we help define and create our respective futures, no magic bullet or easy path is available for us. The journey and its inherent activities helps define what constitutes the next series of questions and options—ones that, we hope, will further move us toward campuses that have less substance abuse and violence, and much more of what we determine to be desirable for the lives of our students and ourselves.

REFERENCES

Abbey, A. (1991) Acquaintance rape and alcohol consumption on college campuses: How are they linked? *Journal of American College Health 39*(4), 165–169.

Anderson, D.S., & Gadaleto, A.F. (1997). *College alcohol survey, 1979–1997.* Fairfax, VA: George Mason University. Unpublished data.

Anderson, D.S., & Milgram, G. (1997). *Promising practices: Campus alcohol strategies sourcebook.* Fairfax, VA: George Mason University.

Anderson, D.S., & Pressley, G. (1991). Drug and alcohol survey of community, technical and junior colleges. Unpublished data.

Berkowitz, A., & Perkins, N.W. (1996). *From reactive to proactive prevention: Correcting misperceived norms of substance use on campus.* Baltimore, MD.

Campus crime linked to students' use of drugs and alcohol. (1990, November-December). *Chemical People Newsletter.*

Chickering, A.W., & Reisser, L. (1993). *Education and identity* (2nd ed.). San Francisco: Jossey-Bass.

Cockey, M., Sherrill, J.M., & Cave, R. (1989). Towson State University's research on campus violence. In J.M. Sherrill & D. Siegel (Eds.), *Responding to violence on campus* (pp. 17–27). San Francisco: Jossey-Bass.

Coleman, S., & Anderson, D.S. (in press). *Charting your course: A lifelong guide to health and compassion.* South Bend, IN: University of Notre Dame Press.

Coleman, S., Kigar, G., & Shoup, J. (1995). *Challenge 2000: Shared visions.* South Bend, IN: University of Notre Dame.

Collins, J., & Messerschmidt, P. (1993). Epidemiology of alcohol-related violence. *Alcohol Health and Research World, 17*(2), 93–100.

Collins, J., & Schlenger, W. (1988). Acute and chronic effects of alcohol use on violence. *Journal of Studies on Alcohol, 49,* 516–521.

Johnson Institute (1993). *The effect of alcohol and other drug use problems on the school environment.* Minneapolis, MN: Author.

Johnston, L., O'Malley, P., & Bachman, J. (1995). *Drug use among American high school seniors, college students and young adults, 1975–1994.* Washington, DC: National Institute on Drug Abuse.

Marlatt, A.G., Baer, J.S., & Larimer, M. (1995). Preventing alcohol abuse in college students: A harm reduction approach. In *Alcohol problems among adolescents.* Northvale, NJ: Lawrence Erlbaum.

The Metro Toronto Research Group on Drug Use. (1996). *Facts on Rohypnol (flunitrazepam)* [On-line]. Available: http://sano.arf.org/geninfo/rohypnol.htm

Napierkowski, C.M., & Gadaleto, A.F. (1996). *Developing campus strategies for violence prevention.* Paper presented at American College Personnel Association Convention, Baltimore, MD.

Pernanen, K. (1993). Research approaches in the study of alcohol-related violence. *Alcohol Health & Research World, 17*(2), 101–107.

Presley, C., Meilman, P., & Lyerla, R. (1996). *Alcohol and drugs on American college campuses: Use, consequences, and perceptions of the campus environment: Vol.3. 1991–1993.* Carbondale, IL: The Core Institute, Student Health Programs, Southern Illinois University at Carbondale.

Rickgarn, R.L. (1989). Violence in residence halls: Campus domestic violence. In J.M. Sherrill and D.G. Siegel (Eds.), *Responding to violence on campus* (pp. 29–40). San Francisco: Jossey-Bass.

Roark, M. (1995). Conceptualizing campus violence: Definitions, underlying factors, and effects. In L.C. Whitaker and J. W. Pollard (Eds.), *Campus violence: Kinds, causes and cures* (pp. 1–27). New York: Haworth Press.

Sherrill, J.M. (1989). Models of response to campus violence. In J.M. Sherrill & D.G. Siegel (Eds.), *Responding to violence on campus* (pp. 77–88). San Francisco: Jossey-Bass.

Smith, M. (1989). The ancestry of campus violence. In J.M. Sherrill & D.G. Siegel (Eds.), *Responding to violence on campus* (pp. 5–15). San Francisco: Jossey-Bass.

Staaten, C. (1998, February 10). "Roofies": The new date rape drug of choice. Obtained through electronic message from G. Deisinger, February 10, 1998.

U.S. Drug Enforcement Administration. (1998). *Flunitrazepam (Rohypnol)* [On-line]. Available: http://www.usdoj.gov/dea/pubs/rohypnol/rohypnol.htm

Wechsler, H., Dowdall, G., Davenport, A., & Castillo, S. (1995a). Correlates of college student binge drinking. *American Journal of Public Health, 85*(7), 921–926.

Wechsler, H., Moeykens, B., Davenport, A., Castillo, S., & Hansen, J. (1995b). The adverse impact of heavy episodic drinkers on other college students. *Journal of Studies on Alcohol, 56,* 628–634.

White, H., Hansell, S., & Brick, J. (1993). Alcohol use and aggression among youth. *Alcohol Health & Research World, 17*(2), 144–150.

CHAPTER 13

Crisis Management Resulting from Violence on Campus: Will the Same Common Mistakes Be Made Again?

J. Victor Baldridge and Daniel J. Julius

INTRODUCTION

Colleges and universities, like all other organizations, occasionally face serious crisis situations that arise as a result of violence or criminal activity. A union strike paralyzes operations, a student is robbed in the dorm, a faculty member is murdered on campus. While these are rare events, they can seriously hurt the image of an institution or end administrative careers, depending on how the "crisis" is handled. These situations are referred to as "low-probability, high-consequence events." This chapter does not explore the causes of violence on campus or controversy surrounding the reporting of campus crime, but is directed at the development of a management plan and building response capabilities.

This chapter makes three primary assertions. First, institutional leaders are likely to be confronted with violent events on campus. Because of the ever-increasing complexity of life on campus, deterioration of surrounding neighborhoods, pressures to succeed, availability of guns and other weapons, more violence is inevitable. Murders, however gruesome, do occur on campuses. The campus has also become a center of human activity, dependent on computer systems and subject to riots or natural disasters. These are low-probability, high-consequence events. Unfortunately, while rare, they occur. Careers can be ruined, or individuals put at risk, unless adequate preparation is made.

Second, many campus administrators seem to muddle through crisis situations; they make a series of predictable mistakes. In some cases, action is delayed or a cover-up is attempted. Failure to take appropriate action exposes the institution to lawsuits. Mistakes occur when unprepared administrators face low-probability/high-consequence events.

Third, campuses can enhance their crisis management capabilities. Implementing crisis policies and adhering to a strategic action plan will prepare administrators and faculty to perform better in crisis situations. A crisis audit is a series of *actions* to build strength into a campus response. Although it certainly includes a written plan, it may also include exercises, periodic reviews of insurance, and careful analysis of human resources policies. The audit is a reminder of the need to reinforce the campus's crisis management capabilities.

This chapter is a report of a series of interviews that were conducted with college and university officials to determine (1) the manner in which they responded to violence on campus and (2) a methodology to improve crisis management techniques and strategies in response to these events.

IMAGINE THESE SCENARIOS

You are the dean of students at a small liberal arts college in the East. At 3 o'clock on a Tuesday morning, your phone rings. Carl, your residence hall director, is frantically calling to tell you that a women's dorm is on fire. He believes that arson may be involved. Five women, he cries, may have been killed. "Come at once, I really don't know how to handle this!"

You are the director of a research center at Southern state university. The newspaper headline staring at you on the front page of the metropolitan paper is grim: "Student Murders Two Professors in Research Center Scandal!" You are at the breakfast table when the phone starts ringing. "Bob," the strained voice of the president says, "I think you better meet me right away. The press will be here at 9:00 AM."

You are president of a Methodist college in a Midwestern community. As you walk into the Rotary Club breakfast, the pastor of a local church hands you the newspaper, which you have not yet seen this morning. "College Dean Arrested by Undercover Sting Operation. Soliciting and Beating Young Boys in Town Park. Sex Ring Broken Up." You frantically glance around the room. Obviously most of the 100 businessmen *have* read the story. And there sits your least favorite reporter.

You are the vice-president for student affairs at a prestigious West Coast university. On Tuesday morning, you learn that a coed has been murdered after leaving class the night before. As the story unfolds, investigators learn that she had been stalked for weeks and had complained to public safety officials. Her parents are on the phone.

CRISIS EVENTS

These scenarios depict campus leaders confronted by crisis situations. The situations have presented themselves as a result of growing violence in our society. Colleges and universities are not immune. For weeks—even months—these campuses may be in turmoil, careers will be at stake, reputations will be challenged. Unfortunately, presidents and administrators often make common mistakes that others in similar situations have made.

A crisis precipitated by violence or related criminal activity is, by definition, unpredictable. Crisis management, as it is discussed here, concerns events that have a low probability of happening, but high consequences if they occur. All colleges and universities, including the most prestigious universities in the nation, have been affected by crime and violence. Many of these situations have been subjected to intense public scrutiny. Witness the acerbic spotlight thrown on Florida State University when serial murders plagued the university or the attention given to San Diego State University in the aftermath of the murder of three engineering professors!

Parenthetically, inept crisis management can also result in the destruction of political careers. Richard Nixon lost the presidency of the United States because he ineptly managed the aftermath of a dumb burglary. John Kennedy made a debacle of the Bay of Pigs invasion, but rebounded smartly, handling the Cuban Missile Crisis and salvaging his presidency. Campus administrators face the same pressures because of escalating crisis situations, particularly those related to criminal activity. Such crises result from violent murders, rape, sabotage of computer facilities, explosions detonated by "rights" activists; unfortunately, the list is rather long. We argue:

- *More crises precipitated by violence will occur.* Administrators must be trained in crisis management related to violence.
- *Common mistakes are constantly made.* Campus administrators and spokespersons do not always manage crisis situations well. It would seem they consistently repeat the same mistakes.
- *Better managerial performance in crisis precipitated by violence or criminal activity is indeed possible.* Colleges and universities can enhance their management reaction skills, striving to be proactive rather than reactive.

MORE CRISES?

Violence in America is endemic and has been since the landing of the Pilgrims at Plymouth Rock, Massachusetts. A variety of examples illustrate this point. Wars were fought as whites took over the lands of the native American Indians. Violence associated with slavery provided a sordid legacy. Our nation, born out of armed revolution, suffered through one of the most violent civil wars of the 19th century. Workers, attempting to unionize during the mid- to late 19th century and well into the 20th century, were, in many cases, beaten, even murdered, by hired (private) armies; men and women who worked in the coal, steel, lumber and shipping industries were gunned down, sometimes while they were asleep in tents, or hosed with high-powered water cannons. Synagogues are periodically burned and swastikas painted on the doors of Jewish students. Our campuses have also known violent race riots or political demonstrations.

Violence and crime appear to be increasing on campus, even though violent crime rates are declining in major U.S. cities. (This is probably the result of better community policing and the decline of a teenage population traditionally associated with violent crime. It is also the case that until recently, crime on campus was not comprehensively reported.) The focus of this chapter is on the "reaction" to criminal acts and the management of violence through crisis planning. The chapter does not deal with the underlying causes of crime (such as homelessness, the destruction of the family, the proliferation of weapons, growth of low-income pockets in campus neighborhoods, or the deterioration of civility or "values" in America), all of which lead to violence or criminal activity.

Of course, campus managers have always faced crises due to violence. Modern administrators are not alone, or the only ones, who confront tragedies and related campus turmoil. Nevertheless, the complexities facing modern administrators will demand more expertise with crisis management as a result of violence. Consider the following factors and their consequences:

- *College and university campuses are larger and more complex.* They are becoming a target for individuals and groups involved in violent or criminal activities. The thesis is this: the more complexity, the more vulnerability. Large campuses that are held accountable for increased services and responsible for greater numbers of employees are more likely to confront crisis.
- *Political movements historically have been part of campus life.* Civil rights, animal rights, and antiwar movements are only a few examples of group and societal dynamics that disrupt the modern campus. In these cases, the university is seen as an extension of "American policy" for those who object to ideas, policies, or the structure of capitalist society. No longer the Ivory Tower, the university or college has increasingly been at center stage for controversial movements. For example, the recent unionization of teaching assistants is a political and economic trend that has affected many larger campuses—with activism, strikes, campus shut-downs, and the potential for violence.
- *Computer systems are a source of vulnerability.* What campus can function without its lifeline and nerve center, the computer network? Campuses are

uniquely vulnerable, since every academic enterprise continuously uses computers. While these systems provide capability and sophistication, they spawn chaos when tampered with or broken.
- *An aggressive media makes a private crisis a public circus.* The age of "instant news" is here. CNN broadcasts 24 hours a day in more than 60 countries around the world. In a gentler age, the Methodist college president, in our earlier example, would have received a late night call from the chief of police or editor of the local newspaper. The "scandal" would not have leaped to the front page, and the president would not have been blind-sided by an unexpected headline. Today, however, the media bores in. Events that might have been settled in-house are sometimes pushed by aggressive news agencies searching for a gripping story for the 6 o'clock news. Violence, crisis management, and media management have an iron linkage.
- *The legal environment accommodates those who seek litigation as an answer.* We are witnessing a revolution in legal entanglement, a veritable litigation epidemic. University campuses are not alone. Violent incidents spawn publicity and lawsuits. The media attention, insurance policies, and an army of contingency-fee lawyers has produced a crisis in its own right. However, a potential for violence can also occur if administrators fail to take appropriate action because they may fear the possibility of legal action. Consider the hypothetical case of a senior administrator who may be aware of an individual working on campus with a criminal record who lied about his past on an employment application. The employee's behavior becomes erratic. However, the administrator voices "privacy concerns." It is recommended that such employees be questioned if their behavior is suspect. In many cases, the employee will admit to lying on the employment application and can be terminated for just cause. A potentially violent situation is defused. Many lawsuits are perfectly legitimate, an appropriate redress of real harm. Others, however, are frivolous, are expensive, and lead to endless claims. Administrators must be aware of the legal environment but not let it hamper effective action.
- *Environmental concerns have had an impact, particularly in regard to hazardous waste disposal.* Activities that may have been acceptable for a hundred years are now outlawed as environmental "green" movements spread around the world. Although most would agree that the environmental movement is a positive development, it nevertheless leaves many campuses vulnerable to unimagined problems. In the United States, the Environmental Protection Agency has fined numerous organizations for waste disposal and environmental problems. Most managers claim they do not know they are violating the law. Many recent citations involve laws that were nonexistent 10 years ago. Although we normally think of these "environmental prob-

lems" as issues facing business, any campus leader who has ever tried to build a new building, dispose hazardous waste, install an atomic reactor, or simply get a zoning variance for a power plant or laboratory knows that environmental issues, zoning laws, and other regulations stimulate a "community" response.
- *Natural disasters and civil unrest increasingly threaten.* Although administrators rarely incorporate natural disaster or civil disorders into campus plans, the disruption caused by these phenomena is increasing. Hurricanes push across the Southeastern United States, and earthquakes rip California. Major floods paralyzed the central section of the nation during the summer of 1993 and the winter of 1996. Hurricane Andrew disrupted academic institutions across southern Florida in the early 1990s.

Civil unrest is a related consequence and can be a serious factor. The Rodney King riots in Los Angeles played havoc with mid-city campuses for weeks. Every senior administrator who works in a far-flung campus setting should be prepared to deal with natural disasters and accompanying civil disturbances.

The arguments here are straightforward. The social, political, and physical environment inhabited by colleges and universities has grown increasingly complex. Increased complexity produces vulnerability, and vulnerability inevitably leads to crisis.

COMMON MISTAKES: THE BIG NINE

Administrators who have faced major crises have dealt with (1) life-threatening situations, (2) trauma that seriously disrupted their campuses, (3) events that destroyed personal careers, and (4) outside governmental agencies whom (they felt) interfered substantially. Common violence has intruded on the campus in stark, shocking ways. Consider the following events that have occurred during the last few years:

- A student stabbed a roommate and then killed himself at Harvard University.
- Four students were held hostage at gunpoint in a residence hall at the University of Northern Colorado. The gunman, a nonstudent, was killed by a SWAT team.
- At Purdue University, a resident hall assistant reported a student who was using cocaine. The resident assistant then was shot. The individual who shot him barricaded himself in a dorm room, and when police lobbed in tear gas, the man killed himself.
- A student murdered a Spanish teacher and a sophomore in the library at Simon's Rock College in Massachusetts. Four others were wounded in the same attack.

- At San Diego State University, an engineering student, angry about the manner in which his dissertation was handled, killed three engineering faculty who served on his committee.
- The University of New Mexico canceled its summer program in Ecuador after a group of students and advisors in a shuttle bus were ambushed by a group of armed bandits. The wife of the group leader was shot and killed. University officials later stated they were aware of the dangers but said there was no choice but to travel at night. They thought their route was safe.
- A white dean at predominantly black Albany State College in Georgia was hospitalized after being stoned for a second time. In an earlier attack, the dean was surrounded by 40 students. A complaint has been filed with the U.S. Equal Employment Opportunity Commission.
- The Unabomber sent bombs to professors at the University of California and Yale University, severely injuring or killing several people (Magner, 1996a, 1996b)

Almost all administrators we interviewed said that they were unprepared for crisis management situations related to violence, and that they had "muddled through." The crisis came as a surprise. Their campus was not ready. Many individuals wished that they were better prepared before the event occurred. On the basis of these interviews, a list of nine common problems and mistakes has been constructed.

Problem 1: A Slow Response

Emergency situations are always a surprise. A crisis does not warn anyone with a fax 3 days ahead. Crises arise in the night when least expected, when staff are away on vacation, when computers are down, when the campus or systems are vulnerable. In these cases, people try to discern what is happening, assess damage, or marshal resources to respond. Chaos slows down the response. Administrators want to make decisions slowly and deliberately based on facts, not guesses. However, the media, the public, and government agencies are present, demanding information and an explanation why the campus did not react faster or in a different way.

Recommendation

It is better to overestimate the threat and to take extremely rapid action than to underestimate and delay a response. To overreact makes individuals look a little foolish; to underreact can lead to disaster. We advise administrators to act swiftly to respond to violence, even at the risk of seeming to overreact.

Problem 2: "Cover-Up" Leading to Scandal

Human beings want to minimize their mistakes. In a crisis, some administrators may endeavor to hide the bad news from the community, alumni, parents, or their supervisor. Their impulse is to shield the news from the media. A "witch hunt" can follow a crisis. Hiding the dimensions of the violence may compound the problem. Inevitably, the details will be pulled out (painfully) by an aggressive newspaper.

Recommendation

Get the bad news out as quickly as possible. It is better to suffer a clean decapitation than death by a thousand cuts!

Problem 3: Muddled Command

In most situations, muddling of authority occurs in response to violence. The press is told two or three different stories by various people, each of whom may claim to have the final authority. Confusing communication results when a campus has no clear plan that specifies lines of authority in the event of violence or criminal activity. The community fire or police chiefs may show up, campus system officials or board members may feel compelled to visit, or the governor may arrive in a publicity-staged event. Many a campus president has had to breathe helicopter dust from the arriving big-shot when time could be better spent addressing the situation. Under these circumstances, who is in charge? And of what? As an example, we recently sat in the command center of a major American city during an emergency drill. The mayor was delayed because of an automobile accident. For more than 40 minutes—while the city was hypothetically paralyzed by a major earthquake—the mayor's cabinet argued. Who was the boss while the mayor's bent fender was examined?

Recommendation

Set up a violence crisis team in advance, and clarify authority lines ahead of time.

Problem 4: Poor Media Management

The media loves violent crimes! Reporters arrive and demand answers. In the interviews we conducted, media management was a major difficulty. At a southern campus in 1997, a sex scandal erupted with several faculty members accused of homosexual relations with young students. The local media smelled a story and dug into the situation. The administration foolishly stonewalled the press and denied the obvious facts. When the local newspaper finally got the whole story, the cover-up became the lead item, not the actual sex scandal. Several senior administrators eventually lost their jobs.

Recommendation

Prepare a media relations plan (Exhibit 13–1) that includes a systematic approach to press relations for crisis situations. If a crisis occurs, it should be "business as usual" rather than a media fiasco.

Problem 5: Failure in Interagency Relations

As a crisis escalates, the circle of activity can move beyond the campus. The press gets involved, government agencies sometimes move in, and administrators can find themselves caught in a web of external agencies. In the interviews we conducted, administrators were reasonably satisfied with how they controlled events inside the campus. The problems resulted from relations with external agents.

For example, when Hurricane Andrew devastated southern Florida in 1992, interagency foul-ups were continual and serious. The National Guard, the U.S. Army, the Federal Emergency Management Agency (FEMA), and countless state agencies were involved. In the aftermath of the disaster, campus administrators were forced to respond to a myriad of agencies in order to rebuild battered campuses; the process was a nightmare of confusing rules, overlapping jurisdictions, and contradictory regulations.

Recommendation

Conduct environmental scans and related analyses of federal and state agencies active in violent crises or disasters in the area. Develop a plan, contact key personnel in external agencies, develop advance rapport.

Problem 6: Failing To Prepare Action Logs

During the initial phases of the crisis, individuals have difficulty remembering what was done. People sometimes forget the details—what they did, when they did it, and what impact their actions may have had. To the extent possible, write down the sequence of events, keep tape recordings of messages or—if possible—telephone conversations. Senior administrators should immediately write an "after action report." At one Midwestern college, a suspicious dorm fire caused several injuries and damage to an expensive building. Campus administrators failed to keep adequate records of costs and careful logs of their actions. This resulted in

Exhibit 13–1 Guidelines for Media Management

Meet the Press: The Do's
- Be accessible, act fast. Express care and sympathy.
- Get the facts as they are known and keep them up to date.
- Prepare press releases in advance. Tell *your* story.
- Brainstorm possible questions. Who? What? When? Why? What are you doing? Why weren't you better prepared?
- "Keep It Simple, Stupid" (KISS). Use plain English, no technical jargon.
- Explain your containment strategy. People need confidence that the situation is under control. What are you *doing* about it?
- Rehearse. Then rehearse some more.
- Try to plan a *positive* message. Always bridge back to it.

Meet the Press: The Don'ts
- Do not assume the press is out to get you.
- Do not say anything "off the record." Everything is *always* on the record.
- Do not speculate. Your speculation is often wrong. Give only the known facts.
- Do not characterize other people's action (e.g., attributing motive or blame), and do not give praise unless you are sure.
- Do not pretend to have secrets. "If you only knew" always gets someone in trouble.
- Do not try to be humorous. Violence is never funny to the people who are hurt.
- Do not give instant interviews. Prepare, prepare, prepare.
- Do not give a response without coordinating actions with senior administrators or those normally responsible for media relations.

---

the loss of hundreds of thousands of dollars as a result of rejected insurance claims and filed lawsuits where the campus was portrayed as led by "bumbling managers" who could not get simple facts straight.

Recommendation

It is essential to keep careful logs of actions taken. These logs will be invaluable if lawsuits commence.

Problem 7: Recovering Expenses from Insurance or Government Agencies

Virtually every college or university carries insurance for situations relating to violence or disaster. A common complaint among the people we interviewed concerned inadequate insurance. Written policies can contain a variety of fine points that limit the liability of insurance companies.

After the destruction from Hurricane Andrew, many Florida colleges and universities discovered yet another problem. Claims worth millions of dollars were disallowed by insurance companies or by public agencies such as FEMA. Most were challenged because of poor records and the inability to justify claims. No one is sure of the exact amount, but it was easily a "hundred-million dollar misunderstanding."

Parenthetically, in many natural disasters (such as earthquakes and hurricanes), government agencies may help with cost recovery. Any recovery of cost, however, depends on documentation of losses.

Recommendation

Conduct an insurance review, establish a system for documenting losses, and enforce its use in case of an emergency. Additionally, review the potential assistance that is available from governmental agencies if losses are incurred as the result of a disaster.

Problem 8: A Comprehensive Crisis Management Plan

Many administrators we interviewed worked at public campuses. Under most state laws, public entities are often required to have disaster plans. In some instances, the plans were current, comprehensive, and useful. In other cases, however, the plans were "fat paper doorstops," as one president derisively commented about his university's crisis plan. One exception to the rule concerns data processing and information systems. Here, planning for violence, criminal activity, or disaster is a major enterprise. In most other divisions on campuses, planning (which should include how the plan will be communicated and implemented) for violence or criminal activity is often inadequate.

Recommendation

Every campus should develop a comprehensive plan to accommodate violent acts or, to the extent possible, criminal activity and natural disasters. The plan should be current and its contents known to faculty, staff, and administrators. To the extent required, individuals should practice following the plan in the event of an emergency. Specific responsibilities should be assigned to various offices (student affairs, public safety, etc.). People must be held accountable in the event that it becomes necessary to institute the plan.

Problem 9: The Small Human Steps

Compare two airlines' handling of separate plane crashes. When one of its planes crashed at Boston Airport several years ago spilling passengers into the icy harbor, World Airways took little action to express humane sympathy for victims. It was, in fact, a public relations disaster; the airline refused to admit that anyone had been killed. In this instance, the media depicted a frantic woman pounding at the World Airways counters, claiming her husband and son were missing. The airline callously denied that they were on the plane until the wreckage was raised and bodies were discovered.

In contrast, United Airlines suffered a crash in Sioux City, Iowa. A number of people were killed and injured. The company flew in a task force of psychological counselors, clergy, and company officials. They contacted family members. United officials visited the injured in hospitals, distributed flowers, and assured victims the airline would reimburse expenses. Psychological counseling was offered to relatives and loved ones.

Sometimes the same callous behavior infects campuses. At one large Midwestern university, the director of the counseling center recounted the story of an on-campus rape where the victim was treated (in his opinion) like a criminal by local police. To the director's dismay, the counseling center staff were not used in the debriefing and recovery process for this unfortunate woman. Later that semester, possibly due to the insensitive and crude handling of her tragic case, she withdrew from the university. The simple human steps were ignored.

Recommendation

In situations related to violence, criminal activity, and related disasters, people are injured or financially damaged and careers are disrupted. There are always human costs. Administrators, faculty, and staff must be sensitive and act humanely and, to the extent possible, ensure that nonuniversity employees act similarly. Those responsible must pay attention to the feelings, emotions, and personal lives of people affected.

BUILDING CRISIS-RESPONSE CAPABILITIES

Campus administrators can learn how to approach and manage violent or criminal crisis situations by using a "crisis audit"—a set of actions to guide campus administrators before, during, and after the act or event. Enhancing crisis-response capabilities requires an understanding of the "elements" involved. Responding to crises precipitated by violent events, acts, disasters, or related phenomena has cyclical components. Crisis response comes in four stages:

1. *Prevention.* The best way to handle a crisis is to avoid it. Efforts are directed toward preventing crises from happening. For example, if sprinkler systems are in place, the likelihood of a major fire is low; if a college has strict controls on behavior in residence halls, the school is less likely to be scandalized with dormitory violence.

2. *Preparation*. Preparation involves the specific planning for the crisis itself. Preparation might involve setting up a command center in advance, asking legal counsel to review the crisis plan, acquiring sophisticated communication or surveillance systems, or designating a counseling or crime reaction team.
3. *Response*. Response is the activity focused directly on the event. A fire must be extinguished, a poisoned product withdrawn by the food service, a faculty strike settled, a rape case managed properly.
4. *Recovery*. Resumption of normal campus activity is the key focus of the recovery cycle. Once the immediate crisis has been managed, recovery can take months, even years!

Managing violence or criminal activities is synonymous with the response phase. It is extremely important, however, to understand the full cycle, because the response comprises only part of the overall cycle. If a campus has excellent mitigation or deterrent programs, it may avoid the difficult, agonizing experience of a crisis response. However, regardless of how large and skilled the public safety force, violence inevitably occurs. Campuses with adequate recovery plans return to normal operation. Where recovery has not been contemplated, the result may be a campus that encounters reduced effectiveness and public relations problems. Innocent people may suffer.

A crisis audit should address the entire cycle: prevention, preparation, response, and recovery. Table 13–1 provides crisis phases and activities for a crisis audit.

Crisis Audit: Prevention Phase

1. *Assign the crisis management function to an office*. However, the management function must cross jurisdictional lines of authority. Many colleges and universities have an office of "contingency planning." Contingency planning rarely moves beyond security or safety activities. The real essence of crisis management is that instant *response* to a violent emergency is not the sole problem. Another primary problem is maintenance of *the totality of campus life* and reaction to what occurs after the act is committed. Senior administrators are responsible for the overall management of this phase.
2. *Data processing plan*. Of course, every major data processing operation has plans for backup and recovery, but data systems are nevertheless extremely vulnerable. Even an operation as sophisticated as Bank of America, which has one of the largest data processing operations in the banking industry, found itself in trouble during the Los Angeles riots in 1991. The odds are high that data processing operations are not as immune to criminal activity or riots as one might believe. Recall the virus initiated by a gradu-

Table 13–1 Crisis Audit

Crisis Phase	Audit Activities
Prevention	Response plan Staffing for deterrent Data processing plan Human resources policies Event and facility assessment Insurance and legal review
Preparation	Crisis team Operations center Media kit
Reponse	Response plan implementation Action logs and crisis drills Media management
Recovery	Campus resumption plan (counseling, interface with police, etc.) Cost recovery plan Legal containment plan Strategic debriefing

ate student at one Ivy League institution several years ago. "Computer geniuses" sometimes experiment with viruses, almost to the extent of playing a game. Potentially disastrous and criminal activities can result.

3. *Human resources regulatory review.* After any injury in a crisis or violent crime, government regulatory agencies may examine how closely the campus adhered to criminal, occupational, and/or safety regulations. Individuals may inadvertently violate serious regulatory rules. An assigned office (perhaps human resources, public safety, or facilities management) should conduct periodic reviews, updating or reacting to changing government regulations. Compliance should be rigid. In a crisis situation, scrutiny can be intense after injury, death, or a violent event. Subsequent to a balcony falling during a graduation ceremony at the University of Virginia in 1997, local officials began a painstaking review of building codes.

4. *Event probability assessment.* During the crisis audit, it is wise to conduct an event probability assessment. It is often possible to identify the most likely violent events and potential criminal activities. The impact on cam-

pus should be discussed and incorporated into a response plan. There is no substitute for advance planning.
5. *Insurance and legal review.* Campuses may not have adequate coverage for insurance purposes. Violence is synonymous with legal liability! Insurance policies and products change rapidly; threats to the institution also shift over time. A comprehensive review of legal and insurance requirements is recommended.

Crisis Audit: Preparation Phase

The preparation phase moves closer to the actual crisis situation; from broad and general mitigation activities, to specific preparation actions. This phase involves at least three activities.

1. *Establishing a crisis team.* To some extent, the nature of the event will determine who will lead the campus in case of a crisis. Teams should be small and involve senior administrators. Here is a list of potential team members in the event a major violent act occurs.
   - chief executive officer
   - chief academic officer and financial officer
   - chief student affairs officer
   - counseling, public safety, or human resources director
   - dean or director of the division or school involved
   - public relations specialist
   - legal counsel
2. *Establishing a crisis center.* Of course this activity will depend on the nature and severity of the event. In some cases, the president's office will be adequate. However, in the event of a serious problem that draws substantial media attention, space to brief the press may be arranged. Subsequent to many acts of this nature, the physical communications systems can be overloaded. (In major disasters, such as earthquakes, the entire phone system of a region can be out of commission.) Usually the problem is not communications failures (phone companies build elaborate sophisticated backups) but simple overload. In a crisis situation, campus communications will be overused, with calls from the press, concerned parents, employees, or managerial staff. A crisis center may need additional phone lines or fax facilities.
3. *Developing a media plan.* Activities with the press or the media can be anticipated. A media relations plan should specify the spokesperson, information to be released, whether press conferences are allowed, and what facilities will be offered to the press. In addition, a "media kit" is recom-

mended. When a violent act occurs, the campus spokesperson must present an accurate and positive view of the organization to the community. The kit should emphasize positive institutional and demographic data such as how many people are employed, the history of the institution, background of key administrators, etc. The first press releases about the event should begin an honest relationship with the press that continues through the recovery.

Crisis Audit: The Response Phase

Every event and crisis will turn on its own facts; however, enough common themes exist to permit discussion of plan implementation.

1. *Resource lists.* When crisis strikes, campus administrators should not stumble through the yellow pages to obtain resources. Will the campus need specialized equipment? Prepare a list of vendors in advance. Will college leaders need to confer or coordinate with regulatory officials or the law enforcement community? Consider contacting these individuals on occasion to establish a pattern of interaction. Set forth the resources needed and actions necessary to obtain these resources.
2. *Action logs and crisis drills.* It may be imperative to keep concise action logs. In a serious disaster or crime, government officials and regulators will surface. Lawyers will approach the parties and possibly see individuals who are employed. Insurance companies need information—what was done, when, and why. For purposes of legal liability and cost recovery, it is important to keep accurate logs of events. This includes, in addition to recording actions during the event, "after-action" reports from senior personnel. Unfortunately, drills and exercises are virtually nonexistent on college and university campuses. Drills and exercises take their cue from law enforcement and the military, so campuses may not feel comfortable with the process. Nevertheless, there are important lessons to be learned. Law enforcement and military also face low-probability, high-consequence events. Like crisis managers everywhere, these organizations spend time waiting and preparing for crisis events. Over the years, they have learned that drills and exercises are mandatory.

In the arena of crisis management, academic organizations can learn from the public sector. Although the logistics of drill actions are difficult in environments with diffuse authority and command structures, "practice" for crises is recommended.

Crisis Audit: The Recovery Phase

When the immediate event is over, many will breathe a sigh of relief. However, the crisis may not be over, certainly not for victims whose lives are irreparably shattered. In the aftermath, there will be a fallout from public relations, legal liabilities, employee stress, and possible career destruction. In cases where a crisis is managed well, public confidence will be maintained in the president and in the institution, senior administrators will retain their jobs, and legal liability will be minimized.

The campus's first task, of course, is to address the emergency. If lives are endangered or property is threatened, the initial act must be a response. In the larger perspective, a goal is to resume normal activity on campus as well as to enable victims to reconstruct lives. In every case, administrators, faculty, students, and staff must move from initial containment and emergency response to campus recovery. The recovery phase normally involves four actions:

1. *campus resumption plan,* which is a written document outlining how key institutional services and functions will be continued, and, if necessary, improved
2. *legal containment plan,* which defines the key strategies and steps in responding to inevitable legal charges
3. *cost recovery plan* that identifies and documents costs that may be recovered from insurance or from public relief laws
4. *strategic debriefing session* in which the crisis management team reviews what occurred and how to guard against future recurrences

TRANSFORMING "CRISIS" ACTIVITIES TO "ROUTINE" MANAGEMENT

No one enjoys planning for crisis management due to violence, criminal acts, or disaster. These are low-probability events, but they have extremely high consequences. The general reaction by most institutions is, unfortunately, simple apathy. "We have too much to do to keep the normal campus going to worry about unlikely events like this," or "I'll wait to hear from the legal office." One way to overcome inertia is to weave the emergency activities into the fabric of routine management. This can happen in several ways:

- A "hypothetical crisis situation" can be put on a regular agenda. The crisis event team will undoubtedly include many senior managers at the campus.
- Be sure that resources lists for crisis management dovetail with normal acquisition procedures.

- Install and constantly test sophisticated communications and surveillance systems.
- Develop media kits and press relations strategies.

In other words, endeavor to make activities of the crisis audit an offshoot of normal administrative processes. In this manner, it may be possible that routine procedures will be consolidated and focused when a crisis event occurs.

REFERENCES

Magner, D. (1996a, April 12). FBI nabs Unabom suspect who studied and taught at top colleges. *The Chronicle of Higher Education,* p. A24.

Magner, D. (1996b, April 19). Colleges keep security measures despite arrest in Unabomer case. *The Chronicle of Higher Education,* p. A25

SUGGESTED READINGS

Bausell, R.B., Bausell, C.R., & Siegel, D.E. (1990). *The links among alcohol, drugs and crime on American college campuses: A national follow-up study.* Silver Spring, MD: Business Publishers.

Council for Advancement and Support of Education. (1990, December). *Crime incidents on campus* (CASE Issues Paper for Communications Professionals, Public Affairs Services, No. 3), Washington, DC: Author.

Rickgarn, R.I.V. (1987, January 7). *Violence and suicide: Deadly connections.* Paper presented at First National Conference on Campus Violence, Baltimore, MD.

Schneidman, E. (Ed.). (1972). *Death and the college student: A collection of brief essays on death and suicide by Harvard youth.* New York: Human Sciences Press.

Shoemaker, E. (1993, January-February). When the wind blew through "cane country." *Campus Law Enforcement Journal, 23*(1), 27–30.

Siegel, D. (1994). *Campuses respond to violent tragedy.* Phoenix, AZ: Oryx Press/American Council on Education.

Zinner, E. (1985). *Coping with death on campus.* San Francisco: Jossey-Bass.

CHAPTER 14

Communicating about Violence on Campus: An Integrated Approach

Larry D. Lauer and Carolyn N. Barnes

WHAT IS INTEGRATED COMMUNICATION?

For years, the communications function of a university was thought of as little more than the mass distribution of press releases by the campus news and information office. Few people considered communications to be the responsibility of everyone at the university. Fewer still considered the idea that to communicate effectively, a university needed to have a strategy for doing so. In the 1990s, however, universities are discovering, and beginning to implement, the concept of integrated marketing.

Everyone at a university engages in communication, whether it is the university president speaking before the legislature, an admissions counselor recruiting a prospective student, a fund-raiser asking for a six-figure gift, or a media relations specialist pitching a story idea to a reporter. Integrated marketing, at its most basic level, acknowledges this reality. But, it takes that reality a step further by advocating a holistic approach to organizational communication. Combining marketing, public relations, and advertising, integrated marketing involves the entire organization in communicating consistent messages.

Nowhere is the need for integrated communication stronger, nor its value greater, than in communicating about a crisis, such as an incident of campus violence. In times of campus violence, victims as well as the public at large are looking for the comfort and reassurance that come from believing that the people in charge are capable and effective. People generally understand that life is full of mishaps. They accept that a university is a community and, like any other, it will have its share of violence. What people, especially parents, won't understand or

Source: Adapted from L.D. Lauer, *Communication Power: Energizing Your Nonprofit Organization*, pp. 149–165, © 1997, Aspen Publishers, Inc.

accept is the mishandling of campus violence by the professionals in whom they have placed their trust—and the welfare of their children. At a time like this, integrated communication means calling upon all the appropriate media and involving all the appropriate people to manage the crisis effectively.

Admittedly, taking charge decisively and quickly in the midst an incident of campus violence is more easily said than done. The aftermath of a violent act can be emotionally charged and intellectually confusing. Consideration must be given to the rights and needs of victims, to the rights of anyone accused, and to the integrity of any police investigation or disciplinary proceeding. This must be accomplished even as the media, who may know about the incident before you do, are demanding full disclosure of all "the facts."

UNDERSTANDING THE NEWS MEDIA

The News Business

The most important thing to remember about the news business is that it is a business. News organizations—print and broadcast—do not operate as a public service, even though one of their much-stated objectives may be to fulfill the public's "right to know." They are, in fact, corporate entities with the organizational characteristics and goals common to the corporate culture. Understanding how news organizations work, and how they do not, is critical to the handling of any incident of violent crime.

- The news business is a business, not a public service.
- The media's goals, not the university's, determine news coverage.
- Large profits require large audiences.
- Different perspectives can lead to conflict over what is "the truth."

The Media's Goals

A news organization exists to meet its own goals, not yours. It may seem to you that the less said about a violent crime on campus the better, for everyone involved. Many, perhaps all, of your peer opinion and community leaders may agree with you. But newspaper publishers and television and radio news editors will see it differently. And it is their goals, not yours, that will determine their coverage of your violent crime.

The Profit Factor

News organizations are commercial enterprises. They operate in a highly competitive environment. They succeed only when profitable, and their profits are produced by bringing large audiences to advertisers. Large profits require extremely large audiences. People working in the news business discover over time what attracts and what does not attract audiences. A television producer once said that he really got into the business to educate society. His aim was to bring interesting, insightful stories about complex problems to large audiences. But the more complex his stories became, the smaller his audiences became. And he could not survive with small audiences. The nature of the medium dictated what worked and what didn't. Attracting audiences, more often that not, determines when, how, and for how long the news media cover a story. The business of the news media requires it to overreport a crisis. There is a lot of mass-audience interest in emotional events involving any kind of conflict, such as violent crime on campus, and a lot of competition among media outlets to beat one another to the latest news. This combination of public interest and professional competition can lengthen media coverage of a story far beyond what may seem appropriate to you.

Your Perspective versus Theirs

It is also important to remember that you and the media have different perspectives on the story. From your perspective, it is best to tell the story, get it over with. From the media's perspective, the best course of action may be exactly the opposite—to keep the story on the front page or at the top of the newscast for as long as it draws a large audience, and the longer the better. These different perspectives and goals can lead to conflict between you and the media over what is the truth and what is not. Each interpretation may be correct in its own way, with the conflict coming not over the facts themselves but how each of you interprets them based on your different perspectives and goals. When these situations arise, and they will, it is critical to maintain mutual respect despite your differences.

CULTIVATING MEDIA RELATIONSHIPS

How effective you are in communicating about a crisis, such as campus violence, will depend in large part on how effective you have been in establishing good relations with the media long before the incident occurs. Taking the time to provide the media with background on your policies and procedures regarding campus violence and to introduce them to key players on campus will pay big dividends when you need them the most. Providing background information gives the media a head start they will appreciate. It also saves you from having to explain your policies and procedures, and to justify their use, over and over. To cultivate the media, take the following steps:

- Begin establishing relationships long before a crime occurs.
- Use an explanation of the Campus Crime and Campus Security Act as a starting point.

- Explain the differences between the criminal justice system and a disciplinary proceeding.
- Explain the confidentiality provisions of the Buckley Amendment.
- Introduce key players.

CAMPUS AWARENESS AND CAMPUS SECURITY ACT

A logical starting point for background information is the Campus Awareness and Campus Security Act of 1990. Although some reporters are aware of this federal law, which requires colleges and universities to report annually the incidence of certain crimes and arrests on their campuses, many are not. Explaining the act and the categories of crimes and arrests that must be reported provides a framework in which to present your specific policies and procedures.

The Campus Awareness and Campus Security Act requires that colleges and universities covered by the act—virtually all higher education institutions in the nation—must report incidents of murder, rape, robbery, aggravated assault, burglary, and motor-vehicle theft occurring on campus. Statistics must be reported for the most recent school year and the two preceding school years for which data are available. In addition, colleges and universities must report the number of arrests on campus during the same periods for liquor law violations, drug abuse violations, and weapons possessions. The act further requires that colleges and universities annually issue descriptions of campus crime prevention programs, as well as programs to inform students and employees about campus security practices and how to be responsible for their own security.

Presenting this information to the media as background gives you many opportunities to explain your crime statistics and level of preparedness under the best possible circumstances—when there is no crime demanding your attention and that of the media. This is the time to explain your ongoing efforts to prevent crime, how the campus police and the student affairs office handle reports of crime, as well as how your university assists the victims of crime.

THE DISCIPLINARY PROCESS AND THE BUCKLEY AMENDMENT

A comprehensive backgrounding program includes information on two items that frequently are misunderstood not only by the media, but also by many within the university community. One is the student disciplinary process, and the other is the Family Educational Rights and Privacy Act, also known as the Buckley Amendment. The media often think of the disciplinary process in terms of the criminal justice system, when in fact the two are very different. They also expect disclosure of information that you are legally prohibited from releasing if the individuals involved are students.

In a criminal trial, the defendant appears, evidence is presented, and the verdict is announced—all in open court. In a disciplinary proceeding, it is usually just the opposite. The students involved are not identified publicly, the proceeding is held behind closed doors, and specific disciplinary actions are not disclosed. The differences between the criminal justice system and the disciplinary process, if not explained thoroughly, can lead to accusations by the news media (and others) that the university is attempting to cover up the real story.

Any explanation of the disciplinary process by necessity will include discussion of the Buckley Amendment. Enacted in 1974, the Buckley Amendment protects the confidentiality of student educational records, which generally are defined as any information maintained by the institution about a student. Universities may release directory information—name, address, telephone, date of birth, major, dates of attendance, awards, etc.—about students without their permission. Other information of a "personally identifiable" nature may not be released. Campus law enforcement records are considered exempt from the Buckley Amendment, meaning they can be released. Disciplinary records always have been considered educational records and therefore confidential. But recent court decisions have clouded the issue of whether disciplinary records are confidential. The Ohio Supreme Court has ruled that disciplinary records are not protected from release by the Buckley Amendment. Meanwhile, the U.S. Department of Education has notified Ohio public universities that they must comply with the Buckley Amendment, which defines disciplinary records as educational records. The case is being appealed to the U.S. Supreme Court. While awaiting the court's decision, most institutions continue to consider disciplinary records confidential.

It is your responsibility to clarify the differences and distinctions for the media. Provide reporters and editors with copies of your disciplinary process and with the Buckley Amendment, and walk them through the information. You will be much more effective at this if the discussion takes place on a weekday morning over a cup of coffee than at midnight outside the scene of a fraternity brawl.

BRINGING TOGETHER KEY PLAYERS

Providing information is only one aspect of backgrounding. The other is to arrange meetings with key players at the university who likely will be part of the communication loop when a violent crime occurs on campus. These individuals might include the campus police chief and his or her assistant, the vice president for student affairs and/or the dean of students, the counseling services director, the health services director, and others. These meetings help reporters and administrators match faces to names, and establish professional identities and connections. Questions can be asked and answered thoroughly, policies and procedures explained in detail, without the pressure of deadlines or the glare of the television cameras.

PREPARING FOR A CRISIS

When a violent crime occurs on your campus, it's a crisis. It may seem impossible to be prepared for a crisis, but it isn't. The first step in becoming prepared is to acknowledge that sooner or later it will happen to you and your university. It is surprising how many university administrators adopt the "it will never happen to us" attitude. Not surprisingly, these are the people whose institutions suffer when the impossible happens and they are totally unprepared. The second step to preparedness, is to ask the question, "How will we handle a crisis?" This question becomes the framework for discussions among key staff. Out of these discussions will come the policies and procedures that will guide your actions.

YOUR FIRST RESPONSE

Crime, and crises in general, always seem to occur at the most inconvenient times. But you must respond quickly, because hesitation can cause you real trouble. Every minute that you delay puts you further behind in achieving your objective—a timely response and the appearance of control. The reporter who only last week wrote so glowingly about the success of your crime prevention programs is calling at 6 AM for information on the two students who were shot and killed outside the university library during a botched robbery. And this is the first you've heard about it.

In this kind of situation you also can anticipate that you will soon be receiving more media calls. A local television reporter may call to say that he or she wants to interview you on the steps of the library explaining how such a crime could have been "allowed" to happen on a supposedly safe campus. Are parents calling to express their concern? Have any students left the campus? If you refuse the interview request, you are informed that the story will be told anyway, just not your side of it.

Now the stakes are raised even more. Radio stations will call for on-air comments. More television stations will want interviews. And the editors of the newspaper that broke the story will be planning to talk to everyone they can find before deadline. They will be scrambling to outdo each other with the latest revelations.

It is at this point that the situation becomes dangerous. With the news media treating the story as "breaking news," misinformation and speculation will be everywhere. Focusing on the truth will be difficult as the objective shifts from reporting the facts to filling the public's and the media's appetite for emotional, dramatic stories by keeping the story alive as long as possible.

Into this circus atmosphere comes the university's communications staff and administration. Contrary to the notion that in a crisis the news media report the facts and the public relations staff exists to cover them up or spin them positively, the situation really is reversed. The media usually overreport and confuse the situation, while the university becomes responsible for bringing everyone back to the essential facts of what happened and what is being done about it.

A CRISIS COMMUNICATIONS PLAN

Whether the crisis is a violent crime or an incident of administrative misconduct, your management of the situation will be enhanced if you have a crisis communications plan.

No two crises are ever exactly alike. Crisis plans that are too specific often become unusable. Those trying to following the plans become frustrated as they go through lists that don't relate to the situation and end up putting aside the so-called plan. When policies are clear and procedures are basic, a crisis communications plan can prepare you for the approaches and priorities that will remain unchanged no matter the crisis. At the same time, it gives you the flexibility needed to address the aspects of the situation that make it different from any other. When developing a communications plan:

- Identify the person responsible for managing the situation, usually the chief communications officer, with the close involvement of the president.
- Outline procedures for gathering the facts.
- Define public versus private information. State the university's views on privacy, open records, and confidentiality.
- Decide how spokespersons will be identified. Who will be the chief spokesperson?
- Determine how the university will handle overly aggressive reporters.
- Decide what the organization will say when it cannot, or does not want to, say anything.

Managing the Crisis

The chief communications officer of the university usually is the most appropriate person to manage a crisis. This depends on that person's experience and ability to organize tasks under pressure. If possible, the president delegates this responsibility to the communications officer and then assumes a strong participatory role. This enables the president to gain perspective and prepare for his or her performance. If the communications officer is unavailable, then the president or another officer with the right skills assumes the role of manager.

Gathering the Facts

An organization must never rely on the news media to tell it the facts. The organization must be prepared to do its own investigation, and do it quickly and thoroughly. The communications officer and staff coordinate the fact gathering, but the actual investigations should be done by those closest to the situation. If a communication plan makes this process clear, the right people can go to work at a moment's notice. The communication plan also should stipulate that information needs to come in the form of timelines and fact outlines. Long narratives leave too much room for misinterpretation and take too long to digest.

Releasing Information

Determining what is public information is an area where the university and the news media likely will be in disagreement. The media will think and argue that the public has a "right to know" virtually everything about the crime. Talking to everyone involved gives more human interest and emotion to the story. Getting every fact leads to more follow-up stories and adds intensity.

The university will see the situation differently. You will want to protect the rights of the victim and preserve the integrity of any investigation, and you may be legally prohibited from releasing information about potential suspects if they are students. Recognizing this, the communication plan should spell out the university's views on confidentiality, open records, and personal privacy. Then, when a crisis occurs, it can give the media the policy, and its rationale, independent of the situation at hand.

Designating a Spokesperson

The trend in designating spokespersons seems to be toward a single person to speak for the university. But, the media may find speaking to one person unsatisfactory, and you might find that it does not work to your advantage, either.

First, you should think through those situations in which your president will and will not speak. If the crime is isolated or limited in nature, it probably is not necessary for the president to speak at all, nor is it a good idea. If it is a crime whose repercussions extend to or alarm the entire university, then the president will have to speak and do so fairly quickly. This should be spelled out in the communications plan.

A more satisfactory option might be to designate multiple spokespersons. The communications officer could give out facts, while a second (or third) person closer to the situation would be available to go over the details. The president, if necessary, could express the university's concern.

Handling Aggressive Reporters

Putting on paper how you will handle overly aggressive reporters can be quite helpful when those people show up on your doorstep. There are always the few who are trying to make a career at your expense, and to them the drama and emotion inherent in a violent crime seem made to order. They will camp out in your office, stake out victims and suspects, disguise their identity to secretaries, and do virtually anything else to ensure that they beat out everyone else to the "story."

It is completely acceptable for you to define your ground rules for handling these people and to let reporters and editors know in advance what they are. Keep in mind, however, that the ground rules should take into account the news media's legitimate needs as well as your own. For example, you can state that you will not talk to reporters who do not call in advance. But when they call, you need to talk to them. If a reporter has been particularly rude or abrasive, it is possible to tell his or her editor that you will not work with this person anymore, but you need to work with the next one they send.

No Comment?

There may be many times during the aftermath of a violent crime on campus when you cannot discuss certain aspects of the situation with the news media. When this happens, how you decline to talk to the media is very important.

If at all possible, never say "No comment." If you do, you will only sound arrogant and unresponsive, and you may create the impression that you are hiding something. If it's a request for information that you don't have, say you don't have it but you'll get it. If it's an issue that you believe you legally cannot discuss, explain why. Provide a context and a rationale for your decision not to talk.

THE VALUE OF A COMMUNICATIONS PLAN

The value of having a communications plan is that you can share it with the media in advance. Most reporters and editors know how aggressive their business has become and how easily it can get out of hand. They will respect you for taking a stand, if it's considered and fair. It also will demonstrate your commitment to getting out the facts quickly while making it clear that you will not be party to a media frenzy.

HOW TO WORK WITH THE MEDIA WHEN A VIOLENT CRIME HAPPENS ON CAMPUS

It's happened. An attempted robbery gone awry has left two students dead on the steps of your library. The media know it. What do you do? You follow the basic procedures outlined in your crisis communications plan, adapting them to the specifics of the situation and calling upon university resources as needed.

- Gather information and develop a fact sheet, preferably on a timeline.
- Form a communications committee, with the chief communications officer, media relations director, and president as permanent members. Add others as appropriate to the situation.
- Identify priority audiences and the most effective ways to get the information to them.
- Prepare statements for each spokesperson; use the fact sheet.
- Prepare spokespersons.
- Choose the method of release: mail, fax, e-mail, or press conference.
- Set a follow-up procedure.

Developing a Fact Sheet

The first procedure for the crisis manager is to prepare an outline of exactly what happened. To accomplish this, the information-gathering process previously discussed is activated. Since the sequence of events is so important, the facts should be listed in the form of a timeline. The fact sheet should then list and briefly explain every pertinent fact. This first draft should include every fact that can be discovered, and then it can be edited for accuracy and legal, privacy, and security issues. Care should be given to the editing process, because the fact sheet will form the foundation of the integrated communications effort.

Forming a Communications Committee

If you have the time, the next step is to form a committee to manage the university's proactive response. The committee should have several permanent members; others are added as appropriate to the situation. In the case of a violent crime on campus, the permanent members should include the chief communications officer, the president, the internal communications director, the media relations officer, the chief security officer, the dean of students, and usually the chief operations officer. Others are added as it becomes apparent that their involvement will expedite matters and improve communication.

It is always ineffective to exclude key people from committee deliberations. You end up having to say everything twice. But you also cannot allow the committee to become too large and unproductive. One approach is to have two levels of participants in the room: Those at the first level take an active role in planning; those at the second level are there to listen. Before any action is taken, the second group is given the opportunity to comment and ask questions.

What if you don't have the time to follow the procedure and pull together a committee? You may find it necessary to forgo the formality of calling together a committee, at least in the beginning. Instead, you could retain the goal of the committee—to coordinate the collection and dissemination of information—and adapt to a different model. One-to-one meetings, telephone calls, or e-mail may save critical time and be effective. The communications manager must be careful, however, to ensure that everyone who needs to know the latest and most complete details is kept in the information loop.

Preparing Statements

The fact sheet should be the basis of any statement. You simply turn this outline into several brief paragraphs following journalistic style: a lead emphasizing the main point followed by the who, what, where, when, and—if possible—why. You also should consider attaching your fact sheet to the statement in an effort to keep journalists who insist on writing their own stories as accurate as you can.

Preparing the Spokespersons

Once the fact sheet is edited into final form, the spokesperson assumes center stage. If you choose to have more than one spokesperson, it is critical that they communicate essentially from the same page. They should all have the same fact sheet in front of them at all times. Although each spokesperson reflects his or her differing roles and perspectives, they all should ultimately end up reinforcing a point on the fact sheet. The president will reflect his or her overall concern, while the communications officer's perspective is more factual, the security officer's more investigative, the dean of students' more reassuring, and so on. But each will reinforce the central message of the fact sheet.

The key is not to stray toward related facts that are not central to the story but can become pegs for more news stories. The objective is to get the central facts out as soon as possible but to limit the news coverage to a reasonable time frame. Providing these different perspectives on essentially the same facts should meet the needs of the news media and satisfy your objectives as well.

Interviews should be role played in advance between the spokespersons and the media relations person. The primary objective is to state the facts on the fact sheet with clarity no matter what questions are asked. Begin the rehearsal by brainstorming all the questions that might be asked. Having other staff members join in the session will help ensure that all angles are covered. The fact sheet is used as a guide in preparing answers for each question.

At the close of the session, an "interview notes" sheet should be prepared that includes phrases representing each anticipated question and, listed below each phrase, the facts that address the question. These notes can be reviewed as needed.

Releasing the Information

Once you have the fact sheet in place and your spokespersons prepared, then you must decide how you are going to release the information. This will depend to some extent on the circumstances. If the violent crime in question appears to be an isolated incident, if the victim is not seriously injured, or if protecting the safety of others does not seem to be an issue, then you may choose to issue a release based on your fact sheet. This could be hand-delivered, faxed, or e-mailed to the media, perhaps with a telephone call to alert reporters that the release is coming. You also would want to have your spokespersons on alert and include information for the media on how to contact them. This conveys the sense that you are concerned about the incident and take it seriously, but that you do not consider it a crisis.

The news release method will work only if it is based in reality. If the circumstances are other than those described above, or if there already is a demonstrated news media interest in the story, this method will not work. Under those conditions, a press conference may be a better way of getting the information out.

Holding a Press Conference

A press conference gives you a chance to have all the information out, with a defined period for follow-up questions, and all of this is done in a controlled time frame and format. This can be especially important if the facts of the crime are complicated and will require a detailed explanation. When planning a press conference:

- Schedule with newspaper deadlines in mind.
- Make provisions for microphones, photographers, television cameras, and writers.
- Begin with introductions, followed by any statements.
- Open the floor to questions for a specified time.
- Plan for one-on-one follow-up interviews at a later time.

Usually you call a press conference with newspaper deadlines in mind. You will know what they are, if you have cultivated good working relationships with reporters. The protocol typically calls for the people making statements to sit at a table at the front of the room, and for provisions to be made for microphones, still photographers, television cameras, and writers. Telephones should be available nearby. The media relations person welcomes everyone and introduces the spokespersons. Questions are taken for a specified time after the statements. Experienced spokespersons might stay around for a few minutes after the press conference to do one-on-one interviews. Others will leave as soon as the conference is over.

Developing Follow-Up Procedures

You may think that you've said all there is to say at the press conference, but the media will disagree. Anticipate this situation and be prepared for it. For example, one spokesperson may be prepared for one-on-one television interviews and made available later for a limited period of time. Others might be prepared to do telephone interviews at the same time. They all should use the fact sheet and they all should repeat, using different language, what they said at the press conference. Again, clarity, consistency, and containment are the goals here.

Circumstances may dictate follow-up press conferences or briefings. For example, this will be the case with violent crimes that require lengthy investigations or that lead to criminal arrests or disciplinary proceedings.

Working with the Local Police

Other issues that you will need to consider are what the local police are doing or will do to respond to and solve the crime, and how you can coordinate your communication effort with theirs. If the local police are the primary crime investigators, then questions about the investigation should be referred to them. You do not want to jeopardize the investigation by accidentally releasing sensitive or confidential information. If you choose to hold a press conference, it would be advisable to have a local police representative on hand to answer questions that university spokespersons would be uncomfortable answering.

A CASE STUDY IN TAKING AN INTEGRATED APPROACH TO VIOLENT CRIME

This is a case study of how one university used integrated communications in handling violent crime. In this example, Texas Christian University (TCU) involved all the key players in gathering facts, setting goals, deciding strategies, and implementing decisions, and utilized all the appropriate media in their total communications effort.

The Facts

A TCU student was sexually assaulted around 1:30 AM at a private residence not on, but near, the campus. It was the second assault of a TCU student near the campus in 12 days, and the third attempted assault in the area in 5 weeks. Local police initially believed the first two assaults were isolated incidents, but changed their thinking with the third incident. After the second incident (the first involving a TCU student), the university warned students about the assault through fliers placed in campus mailboxes and in a front-page story in the student newspaper.

About the time local police were making the TCU police aware of the third assault, the TCU communications director was learning of it from a newspaper reporter. The reporter had seen the incident on the police blotter and showed up at the communications director's office shortly after 8 AM. After assuring the reporter that information would be made available as soon as possible, the communications director began to adapt and implement the university's crisis communication plan.

The Objectives

In meetings or telephone conversations, the communications director, the vice chancellor for student affairs, and campus police agreed on two primary goals. The first was to ensure the safety of students, faculty, and staff by communicating the facts of the assaults and the need to be especially cautious. The second was to make public the facts about the assaults and the university's response as quickly as possible. A secondary goal was to enlist the support of area neighborhood associations and voluntary "citizens on patrol" to communicate what TCU was doing to address the situation.

The Response: The Campus Community

Throughout the day of the assault, many strategies were implemented in support of the university's goals. A fact sheet (Exhibit 14–1) outlining the circumstances of the assaults and the university's response was completed, as was a flier (Exhibit 14–2) containing the suspect's description and a list of safety DOs and DON'Ts. These were placed in student mailboxes and distributed to faculty and staff, especially those answering the telephones. TCU's campus escort service was expanded to include individuals going to cars, homes, or buildings on nearby off-campus streets. Local police added patrols in the area, while university police

Exhibit 14–1 Sample Fact Sheet

February 9, 1998

FACT SHEET

TCU Steps Up Campus Safety Efforts Following Off-Campus Sexual Assaults

TCU police and student affairs offices are taking several precautionary safety and security measures following the reports of two sexual assaults of TCU students near the campus. The Fort Worth Police Department is investigating the incidents that occurred in neighborhoods adjacent to the campus. The latest incident occurred February 9 at a private residence on the east side of campus. A similar attack occurred January 28 on the north side of campus. Fort Worth police believe the same suspect committed both crimes.

TCU police are working closely with Fort Worth police investigators and have taken the following measures:

- **Increased and expanded patrols of streets adjacent to the campus.** More patrol units have been added throughout the day. TCU officers are also patrolling more streets in close proximity to the campus, in addition to the streets directly bordering the campus that they already patrol on a regular basis.
- **Expanded campus escort services.** In addition to accompanying students, faculty, or staff to residence halls or vehicles parked on campus, campus escorts are now available to accompany individuals to their cars, homes, or apartments on nearby off-campus streets. Faculty, staff, or students may request an escort by dialing campus police at ext. 7777.
- **Assault alert flyers.** Flyers with suspect descriptions and safety measures (see attached) have been distributed to all student mailboxes on campus and hand-delivered to neighborhoods adjacent to the campus. "Citizens on Patrol" volunteers coordinated by the Fort Worth Police Department have assisted in this distribution. Flyers are also being posted in high-traffic areas across campus.

TCU police are urging all students, staff, and faculty to report any suspicious person or activity immediately. If you are on campus dial ext. 7777. Off campus dial 911.

Courtesy of Texas Christian University Office of Communications, Fort Worth, Texas.

---

expanded their patrols to include the streets bordering the campus. Fraternity members volunteered to serve as escorts, and were given caps and tee-shirts to identify themselves. And, the university waived the fee for a self-defense class for students.

Exhibit 14–2 Sample Flyer

ASSAULT ALERT

In the early morning hours of 2/9/98, within two blocks of the campus, a second assault has occurred. The following is a description of the suspect:

> **Sexual Assault Suspect**
> **Black Male**
> **20–30 years old**
> **Approx. 5'10"**
> **Weight approx:160–170**
> **ARMED WITH A HAND GUN AT THE TIME OF INCIDENT**

IF YOU NOTICE *ANY SUSPICIOUS PERSON(S)* CALL THE POLICE!
OFF CAMPUS: DIAL 911
ON CAMPUS: DIAL 7777

DON'Ts

- DO NOT walk alone! Call 7777 for an escort.
- DO NOT jog alone!
- DO NOT enter your residence alone if anything seems suspicious!
- DO NOT prop doors or windows under any circumstances!

DO's

- Leave outside lights on at night and keep lights on in more than one room.
- Secure all doors and windows.
- Before walking several blocks home from campus, call for an escort.
- Always lock your door when going to the rest room.
- CALL CAMPUS POLICE FOR ESCORTS, ON CAMPUS DIAL 7777, OFF CAMPUS DIAL 911.
- Keep car doors locked at all times, even if you are going into a place for a short time.
- *Always* look in your back seat before entering your car.
- *Always* have keys ready to unlock the car door and enter without delay. *Never* walk across the parking lot digging in your purse for your keys. Have them in your hand before leaving a building.
- *Always* park in well-lighted areas.

ABOVE ALL, BE AWARE OF THE PEOPLE AROUND YOU AT ALL TIMES! ALWAYS BE AWARE OF YOUR SURROUNDINGS AND TOTAL ENVIRONMENT!

Courtesy of Texas Christian University Office of Communications, Fort Worth, Texas.

The evening after the assault, TCU held a "town hall" campus meeting at the student center. More than 200 people, mostly students, attended. Police and university administrators spoke first, providing an update on the status of the case and reinforcing safety measures, and then took questions from the audience.

The Response: The Community

Numerous representatives of neighborhood associations and "citizens on patrol" volunteers with the city police department were asked to come to the campus police station to pick up fact sheets and fliers and distribute them to homes and apartments in the affected neighborhoods. Officers in the nearby city police "storefront" office assisted the university in contacting these representatives. The volunteers discussed the situation with the police department spokesperson, which further enhanced opportunities to pass along accurate information.

The Response: The Media

The university communications director spent most of the day at the police office responding to numerous calls and visits by every major print and broadcast media outlet in the metropolitan area, including a Spanish-language television station. The fact sheet and flier were distributed to the media. The assistant police chief, designated the official spokesperson with support from the communications director and the vice chancellor for student affairs, used the fact sheet and the flier as the basis for his discussions with the press.

The university provided several television crews the opportunity to ride along as TCU police patrolled the affected neighborhoods. The communications director also coordinated media requests to accompany neighborhood volunteers as they distributed fliers off campus.

Questions regarding the investigation of the assaults were referred to the local police. TCU spokespersons were careful not to intrude upon police jurisdiction, talking only about the university's response to the incidents. This duality led to the decision not to hold a press conference, even though media interest in the situation was intense. TCU officials believed this was not solely their story to tell, and were concerned that this distinction would be lost in the questions and answers of a press conference.

The media were invited to cover the town hall meeting, and turned out in large numbers. Four television stations were represented; two broadcast live from outside the meeting area. TCU officials asked the media not to photograph or tape participants during the question-and-answer session, fearing that it would inhibit the discussion. All complied; many stayed simply to listen to the dialogue.

The Results

TCU officials believe they achieved the best possible results in a bad situation. The campus community was informed and encouraged to be cautious. Safety measures were enhanced. Parents were reassured. The media reported accurately, making frequent use of the fact sheet and flier (Exhibits 14–3 through 14–5), and acted responsibly. Although no arrests were made, it can be said that awareness throughout the university community has been heightened. Follow-up fliers were distributed to ensure this awareness (Exhibit 14–6).

Exhibit 14–3 Media Report, February 10, 1998

Rapist Strikes Again Near TCU Campus

By Kathy Sanders
Star-Telegram Staff Writer

A masked rapist, possible the same man who has taunted women with frightening phone calls, struck again near the Texas Christian University campus early yesterday, attacking a student as she returned home, police said.

The assailant has targeted young women who live alone in apartments near the university and has sexually assaulted his victims at gunpoint, police said. Two women have been raped and a third woman escaped as the man broke into her home, police said.

"It's what we would classify as a serial rapist," Fort Worth police Sgt. Judy Jones said. "We know of three victims in the same area with the same general [mode of operation] with the same victim characteristics."

Two victims, including the 22-year-old attacked when she entered her apartment about 1:30 a.m. yesterday, are TCU students, and all three are young women who live alone north or east of the campus, investigators said. All were attacked in the early morning hours—on Jan. 4, Jan. 28 and yesterday.

The rapist wore all-black clothing, masked his face with dark pantyhose or a leotard, and brandished a handgun. He was described in two of the attacks as being African-American, in his 20s or 30s, with an athletic build and standing between 5 foot 6 inches and 5 foot 10 inches tall, police said.

"We have a male who has knowledge of the area who is preying upon TCU students or single women who are living alone in the area," said Lt. M.C. Slayton.

The first attack, police said, was apparently foiled when the 31-year-old woman awoke to footsteps outside her home, found a window broken and discovered that her

continues

Exhibit 14–3 continued

phone lines had been cut. When she fled from the apartment, she said yesterday, a man ran around the side of the building and into her open door.

"Oh, my God, my stomach goes in circles thinking about it," she said about the Jan. 4 incident. "I was so scared."

The recent attacks prompted Fort Worth police and university officials to mount an information assault on the campus and surrounding areas, alerting women to the rapist and passing on tips to reduce chances that they would become victims.

Campus officials have scheduled a meeting at 6 tonight in the student center to answer questions and address concerns, TCU Police Chief Steve McGee said.

TCU police are working with Fort Worth officers because "we have jurisdiction on campus only," McGee said.

"But since this has happened, we've added extra officers working details on late-night shifts, and we've beefed up our patrol on the perimeter of campus."

McGee said he doesn't believe the measures are premature.

"If we can prevent any sexual assault, we're willing to take any measure to protect students and neighbors of the university," he said.

The latest attack occurred when the woman arrived home after studying with her boyfriend, Slayton and Detective Don Hanlon said. The woman, who like the other victims is not being named because of the nature of the crime, was vaguely aware of the other attacks from pamphlets the university had circulated, Hanlon said.

"She's relieved she didn't get killed," Hanlon said. "She was worried she was going to be killed. She is being very cooperative. She managed to escape from him, and the assault was not fully completed."

The woman told police that when she arrived home, her cat began acting strangely and would not go into the apartment with her. She walked into the apartment, leaving the door open, and was putting her things down when she saw an armed masked man standing in the kitchen, Slayton said.

She fled from the apartment but tripped on an incline and fell, he said.

"She looks around and sees him coming. He grabs her by the hair, pulls her up and the first thing he said, like a robbery, if she had any money," Slayton said.

With an arm around her neck and a gun first to her head and then her side, the attacker walked her down the street to a vacant field, the woman told police. Trying to scare him off, she said her boyfriend was right behind her and would call police when he found her open door and a shoe that had fallen off, Slayton said.

In the field, after covering the woman's head with her jacket, he sexually assaulted the woman, police quoted the woman as saying. Twice, she said, he went back to look down the street. The second time, the woman ran, hiding in bushes until she felt safe to go to a neighbor's house, police said.

About the same time, a neighbor pulling into his driveway saw a man clad all in black, Slayton said. The neighbor yelled at the man, who then vaulted a fence. Police

continues

Exhibit 14–3 continued

> summoned by the neighbor were looking for the prowler when the call was received about the sexual assault.
>
> Police said they found pry marks on the woman's apartment door and a window unlocked.
>
> In the other rape, which occurred Jan. 28, a 20-year-old woman was awakened by a man whispering in her ear and holding a gun to her head, police said. Officers said the rapist got inside the woman's apartment through an unlocked window.
>
> At least one of the victims has received harassing phone calls from a man who said he had been watching her change clothes, Jones said. Other women have reported similar phone calls, Jones added.
>
> Anyone who may have received similar or strange phone calls should call police, she said, because phone calls can be a step in a progression toward rape, she said.
>
> In watching their targets, "they will resort to window peeping and obscene phone calls," Jones said, referring to characteristics of most rapists.
>
> Students questioned on the campus yesterday were unaware of the most recent attack. Some were also unaware of the previous sexual assault, although 4,000 fliers on personal safety and a description of the assailant were recently put in student mailboxes and posted in dormitories, a university official said.
>
> Don Mills, vice chancellor for student affairs, said another round of fliers was to be distributed yesterday evening and today. He said the neighborhood Citizens on Patrol group has been asked to distribute them off campus.
>
> Campus police will also make more officers available to escort students to their dorms and will expand the service to some off-campus areas in the neighborhood.
>
> "Money is not the issue," Mills said. "We want to help make sure students on and off the campus are as safe as they can be."
>
> Staff writer Paul Bourgeois contributed to this report.
>
> *Source:* Reprint courtesy of the *Fort Worth Star-Telegram*.

Exhibit 14–4 Media Report, February 11, 1998

> **Forum Addresses Assaults**
>
> by Ellen Miller and Melanie Rodriguez
> Skiff Staff
>
> Following instincts and being alert are the best weapons for protection against possible attack, police officials told about 200 students Tuesday at a safety meeting called to address concerns about recent sexual assaults near the TCU campus.

continues

Exhibit 14-4 continued

TCU administrators and both Campus Police officers and Fort Worth Police officers discussed safety issues and the heightened campus security to help calm fears and quench rumors about the aggravated sexual assaults of two TCU students within the past two weeks.

Both students who were assaulted lived in apartments within a few blocks of the campus.

Vice Chancellor for Student Affairs Don Mills, Fort Worth Detective Don Hanlon and TCU Police Chief Steve McGee answered questions and addressed the concerns of students. TCU and Fort Worth police officers, neighborhood patrol officers and representatives from the Fort Worth Rape Crisis Center and TCU Victims Advocate Program also participated in the program.

TCU Police Sgt. Connie Villela said the best guard against an attack is to be aware of the surroundings.

"The best thing to do is to always stay alert," she said. "Be alert of your area and walk in pairs. Look out for each other."

The university has increased security measures by providing escort service to both on- and off-campus students.

McGee said 24-hour TCU police officer escorts are available to students by calling 921-7777. The campus escort service, Froggie 5-0, is trying to provide service on campus beginning at about 6 p.m., two hours earlier than the normal 8 p.m. starting time.

Intrafraternity Council President Chad Cook said that TCU fraternities will volunteer time as escorts.

Beginning tonight, participating fraternity members will wear white baseball caps with "Froggie 5-0" printed on the front and "Security" on the back. Members of fraternities can sign up at the TCU Campus Police Office at 3025 Lubbock St.

Manuel Rodriguez, a freshman premed major who works as an escort for Froggie 5-0, said a lot of women on campus are scared because the suspect has not been caught.

"We can't be too careful," Rodriguez said. "I still see girls on the first floor with windows open. Do we need to go out there and hit them on the head to get them to be careful?"

McGee said students should be suspicious of someone doing anything unusual, like driving around a parking lot without a parking sticker.

Hanlon said students should report any suspicious behavior.

"Follow your gut feeling." Hanlon said. "It is a safety measure."

Mills emphasized that because TCU is a community, students should treat each other as family.

"This is not a Fort Worth problem and not a TCU problem," Mills said. "This is our problem."

continues

Exhibit 14–4 continued

> Hanlon and McGee would not comment as to whether the assailant may be a TCU student or if phone calls or stalking preceded the attacks.
>
> Jessica Leonardo, a freshman premajor who lives on campus, said news of the assaults is frightening.
>
> "I'm a bit wary, but I respect the university putting so much effort in letting us know," she said. "I don't feel too much fear, but it's scary."
>
> Davreen Dixon, a senior sociology major, said the assaults have been the most upsetting incident to occur since she began attending TCU in 1994.
>
> "It's a serious situation that should be given the serious attention it's getting," Dixon said.
>
> The suspect in both assaults has been described as a black male, 20 to 30 years old, 5 feet 10 inches tall and weighing about 160 to 170 pounds. He carried a handgun in both incidents.
>
> Due to a general suspect description and a composite that has not yet been released by police officials, some black male students have expressed concern.
>
> One black male student, who asked not to be identified, said he didn't like the impact the assaults have had on his life.
>
> "It's already hard on us black athletes in white schools," he said. "(Fellow students) are looking at us like we're all bad."
>
> *Source:* Reprint courtesy of the *Fort Worth Star-Telegram*.

Exhibit 14–5 Media Report, February 14, 1998

> **Harassing Calls, Suspicious Men Reported Following TCU Rapes**
>
> by Kathy Sanders
> Star-Telegram Staff Writer
>
> Since police disclosed that a serial rapist is in their midst, young women attending TCU or living near the campus have inundated officials with reports of harassing telephone calls, "peeping Toms," suspicious men and stalkers.
>
> Investigators say a deluge of information about the attacks and personal safety tips is responsible for many of the reports, although none has led to an arrest.
>
> Some of the reports are from women who called police this week after receiving an inordinate number of hang-up calls or harassing calls or who have seen men looking in their windows.
>
> "Our list of possible suspects just grows, but there's nothing real definitive," said Detective Don Hanlon. "We are just prioritizing our leads and focusing on the hot leads."

continues

Exhibit 14–5 continued

> University police, city patrol officers, detectives, and citizens on patrol descended on the neighborhood around TCU after two students were raped off campus and an attempt was made on another woman in the area.
>
> The first attack was Jan. 4, the next Jan. 28 and the latest early Monday, when a 22-year-old student walked into her home about 1:30 a.m. and found an armed, masked man clad in black standing in her kitchen, police said.
>
> The student fled but was caught by the man when she tripped outside. She told police that she was forced at gunpoint to a nearby field, where she was sexually assaulted. She escaped when the man went to see if anyone was coming down the street, she said.
>
> The *Star-Telegram* is not publishing the names of the women because of the nature of the crimes.
>
> TCU police said the students' concern for safety has heightened since reports of the attacks.
>
> "We've had a substantial increase in the number of requests for escorts around campus," TCU Assistant Police Chief J.C. Williams said.
>
> Police said they are pursuing several leads, including one woman's report that she was followed a long distance by a man who matched the description of the rapist. Another woman reported that in checking an acquaintance's belongings, she found a book of "strange writing that related to sexual activities," a police report says.
>
> The woman's acquaintance, the report continues, is known to hang out in the area and would leave home in the early morning. He also matched the vague physical description released by police of the rapist, the report says.
>
> In another report, a 21-year-old TCU student who lives in southwest Fort Worth told police that a man was standing outside her apartment window after she returned home late Thursday. The man fled when she banged on the window, police said.
>
> The woman told police that "she had just heard about sexual assaults that were occurring in the TCU area and was very concerned about them," the report says.
>
> The women described the rapist as wearing all black with dark pantyhose or a leotard pulled over his face. He is described in two of the attacks as being an African-American in his 20s or 30s, with an athletic build and standing between 5-foot-6 and 5-foot-10, police said.
>
> Staff writer Yvette Craig contributed to this report.
>
> *Source:* Reprint courtesy of the *Fort Worth Star-Telegram.*

Why It Worked

The communications director believes his good working relationship with reporters and editors was crucial. Reporters knew, from having worked with him before, that he would not attempt to get in the way of the story. They trusted that if

Exhibit 14–6 Sample Follow-Up Flier

ASSAULT UPDATE

February 27, 1998

There have been no new assaults since 2/9/98.

However please continue to be mindful of all the safety tips below!

IF YOU NOTICE *ANY SUSPICIOUS PERSON(S)* CALL THE POLICE!
OFF CAMPUS: DIAL 911
ON CAMPUS: DIAL 7777

DON'Ts

- DO NOT walk alone! Call 7777 for an escort.
- DO NOT jog alone!
- DO NOT enter your residence alone if anything seems suspicious!
- DO NOT prop doors or windows under any circumstances!

DO's

- Leave outside lights on at night and keep lights on in more than one room.
- Secure all doors and windows.
- Before walking several blocks home from campus, call for an escort.
- CALL CAMPUS POLICE FOR ESCORTS, ON CAMPUS DIAL 7777, OFF CAMPUS DIAL 911.
- Keep car doors locked at all times.
- *Always* look in your back seat before entering your car.
- *Always* have keys ready to unlock the car door and enter without delay.
- *Always* park in well-lighted areas.

Sexual Assault Suspect
Black Male
20–30 years old
Approx. 5'10"
Weight approx: 160–170
ARMED WITH A HAND GUN AT THE TIME OF INCIDENT

ALWAYS BE AWARE OF YOUR
SURROUNDINGS AND TOTAL ENVIRONMENT!

Courtesy of Texas Christian University Office of Communications, Fort Worth, Texas.

he said he would get back to them with information as soon as possible, he would. This comfort level kept reporters coming to him, rather than trying to circumvent him, for information.

The communications director also was able to correct misinformation or misinterpretations of the situation by the media, without causing anger, before they were reported. One news editor called for an update on the "rape on campus." The communications director was able to correct the misstatement without causing offense—before it was repeated on the air.

Having a flexible crisis communications plan also helped jump start the communications process at a time when minutes were critical. The urgency and "breaking-news" aspect of the situation made it impossible to put together a formal crisis communication committee. However, the communications director, vice chancellor for student affairs, and campus police were able to function as an informal committee—meeting one on one or talking on the telephone—because of the flexibility of the plan and their confidence in it.

CHAPTER 15

The Legal Response to Violence on Campus

Kay Hartwell Hunnicutt and Peter Kushibab

INTRODUCTION

This book has focused on violence occurring primarily in the 1990s on postsecondary campuses. During the 1990s, the courts have held colleges and universities to greater responsibilities in overseeing and protecting the lives and welfare of students (Wood, 1996). In addition, society has become more concerned about campus crime and the prevalence and effects of alcohol and drug use, hazing by fraternities and other groups, "hate crimes," rape, and assaults of students by third parties and peers on campus.

State legislatures have responded, for example, by increasing the drinking age and implementing state laws related to hate crimes, antihazing, and "duty to warn." Universities have strengthened student codes of conduct; provided escort services; increased security; and developed educational programs regarding a variety of topics addressing violence on campus including "date rape," sexual harassment, domestic violence, and "hate-speech" activities.

Public concern with campus crime has mandated that Congress, state legislatures, and postsecondary institutions take affirmative steps in dealing with violence against students. The purpose of this chapter is to review the legal response to violence on campus and to suggest guidelines in promoting safety on campus and preventive law activities to reduce liability.

The authors wish to acknowledge the contributions of Holly Carpenter, Jann Contento, and Richard R. Sines to the preparation of this chapter. They provided valuable help in bibliographic research and editing assistance. In addition, Janice Bradshaw, General Counsel, Maricopa County Community College District, provided guidance in the development of the chapter in the initial stages.

Griffaton (1993) states that campus crime leaves an indelible mark on its victims, their families, and college communities. College and university campuses are no longer "safe, bucolic havens, academic groves where the pursuit of knowledge and the cultivation of fellowship shut out many of the threats and fears of everyday life" (Purdon, 1988).

THE FEDERAL RESPONSE TO VIOLENCE ON CAMPUS

In response to media attention (Kalesse, 1990; "Student Kills 4," 1991) and what appeared to be an increase in violent crime on campus, several states and the U.S. Congress passed laws requiring postsecondary institutions to provide information on the number and types of crimes on and near campus (Kaplin & Lee, 1995).

Griffaton (1993) found that as of 1990, only 13 states had enacted campus crime reporting laws and that there were considerable differences in the reporting and dissemination of crime statistics, security policies and procedures, and enforcement procedures for noncompliance. Members of Congress recognized that state laws and individual institutional initiatives were not preventing or resolving the problem of campus crime and not providing the information students, prospective students, and faculty needed to avoid becoming victims of crime ("Statement of Rep. Goodling," 1990; "Statement of Sen. Gore," 1990).

Campus Security Act

The first of several congressional actions to protect students from violence on campus and to warn potential victims was the Student Right-To-Know and Campus Security Act (Public Law 101-542). Commonly known as the Campus Security Act, this law was signed in November 1990 and amended Section 485 of the Higher Education Act (20 U.S.C.§ 1092). The amendment imposes requirements on colleges and universities for preventing, reporting, and investigating sex offenses that occur on campuses. This amendment also requires institutions to provide information on policies related to the reporting of other criminal actions, campus security, and law enforcement. Additionally, postsecondary institutions are required to disclose the results of campus disciplinary proceedings to victims of violent crime. Institutions are also required to provide information describing the type and frequency of programs designed to inform students and employees about campus security. Title II of this act, known as the Crime Awareness and Campus Security Act of 1990, requires institutions participating in student financial programs under Title IV of the Higher Education Act of 1965 to disclose information about campus safety policies and procedures and to provide crime statistics related to on-campus crime.

The Campus Security Act was intended, in part, to encourage postsecondary institutions to put more emphasis on campus safety and on crime prevention services and programs (Lewis & Farris, 1997). Specifically, under the act, postsecondary institutions are required to report statistics on an annual basis concerning criminal offenses reported to campus security authorities, local police agencies, or any official of the institution with significant responsibility for student and campus activities. The act also specifies that institutions are to compile their crime statistics in accordance with the definitions used in the Federal Bureau of Investigation's (FBI) Uniform Crime Reporting program. Criminal offenses to be reported under the act include murder, sex offenses (forcible or nonforcible), robbery, aggravated assault, burglary, and motor-vehicle theft. The number of arrests for liquor law, drug abuse, and weapons violations must also be reported.

A major provision of the act requires institutions to develop and distribute to students, parents, prospective students, and the Secretary of Education a statement of policy detailing the institution's (1) campus sexual assault programs aimed at prevention of sex offenses; and (2) the procedures followed once a sex offense has occurred (Campus Security Act, 1994).

One of the controversial provisions of the act was how the law defines campus as "any building or property owned or controlled by student organizations recognized by the institution," or any "building or property controlled by the institution but covered by a third party." According to Kaplin and Lee (1995), this definition would, arguably, make fraternity and sorority houses part of the campus, even if they are not owned by the college and are not on land owned by the college.

Other controversial provisions of the act include the following:

- There are no reporting requirements of crime information related to off-campus victimization of students. As a result, campus crime statistics may appear to be skewed and alarming, when in fact, campus crime rates may be far lower than crime rates in the surrounding community. Thus, the act does not give students information about crime patterns in the neighborhoods around colleges—neighborhoods where many college students live (Smith & Fossey, 1995).
- There is confusion in reporting statistics referencing "forcible" and "nonforcible" sex offenses. Some institutions continue to provide statistics only for rape (Lewis & Farris, 1997).
- Many institutions withhold or underreport crime statistics to promote an image of a safe, secure campus to the public, prospective students, and parents.
- Crime reporting obligations of student service personnel and counselors could jeopardize the confidentiality of communications between counselors and student-clients (Lederman, 1994).

- The annual crime reports do not have to account for the nature of the institution (rural or urban, residential or commuter) (Lewis & Farris, 1997).
- Colleges with professional police departments may pursue crime more aggressively than do colleges that handle most incidents through a campus judicial system, resulting in a higher reported incidence of crime.

The final regulations were published at 59 Fed. Reg. 22314 (April 29, 1994). When the regulations were published, the annual crime reports were to include not only crimes reported to law enforcement authorities, but also crimes reported to campus officials "who have significant responsibility for student and campus activities" (Smith & Fossey, 1995, p. 221). According to Smith and Fossey, crimes reported to deans, residence hall directors, or other staff members involved in student affairs, including student counselors, must be included in the institution's annual crime report. However, institutions are apparently not required to make timely reports to the campus community about crimes that may pose a threat to staff or students.

The U.S. Department of Education (DOE) has been criticized for the delay in publishing the final regulations telling colleges how to implement the act and how to publish their final reports. To date, no colleges have been punished for violating the law. In a 1996 article in *The Chronicle of Higher Education,* DOE officials are quoted as stating that colleges are monitored only as to whether the annual reports were prepared and that the DOE does not check for accuracy unless it receives a formal complaint (Lively, 1996). The practice of the DOE is to confirm the publication of crime information by a college when outside auditors conduct annual reviews of compliance with federal regulations governing student-aid programs.

Confirming the accuracy of college crime statistics is rare. In 1996, Moorhead State University's crime statistics were investigated after a former student filed a complaint alleging the college was not following federal definitions of some crimes and that the numbers reported were not consistent with figures elsewhere (Kaplin & Lee, 1995).

The Student-Right-to-Know and Campus Security Act of 1990 authorizes a range of sanctions for violations including the withdrawal of federal funds. Accuracy in reporting crime statistics continues to be a significant issue related to the overall effectiveness of this law.

Federal Efforts To Collect Data on Campus Crime

The National Center for Education Statistics (NCES) conducted a survey of on-campus crime and security at 1,522 postsecondary institutions. The survey was conducted in response to the requirement of the Crime Awareness and Security Act of 1990 that the Secretary of Education make a one-time report to Congress on campus crime statistics. The survey collected information from institutions about campus crime statistics for 1992, 1993, and 1994; annual security reports compiled by institutions; and campus security procedures and programs. While the results of this important study are beyond the scope of this chapter, the study confirmed problems of accuracy in reporting on-campus crime statistics. However, this survey and report to Congress provides the first national estimates about campus crime and security, allows comparisons to be made between various kinds of institutions, and provides the context for interpreting the campus crime and security information provided to the public by individual institutions (Lewis & Farris, 1997).

As the result of widespread criticism of inaccurate reporting of crime statistics by postsecondary institutions, a bill was introduced on February 12, 1997, in the U.S. House of Representatives by Representatives John Duncan, Jr., and Charles E. Schumer, titled the "Accuracy in Campus Crime Reporting Act" (ACCRA), H.R. 715. As of February 1998, the bill had 60 cosponsors. Sponsors of the ACCRA bill believe that without access to accurate campus crime data, students are put at risk. The ACCRA bill is currently pending in the Postsecondary Education Subcommittee of the U.S. House Committee on Education and the Workforce. The future status of the bill is uncertain. On March 5, 1998, Stanley Ikenberry, president of the American Council on Education, presented testimony before the Senate Subcommittee on Labor, Health, and Human Services and Education Committee on Appropriations, criticizing some of the proposed changes in the bill including reporting requirements by campus administrators, deans' offices, athletic department officials, housing officials, and counselors. He criticized this proposed version of H.R. 715 because it would create a complex and expensive training burden, it would require all the proposed reporting individuals to be trained in criminal law, and it would violate confidential counseling relationships with victims (Ikenberry, 1998).

Ramstad Amendment

Another federal law regarding violence on campus is the 1991 Campus Sexual Assault Victim's Bill of Rights; this bill was sponsored by Representative Ramstad of Minnesota and signed into law as Public Law 102-325 in 1992, amending the Higher Education Act of 1992. This law is commonly known as the Ramstad Amendment. The law requires postsecondary institutions to adopt policies to prevent sex offenses and procedures to deal with sex offenses once they have occurred. The Ramstad Amendment requires institutions of higher education to develop and distribute a campus sexual assault policy that describes (1) education programs to promote awareness of rape, acquaintance rape, and other sex crimes; (2) institutional sanctions for sex offenses, both forcible and nonforcible;

and (3) procedures informing victims of the name of the person who should be contacted and the importance of retaining evidence.

The Ramstad Amendment requires sex offense policies to state that the victim has the same right as the accused to have others present during a disciplinary hearing and that both the accused and the accuser will be informed as to the outcome of any on-campus disciplinary proceeding. An additional provision requires institutions to notify the victim of the option of reporting sexual assault to law enforcement authorities and to provide assistance in the reporting process. Finally, campus authorities must notify victims about available counseling services and options for changing academic schedules and reasonable accommodations in changing living arrangements necessitated by sexual assault. (Smith & Fossey, 1995).

Violence Against Women Act

The Violence Against Women Act (VAWA; 20 U.S.C. § 1145h[b]), signed into law by President Clinton in 1994, established a federal cause of action for gender-motivated violence. Victims of a felony crime of violence motivated at least in part by gender may bring a civil suit for damages or equitable relief in federal or state court. The act provides a comprehensive approach to domestic violence and sexual assault, requires sex offenders to pay restitution to their victims, increases funding for battered women's shelters, and also includes a ban on firearm possession for domestic abusers. The act also assists college campuses, and its sexual assault provisions were incorporated into the Higher Education Act under Part D (Grants for Sexual Offenses Education). The act provides for the following:

- training campus security and college personnel, including campus disciplinary or judicial boards that address the issue of sexual offenses
- developing, disseminating, and implementing campus security and student disciplinary policies to prevent and discipline sexual offense crimes
- developing, enlarging, or strengthening support service programs including medical or psychological counseling to assist victims' recovery from sexual offense crimes
- creating, disseminating, or otherwise providing assistance and information about victims' options on and off campus to bring disciplinary or other legal action
- implementing, operating, or improving other sexual offense education and prevention programs, including programs making use of peer-to-peer education.

The act stipulates that no institution of higher education or consortium of such institutions shall be eligible for funds unless:

- Their student code of conduct or other written policy governing student behavior explicitly prohibits all forms of sexual offenses.
- Victims and offenders will both be notified of the outcome of any sexual assault hearing.
- They currently have a campus sexual assault education program (Bohmer & Parrot, 1993).

In *Doe v. Doe,* a wife sought damages against her husband under the civil rights provision of VAWA, based on alleged deprivation of the wife's right to be free from gender-based violence. The husband's motion to dismiss was based on the grounds that the remedy provision was unconstitutional. The federal district court in Connecticut held that (1) the remedy provision had a rational basis supporting a finding that gender-based violence substantially affected interstate commerce; (2) the remedy did not impermissibly encroach on traditional state police powers; and (3) the remedy was reasonably adapted to its intended end of deterring gender-based violence (*Doe v. Doe,* 1996).

In *Brzonkala v. Virginia Polytechnic and State University* (1996), a former university student sued other students who allegedly raped her, under a cause of action under VAWA. The federal district court in *Brzonkala* held that the VAWA was not a proper use of Congress's power under the test of substantial effects on commerce, even though it provides for a civil rather than a criminal remedy. That holding was later reversed, however, by the U.S. Court of Appeals for the Fourth Circuit. Relying in part on the *Doe v. Doe* decision, the Fourth Circuit found a rational basis for Congress's finding that violence against women substantially affects interstate commerce (*Brzonkala v. Virginia Polytechnic and State University* (1997).

Further case decisions at the appellate level will determine the force and validity of this act as a valid civil remedy for all persons within the United States to have the right to be free from crimes of violence motivated by gender. These court decisions do not affect the sections of the act incorporated into the Higher Education Act in 1994.

Title VII and Title IX

Title VII of the Civil Rights Act of 1964 and Title IX of the 1972 Education Amendments represent two bodies of federal law prohibiting sexual harassment (a form of sex discrimination) in postsecondary institutions. Title VII prohibits employers (including colleges and universities) from engaging in sexual harassment against employees. Title IX prohibits postsecondary educational institutions that receive federal funds from engaging in sexual harassment against either employees or students. Because students are often employees of a college or university, educational institutions may be subject to cause of action under both statutes (see, e.g., *Karibian v. Columbia University,* 1993).

Sexual harassment in higher education is a serious matter and may constitute a criminal as well as a civil offense. Obtaining sexual favors through some type of coercion in the form of grades, jobs, or promotions falls within the legal definition of force or threatened force in many state sexual assault statutes (Smith & Fossey, 1995).

Title VII makes it an unlawful employment practice for an employee "to fail or refuse to hire or to discharge any individual, or otherwise to discriminate against any individual with respect to terms, conditions, or privileges of employment, because of such individual's race, color, religion, sex, or national origin" (42 U.S.C.A. § 2000e-2[a][1]).

Title IX is the statutory corollary to Title VII in the context of public education or education financed in part by public funds. Title IX states that "No person in the United States shall, on the basis of sex, be excluded from participation in, be denied the benefits of, or be subjected to discrimination under any program or activity receiving financial assistance" (20 U.S.C. § 1681[a]).

Under Title VII, the Equal Employment Opportunity Commission (EEOC) is charged with the responsibility of ensuring compliance. Guidelines developed by the EEOC to assist in the investigation and prosecution of discrimination based on sex are found in the *Code of Federal Regulations* (29 CFR § 1604, et seq.).

The DOE's Office of Civil Rights (OCR) is the regulatory agency charged with enforcement powers under Title IX. The regulations found in the *Code of Federal Regulations* (34 CFR § 106, *et seq.*) prohibit discrimination on the basis of sex in admission and recruitment, in education programs and actions, and in employment in education programs and activities.

Two types of sexual harassment are prohibited by these discrimination statutes: (1) *quid pro quo* and (2) "abusive" or "hostile" environment harassment. Judicial interpretations of these two statutes have defined discrimination and sexual harassment to include unwelcome conduct offensive to a reasonable woman such as touching; kissing; solicitation; name calling based on gender; derogatory terms; offensive sexual jokes; and conditioning pay raises, promotions, and job retention in exchange for sexual favors.

Sparse legislative history exists regarding legislative intent to address sexual discrimination by the specific enactment of Title VII. In 1986, the U.S. Supreme Court in *Meritor Savings Bank v. Vinson* recognized a course of action for sexual harassment in the workplace. The opinion addressed one actionable form of sexual discrimination in the workplace as sexual harassment in a discriminatory environment that affects the victim's working conditions.

However, Title VII, while expressly addressing the problems of gender discrimination in *Meritor* and its more recent cases, does not extend to student victims except in the employer/employee context and would protect those students discriminated against in their capacity as student employees, such as research assistants or teaching assistants (Mango, 1991; see also *Pallet v. Palma,* 1996).

Litigation under Title IX has addressed alleged sex discrimination in the funding of women's athletics, the employment of women faculty, and sexual harassment of women faculty and students. In *Franklin v. Gwinett County Public Schools* (1992), the U.S. Supreme Court unanimously held that plaintiffs suing under Title IX were not limited to equitable relief but could claim monetary damages as well. The *Franklin* case settled a conflict among federal appellate courts. In *Franklin* the federal district court had ruled that monetary damages were not available; however, the U.S. Court of Appeals had ruled in *Pfeiffer v. Marion Center Area School District* (1990) that monetary damages were available.

In *Franklin,* a high school student in Georgia sued the school board under Title IX and sought relief from both *quid pro quo* harassment and hostile environment harassment by her sports coach. The coach engaged in sexually oriented conversation, unwelcome and forcible kissing on the mouth on school property, harassing calls to her home, and requests to see her socially. During her junior year the coach asked one of Franklin's teachers to excuse her from class. He then took her to a private office and raped her. Franklin alleged that the school officials and teachers were aware of the coach's conduct and took no action to stop the harassment. The school board agreed to let the teacher resign. After investigating Franklin's complaint, OCR determined that the district had violated her rights under Title IX and that sexual harassment had occurred. OCR concluded that the district was in compliance with Title IX due to the resignation of the coach and the school principal and because the school had implemented a grievance procedure.

The U.S. Supreme Court ruled in favor of Franklin and against the district for failing to stop the harassment and held that the district intentionally discriminated against her. The Court did not discuss recovery of damages for unintentional discrimination, or whether the district could be found liable because it had actual knowledge of the misconduct but failed to stop it, or whether the school could be liable even absent such knowledge because the teacher's or employee's intentional discrimination may be imputed to the school. Kaplin and Lee (1995) state that if courts apply agency principles to Title IX sexual harassment claims, it may be easier to hold colleges liable for harassment by their employees because plaintiffs will not need to show intentional discrimination on the part of the college. As a result of the *Franklin* decision in 1992, it is likely that an increasing number of students and faculty will use Title IX to challenge alleged sexual discrimination on postsecondary campuses (Kaplin & Lee, 1995). For example in *Slater v. Marshall* (1995), a federal district court in Pennsylvania refused to dismiss the Title IX *quid pro quo* charge because the student, who graduated with a 4.00 grade point average, raised material issues of fact to whether the harassment limited the variety of course work she might have taken while enrolled in a community college.

In a 1996 case against Temple University, a federal district court found that Title IX did not support a student's course of action against the university by a student alleging retaliation for filing a sexual harassment complaint; however, the court did not dismiss the claim against an administrator (*Nelson v. Temple University*, 1996).

A final issue addressed by Title IX relates to peer-to-peer sexual harassment. Liability in the form of monetary damages can be found against public schools when school officials have notice of a sexually hostile environment and fail to take immediate and appropriate corrective action. According to OCR's recent Sexual Harassment Guidance (1997), a school has notice if it actually "knew, or in the exercise of reasonable care, should have known" about the harassment. In addition, as long as an agent or responsible employee of the school received notice, the school has notice. Constructive notice exists if the school "should have" found out about the harassment through a "reasonably diligent inquiry" (OCR, Sexual Harassment Guidance, 1997). Given the result in *Franklin* and the deference of courts to guidelines developed by both the EEOC and the OCR related to sexual harassment, postsecondary institutions may wish to review their policies on sexual harassment of students by faculty, other employees, and peers.

Family Educational Rights and Privacy Act

The Family Educational Rights and Privacy Act (FERPA), often called the Buckley Amendment, prevents postsecondary institutions from releasing personally identifiable information in a student's educational records to third parties without the student's permission (20 U.S.C. § 1232[g]). There are two exceptions in which student records can be released without the student's permission. First, information must be released in educational records when subpoenaed as part of a lawsuit or upon a court order. Second, information must be released in an emergency situation that presents danger to other persons. FERPA regulations protecting student privacy and confidentiality of educational records can conflict with provisions of other federal laws such as the Student Right-To-Know and Campus Security Act and the Ramstad Amendment discussed previously in this chapter.

Student Right-To-Know and Campus Security Act

The Student-Right-To-Know and Campus Security Act of 1990 contains provisions permitting postsecondary institutions to release the results of campus disciplinary proceedings to the victim. And the Ramstad Amendment requires both the accuser and the accused to be notified of the results of disciplinary proceedings that allege a sexual assault. Crime reports in school newspapers naming student victims and the perpetrator have been litigated. In a 1991 federal district court decision, the judge held that campus crime records were not educational records under FERPA and ordered Southwest Missouri University to make the campus crime records available to the editor of the university's student newspapers in compliance with Missouri's public records law (*Bauer v. Kinkaid*, 1991).

Following the *Bauer* decision, Congress amended FERPA to state that law enforcement records are not educational records. Thus, apparently no conflict exists with the Campus Security Act, which expresses a policy view that students are entitled to information about the incidence of crime on their campus to better protect themselves from becoming victims (Smith & Fossey, 1995). However, Ikenberry (1998) challenged provisions in laws where the federal government dictates that campus disciplinary proceedings be open to the public. According to Ikenberry, this would have a chilling and counterproductive effect on the reporting of crime on campus; victims of campus crime—especially sexual assault—often use disciplinary proceedings as "justice" but are not willing to press criminal charges. He also stated that victims will be less likely to report crimes and seek a resolution if the incident is discussed in a public forum.

The controversy of disclosing the results from records of campus disciplinary proceedings (required by the Ramstad Amendment if involving a sexual assault) was illustrated in *Red and Black Publishing Company v. Board of Regents* (1993). The University of Georgia's student newspaper tried to obtain the records of a campus disciplinary hearing involving hazing complaints against two fraternities. The university argued that the disciplinary records were protected by FERPA, and the newspaper claimed that public universities had to comply with Georgia's public records act. The Georgia Supreme Court ruled that the disciplinary records referenced in this case were public records and the newspaper had access to these documents. The court also ruled that records of campus disciplinary proceedings were not educational records under FERPA.

There is no private right of action to bring a damage suit against an institution or person violating the Buckley Amendment. However, a postsecondary institution might lose its eligibility for federal funding for violations of the law's provisions. A few federal court cases have held that a FERPA violation may be litigated for damages under 42 U.S.C. Section 1983 of the Civil Rights Act of 1871 (*Krebs v.*

Rutgers, 1992; *Norwood v. Slammons*, 1991). This Section provides a private right to bring suit for damages against any person who, acting under color of state law, impairs rights secured by the Federal Constitution and laws. It allows individuals to obtain damages for abridgments of federally protected rights. Postsecondary institutions and their officials have been deemed "persons" under Section 1983.

Campus officials should review current case law addressing FERPA violations and review student records policies in conjunction with federal laws such as the requirements of the Ramstad Amendment and the reporting requirements of the Campus Security Act. State laws protecting student records and state public records laws must be addressed in policy and practice.

STATE EFFORTS TO COMBAT VIOLENCE ON CAMPUS

While the federal response to violence on campus through new laws in the 1990s has not yielded effective enforcement effort, the states have responded to violence on campus by enacting an increased number of state laws dealing with antihazing, hate crimes, and stricter penalties for alcohol and drug violations and possession of weapons. These state laws have resulted in an increased number of cases in the civil and criminal courts for violations of these statutes.

Institutional Liability

An effective way in which the law protects the safety of college and university students is through court decisions in lawsuits filed by student victims of assaultive or similarly injurious behavior. Frequently, if an injury is suffered on campus, or while the student is participating in some school-related activity, the student sues the institution, seeking damages or some other form of relief. The decisions of trial and appellate courts in these actions form a body of common law, the effect of which is to define the respective rights and responsibilities of students and the institutions they attend.

Tort Action

Typically, a student whose injury is alleged to be the fault of a college or university commences an action in tort against the school. In such an action, the student victim must allege that the injury was a direct result of negligence or other wrongful conduct by the institution. In general, to establish a case of negligence, the student must prove that the school was under a duty that is recognized by the law and requires the school to conform to certain standards of conduct (*Robertson v. LeMaster*, 1983).

Sovereign Immunity. If the defendant college or university is a publicly supported institution, a threshold issue for the student plaintiff is whether the state in which the school is located recognizes the doctrine of sovereign immunity. In states recognizing sovereign immunity, an agency of the state, such as a state college or university, may not be held liable for most forms of wrongful conduct. Accordingly, in such states, a student who has suffered an injury due to the school's alleged wrongful conduct would not be allowed to sue the school on the basis of a tort theory, such as negligence (*Cooper v. Delta Chi Housing Corp. of Connecticut*, 1996). Many states, however, have abolished the doctrine of sovereign immunity; a state-run institution in such jurisdictions could be held accountable under a tort theory (*Stone v. Arizona Highway Commission*, 1963).

In loco parentis. In recent years, courts have usually been reluctant to hold postsecondary schools to a high duty of care toward the physical safety of their students. An Ohio court, for example, rejected the claim by the parents of a 17-year-old university student that the university was negligent in allowing their daughter to "become associated with criminals, to be seduced, to become a drug user," and permitting her to be absent from her dormitory. "A university," said the court, "is an institution for the advancement of knowledge and learning. It is not a nursery school, a boarding school or a prison" (*Hegel v. Langsam*, 1971).

Such thinking stems, at least in part, from the courts' acknowledgment that, over the last few decades, the traditional custodial responsibilities of a college or university have diminished. According to one federal appellate court, since the 1960s, "Regulation by the college of student life on and off campus has become limited. Adult students now demand and receive expanded rights of privacy in their college life including, for example, liberal, if not unlimited, partial visiting hours. College administrators no longer control the broad arena of general morals" (*Bradshaw v. Rawlings*, 1979).

Courts have rejected, then, the doctrine of *in loco parentis,* which dictates that, owing to the high degree of control it might exert over its students, a school acts "in the place of a parent" and is therefore responsible for the students' safety. Rather, as "society considers the modern college student an adult, not a child of tender years," a college or university is not obligated to shield its students from the dangerous activities of other students (*Bradshaw v. Rawlings*, 1979).

Special Relationship. The courts' apparent abandonment of the *in loco parentis* doctrine, however, in no way immunizes postsecondary institutions from tort liability for injuries suffered by their students. When an institution undertakes to render services to students in a manner that creates a special relationship, or otherwise willingly assumes a particular responsibility for the student's protection, it might be held liable for injuries suffered by its students.

For example, in Massachusetts, a court upheld a judgment in favor of a student who had been raped in a dormitory, against a college for the latter's failure to afford adequate security. The court recognized a unique situation created by the concentration of young people on a campus and the college's ability to protect its students, and defined the college's duty as including "reasonable care to prevent injury to their students by third persons whether their acts were accidental, negligent, or intentional" (*Mullins v. Pine Manor College,* 1983).

On the other hand, a federal court in West Virginia refused to find a private college liable when a 17-year-old, female student who had lived on campus was raped in her assailant's apartment shortly after she left an off-campus bar. While the court acknowledged that a college has a general duty to its students to maintain a campus environment free of foreseeable harm, it held that such an obligation does not extend to noncurricular activities taking place beyond campus boundaries. Moreover, the court expressly rejected the doctrine of *in loco parentis* despite the fact that the student was a minor (*Hartman v. Bethany College,* 1991).

The Student as Invitee. In measuring a school's liability to its students, some courts have borrowed from laws governing a landowner's responsibilities to visitors who have entered the owner's premises for business purposes (or "invitees"). Generally, a landowner has no duty to protect an invitee on the landowner's property from criminally assaultive conduct by a third party unless the conduct is reasonably foreseeable (Restatement [Second] of Torts, 1965).

A Kansas court adapted this rule to accommodate a lawsuit against a university by a student whom another student had raped in a dormitory room. Specifically, the court held that, while a university is not an insurer of the safety of its students, it has a duty of reasonable care to protect a student against certain dangers, including criminal actions by another student or a third party, if the criminal act is reasonably foreseeable and within the university's control (*Nero v. Kansas State University,* 1993).

State-Based Actions

A college or university, like other public and private institutions, is subject to many state and federal laws that dictate how it may (and may not) conduct its business. These laws often address concerns such as equal employment and confidentiality of student records, as well as matters of health and safety, and frequently threaten fines or other penalties upon an institution for violating these mandates. Moreover, courts, Congress, and state legislatures have ruled that private individuals who are harmed by an institution's failure to comply with many of these laws may themselves sue the institution for damages or other relief.

Student suits in state civil courts against universities have resulted in monetary awards for damages including failure to maintain adequate campus security (*Mullins v. Pine Manor College,* 1983) and failure to exercise reasonable care to protect students from foreseeable harm (*Jesik v. Maricopa Community College,*

1980; *Peterson v. San Francisco Community College District,* 1984; *Tarasoff v. Regents of the University of California,* 1976).

In addition, when an institution attempts to prohibit or to control inherently dangerous activities of students such as hazing during fraternity rush (*Furek v. University of Delaware,* 1991) or alcohol consumption for on-campus functions or school-sponsored, off-campus activities—particularly in states where case or statutory law establishes civil liability for private hosts—a court may find that the institution has a duty to those students (*Hernandez v. Arizona Board of Regents,* 1994; *Kelly v. Gwinnell,* 1984).

Criminal Penalties

Although a college or university may seem an insular environment, its employees and students are generally subject to the laws of the state in which it is located, including the state's criminal code. Frequently, however, those criminal statutes are enforced on campus by the school itself, through a campus security department or university police, rather than by local or state law enforcement officials.

Assault

A state's criminal code invariably includes a series of laws proscribing various types of offensive or harmful physical contact. Typically labeled "assault" or "battery," these statutes prohibit touching a person in an intentional or reckless manner that causes injury, or touching a person with the intent to injure. The offense is deemed more serious, or "aggravated," if the touching in some way causes a serious injury or involves a weapon (Ariz. Rev. Stat. § 13, 1203–1204).

Rape and Other Sex Offenses

The elements of rape generally include compelling a victim to engage in sexual intercourse by force and against the victim's will, or by threat of some bodily injury (Mass. Gen. Laws Ann., Ch. 265, § 22). A criminal code almost always includes sex offenses that do not necessarily involve forced intercourse, but do entail some sort of unlawful sexual contact. Such offenses typically bear labels such as "sexual abuse," or "assault with intent to commit rape" (Ariz. Rev. Stat. § 13, 1401–1404).

Stalking

A relatively new, but increasingly popular, feature of a typical state's criminal statutes is legislation prohibiting stalking or other nonphysical harassment. Stalking laws are distinctive in that they outlaw a series of actions, rather than a particular act. Usually, a state's stalking law proscribes conduct that occurs on more than one occasion and is intended to threaten or harass (Ariz. Rev. Stat. § 13, 2923).

Hate Crimes

During the 1980s, many state legislatures enacted so-called hate crime laws. Typically, such laws prohibit conduct that not only injures (or threatens to injure) a person, but is motivated by some sort of ethnic, racial, or religious animosity.

The elements of the various states' hate crimes vary. In Illinois, a hate crime is any of a relatively short list of offenses (such as assault, battery, misdemeanor trespass to a residence or real property) committed by reason of an individual's or group's "actual or perceived race, color, creed, religion, ancestry, gender, sexual orientation, physical or mental disability, or national origin" (720 Ill. Comp. Stat. Ann. § 5.12–7.3). On the other hand, a hate crime in Massachusetts is "any criminal act coupled with overt actions motivated by bigotry and bias including, but not limited to, a threatened, attempted or completed overt act motivated at least in part by racial, religious, ethnic, handicap, gender or sexual orientation prejudice, or which otherwise deprives another person of his constitutional rights by threats, intimidation or coercion, or which seek to interfere with or disrupt a person's exercise of constitutional rights through harassment or intimidation" (Mass. Gen. Laws Ann., Ch. 22C § 33).

Following their enactment, hate crime laws were subject to numerous court challenges. Generally, these challenges alleged that the statutes prohibiting hate-oriented conduct violated either the U.S. Constitution or the constitution of the state in which the hate legislation was passed.

U.S. Constitution. The first challenge to modern hate crime legislation to reach the U.S. Supreme Court was *R.A.V. v. City of St. Paul, Minnesota* (1992). There, the court reviewed the misdemeanor adjudication, under a St. Paul municipal ordinance, of several juveniles who had burned a cross on the residential yard of a black family. The ordinance proscribed placing "on public or private property a symbol, object, appellation, characterization, or graffiti, including but not limited to, a burning cross or Nazi swastika, which one knows or has reasonable grounds to know arouses anger, alarm or resentment in others on the basis of race, color, creed, religion or gender" (R.A.V., 380).

One of the juveniles claimed that the ordinance violated his right of free speech under the First and Fourteenth Amendments to the U.S. Constitution. Writing on behalf of the majority, Justice Scalia noted that while the First Amendment generally restricts the government from outlawing speech based on its content, prohibitions on certain forms of speech (such as obscenity, defamation, and fighting words) are often permissible. At the same time, however, he warned against an "all or nothing" approach to regulating speech that accompanied conduct: "We have long held, for example, that nonverbal expressive activity can be banned because of the action it entails, but not because of the ideas it expresses—so that burning a flag in violation of an ordinance against outdoor fires could be punishable, whereas burning a flag in violation of an ordinance against dishonoring the flag is not" (*R.A.V., 384–385*).

Ultimately, the Court concluded that the St. Paul ordinance violated constitutional free-speech guarantees in its content-based effect of prohibiting only certain kinds of abusive words. Justice Scalia illustrated such an incongruity with a hypothetical: "One could hold up a sign saying, for example, that all 'anti-Catholic bigots' are misbegotten; but not that all 'papists' are, for that would insult and provoke violence 'on the basis of religion.' St. Paul has no such authority to license one side of a debate to fight freestyle, while requiring the other to follow Marquis of Queensberry rules" (*R.A.V., 391–392*).

One year after it decided *R.A.V.*, the Court upheld Wisconsin's hate crime statute in *Wisconsin v. Mitchell* (1993). Standing outside a Kenosha, Wisconsin, apartment complex, Mitchell had asked several other black individuals (after the group had viewed the motion picture *Mississippi Burning*) whether they felt "hyped up to move on some white people." The group, which included Mitchell, then proceeded to assault a young white boy in the neighborhood, whom they had apparently selected at random.

Mitchell was convicted of aggravated battery under Wisconsin's criminal code. His prison sentence, however, was increased beyond what was normally prescribed for such an offense. The court increased Mitchell's sentence in accordance with the Wisconsin hate crime provision, which called for a longer prison term if the victim of certain crimes was selected because of the victim's race, religion, color, disability, sexual orientation, national origin, or ancestry (*Wisconsin, 479–481*).

Mitchell asserted a claim like that advanced in *R.A.V.*—namely, that Wisconsin's hate crime law violated the First and Fourteenth Amendments to the U.S. Constitution—but the U.S. Supreme Court upheld the Wisconsin statute. While hate-oriented conduct was an element of the offense under the St. Paul ordinance, a hate-based motive according to Wisconsin's provision merely served to enhance criminal punishment. According to Chief Justice Rehnquist (who wrote the opinion on behalf of a unanimous court), the St. Paul ordinance was "explicitly directed at expression," while the Wisconsin statute was "aimed at conduct unprotected by the First Amendment." The court further upheld the rationale underlying the Wisconsin law, noting that it "single[d] out for enhancement bias-inspired conduct because this conduct [was] thought to inflict greater individual and societal harm" (*Wisconsin, 487–488*).

State Constitution. While hate-based motive was a separate element of the Montana statute considered by that state's supreme court, the law nevertheless passed muster under the Montana state constitution. The defendant had been accused of defacing property by attaching, to various county road signs, bumper stickers whose message protested the existence of a religious sect. He was convicted of malicious intimidation, which deemed it unlawful to deface such property "purposely or knowingly, with the intent to terrify, intimidate, threaten, harass, annoy or offend" on the basis of "another person's race, creed, religion, color, national origin, or involvement in civil rights or human rights activities." The Montana court found no violation of the state constitution's free speech provision, holding that the law's purpose was "not to suppress the content of the communication," but rather to "prohibit conduct that violates other criminal laws, such as assault, criminal mischief, and trespass, and that are committed against another person because of that person's race, religion or national origin" (*State v. Nye*, 1997).

Institutional Policies

During the 1980s, many college and university officials created policies prohibiting hate-based activities by students and employees. These policies were likely prompted by the same motivations that led to much of the hate crime legislation that was enacted that same decade. Many of the policies, however, purported to regulate hate-oriented speech, as well as conduct. Many institutional policies were "struck down" by the courts for violation of individual U.S. constitutional rights (Page & Hunnicutt, 1994).

By 1988, officials at the University of Michigan had perceived an increase in incidents of racism and racial harassment at that institution. In response, and following extensive review by university staff and faculty, the university's board of regents adopted a policy on discrimination and discriminatory harassment (*Doe v. University of Michigan,* 1989). The policy prohibited "any behavior, verbal or physical, that stigmatize[d] or victimize[d] an individual on the basis of race, ethnicity, religion, sex, sexual orientation, creed, national origin, ancestry, age, marital status, handicap or Vietnam-era veteran status" that threatened a student or employee, or interfered with a student's or employee's participation in university activities (p. 856).

Courts have long held that a postsecondary institution may impose reasonable and nondiscriminatory time, place, and manner restrictions on all campus speech and conduct (*Heffron v. International Society for Krishna Consciousness,* 1981). Nevertheless, a federal court struck down the University of Michigan's policy as a violation of the First Amendment. The court held that the school could not "establish an anti-discrimination policy which had the effect of prohibiting certain speech because it disagreed with ideas or messages sought to be conveyed," noting that such a rule had "a special significance in the university setting, where the free and unfettered interplay of competing views is essential to the institution's educational mission" (*Doe v. University of Michigan,* 1989, p. 863).

THE COLLEGE AND UNIVERSITY RESPONSE TO VIOLENCE ON CAMPUS

This final section of the chapter deals with the efforts of colleges and universities to educate students about violence and how to prevent it. Policies have been designed to prohibit student conduct involving violence toward others and to provide disciplinary sanctions for students found to have violated student codes of conduct by participating in violent and abusive situations leading to harm of others.

As a result of federal laws affecting colleges and universities in the 1990s (such as the Ramstad Amendment) and the increase of violence on campus, policies related to "hate speech," "hate crimes," assault, sexual harassment, and hazing have been reviewed and revised. Many colleges and universities have updated student codes of conduct regarding procedures to be used when infractions of the code are alleged as well as the type of sanctions for violations.

Although student codes of conduct address both academic and social misconduct, the focus of this discussion is on issues of social misconduct. Social misconduct may include disruption of an institutional function such as teaching, misconduct involving abusive behavior or hazing, as well as more violent types of conduct. Misconduct off campus can be proscribed and students disciplined if the misconduct also violates criminal law. Applying the code to off-campus conduct can pose significant legal issues (*Held v. State University of New York College at Fredonia,* 1995; Kaplin & Lee, 1995; *Ray v. Wilmington College,* 1995). The two cases of *Held* and *Ray* are representative of applying campus discipline to off-campus criminal conduct. In *Held,* a trial court held that a student was entitled to advance notice and a hearing before an impartial body before he could be suspended for arrest and charges for narcotic offenses off campus. In *Ray,* a state appellate court did not find the university's action to be arbitrary and capricious when it suspended a student for sexually assaulting a classmate off campus.

Disciplining students for off-campus misconduct may implicate issues with the Fifth Amendment "double-jeopardy" clause. For example, in the *City of Oshkosh v. Winkler* (1996), the city's attempt to prosecute for disorderly conduct was not barred by the double-jeopardy clause since disciplinary action by a state university for the same misconduct had a nonpunitive purpose. Therefore, student codes of conduct and types of sanctions used for violations should identify aims beyond punishment such as protecting the safety and integrity of the campus community (Pavela, 1997). The attorney general of Maryland addressed this issue in a 1989 opinion upholding the right for a university to discipline "for off-campus conduct detrimental to the interests of the institution subject to the fundamental constitutional safeguards that apply to all disciplinary activities by educational officials" (*Opinions of the Attorney General,* 1989).

Sometimes the courts invalidate a university's discipline for off-campus behavior when the behavior did not pose a threat to the institution. In *Paine v. Board of Regents of the University of Texas System* (1973), the institution disciplined students convicted of drug offenses off campus. The court held that discipline by colleges and universities must be applied equally and cannot be a harsher penalty than penalties given to other code offenders including those with equally serious offenses. The court held that this differential treatment violated the equal-protection and due-process clauses of the Fourteenth Amendment.

In the *Paine* case, the institution automatically suspended students who were put on probation by criminal courts for possession of marijuana. This case also serves a reminder that student codes of conduct must comply with constitutional due process and must be clear enough for students to understand the standards with which their conduct must comply.

Student Code of Conduct Provisions Addressing Issues of Civility

Private and public colleges and universities have revised policies and student codes of conduct to reflect new requirements in federal or state laws, to reflect court decisions regarding student discipline, and to delineate more clearly within the codes the essential values and mission of the institution. Portions of selected codes of conduct are provided as examples directly addressing matters of student conduct on and off campus that may interfere with and endanger the welfare of others or violate the institution's stated essential values and mission.

The University of North Carolina at Chapel Hill enacted a more restrictive policy on alcohol use and possession. The policy, which became effective in January 1996, dealt with individuals and student organizations and their use of alcohol. The policy included the justification for the policy revisions, outlined applicable state statutes, discussed a recent state supreme court ruling on social host liability, clearly delineated rules and enforcement policies for individual students as well as organizations, and outlined requirements for university functions at which alcohol can be served including sorority and fraternity functions. Sanctions for violations of policy related to sorority and fraternity functions include mandatory risk management education seminars for 80% of the chapter, documentation of the incident in the chapter's discipline files, letters to national headquarters, fines on a per member basis, and loss of social privileges.

Deaths of students by acute over-consumption of alcohol, interpersonal violence such as hazing incidents, physical altercations, and date rape as a result of over-consumption of alcohol are well documented. In response to these and similar issues, the policy at the University of North Carolina states:

> We believe that the most productive approach for combating the negative effects of alcohol abuse is through creative alcohol abuse preven-

tion programming, combined with effective early intervention against problem behaviors related to alcohol abuse and a clear statement of University policy related to the use and abuse of alcohol. This revision of the Student Alcohol Policy was undertaken with these principles in mind." (Policy on Student Possession and Consumption of Alcoholic Beverages, 1996).

The *University of Washington Student Conduct Code* (1996) incorporates the terminology from the Federal Crime Awareness and Campus Security Act related to sexual offenses. The policy prohibits conduct on campus constituting a sexual offense whether forcible or nonforcible, such as rape, sexual assault, or sexual harassment.

In response to state and federal laws and the institutional commitment to prevent and educate about violence, Princeton University created a specific office called SHARE (Sexual Harassment/Assault Advising, Resources and Education). The SHARE office provides 24-hour sexual assault services, confidential counseling, and support and advice on legal and disciplinary options. The office emphasizes that the victim does not have to make a report or complaint, but it also provides resources for the victims to "feel safe and in control." Friends of survivors of campus sexual assault also have access to counseling and support. The university's policies provide a range of sexual assault reporting options including the following:

> ... filing a private informal or formal University complaint; reporting to Public Safety; reporting to Police; and/or pressing criminal charges. The following information reviews the offices for reporting and the procedures involved. All forms of sexual assault and all attempts to commit such acts are regarded as serious University offenses that are likely to result in suspension, required withdrawal or expulsion. New Jersey criminal law encompasses the offenses identified in University policy, and prosecution may take place independently of charges under University regulations. (*Princeton University,* 1996)

The SHARE office provides information on Rohypnol and other "date rape" drugs, suggestions of what to do if a victim believes that he or she has been drugged, and ways to reduce the risks of being drugged (*Princeton University,* 1996). The information provided by the SHARE office at Princeton exemplifies the type of information, education, and training needed to prevent sexual assault and how to deal with the aftermath of sexual assault as the victim.

The Wellesley College Campus Police's *Annual Security Report* (1997) details an extensive crime prevention program for students. Among the programs offered by the campus police department are:

- new student orientation programs specifically related to crime prevention
- Rape Aggression Defense System entailing a comprehensive course for women that covers awareness, prevention, risk reduction and avoidance, and basic hands-on defense training
- House Associates Program whereby campus police officers volunteer in various residence halls to participate in hall meetings and offer safety advice
- rape awareness, education, and prevention presentations made by the Sexual Assault Interpersonal Violence Education Committee

The committee includes representatives from the campus police, residence, counseling, health services, and student groups. The campus has a student-run shuttle service and a police escort service when the student shuttle is not in operation.

The University of Michigan's Ann Arbor campus has specific policies on sexual harassment by faculty and staff and includes a consensual relationship policy. Romantic and sexual relationships between supervisor and employee or between faculty or other staff and student are not expressly prohibited by university policy. However, the university and the university senate assembly's position is that sexual relationships between students and faculty, even those involving mutual consent, are a basic violation of professional ethics and responsibility when the faculty member has any professional responsibility for the student's academic performance or professional future.

The University of Michigan's housing program has specific provisions related to sexual assault of women of color and discussion of stereotypes and attitudes that magnify the sexual vulnerability of various groups of women of color that contribute to their victimization.

In addition, the University of Michigan specifically addresses sexual assault in the lesbian, gay, bisexual, and transgender communities. The Sexual Assault Prevention and Awareness Center (SAPAC) works with the campus community to eradicate sexual and physical violence and to provide educational activities related to campus violence. As a part of the university housing program, SAPAC has developed an extensive policy, education, and resource handbook dealing with sexual assault.

Arizona State University (ASU) has an extensive policy statement supporting diversity and free speech. While many university policies regulating student speech have failed due to impermissible infringement of First Amendment rights under the U.S. Constitution, ASU's policy and its Campus Environment Team (CET) have been effective in fostering an environment in which discriminatory harassment is less likely to occur while at the same time safeguarding freedom of speech and academic freedom. Particularly noteworthy is the use of the CET, comprised of 13 members who broadly reflect the campus community. The CET is advisory to the president of the university.

The stated mission of the CET is to: Create and maintain a civil and just campus environment that values diversity, promote respect for all individuals regardless of their status, protect free speech and academic freedom, and promote the pursuit of individual goals without interference from discriminatory harassment. Its specific objectives are: education, monitoring and reporting, referral, and prevention. More detailed information on the CET is provided in Appendix 15–A. ASU's approach in dealing with discriminatory harassment through means other than sanctions and discipline is exemplary (Calleros, 1996–1997).

It is not possible to highlight all the excellent work done by university officials in reviewing and revising policies to meet the mandates of federal and state laws in response to violence on campus. Hundreds of policies are available for review on individual university Web sites and provide a valuable resource for contemplating a revision of student codes of conduct and other institutional policies.

RECOMMENDATIONS FOR PREVENTION OF CRIME ON CAMPUS FROM A LEGAL POINT OF VIEW

This chapter provides a selected overview of federal and state laws regarding crimes and violence on college and university campuses. Not every legal concern is addressed and not every law is discussed. The purpose is to identify the legal response to violence and to provide a brief overview of the law that affects institutional policies and practices. Examples of institutional policies, handbooks, and educational programs provide an insight into the resulting policy and student code of conduct provisions addressing both mandates and violence on campus.

Smith and Fossey (1995) have developed a safety checklist (see Appendix 15–B) for school officials to assist in preventing harm to students on campus.

There are myriad approaches that can be used in preventing violence on campus. Recommendations include the following:

- using the crime statistics that must be reported on an annual basis to the U.S. Department of Education to inform students during orientation of the need for safety, personal safety techniques, and resources for victims
- conducting regularly scheduled violence prevention activities in residence halls
- targeting fraternities, sororities, and athletic teams for additional education programs on violence and liability related to excessive consumption of alcohol, consequences of "date rape," and liability (Case studies of injuries result-

ing from hazing during "rush" activities should be used as discussion materials.)
- establishing strict, unambiguous school policies
- providing clear information on permissible student activities (such as student demonstrations) and designating a public forum area on campus
- developing campus committees composed of broad-based membership of campus representations such as counseling, health services, campus police departments, and student housing
- encouraging faculty and staff to make educational presentations in classrooms and to student organizations
- providing "violence awareness weeks" and self-defense courses
- using multiple prevention approaches concerning acquaintance rape, sexual assault, and sexual harassment that involve all elements of the campus
- using peer facilitators as workshop leaders to attract student interest
- using university counsel during the policy development stage to engage in preventive law (Barr, 1988)
- protecting students' rights when developing and revising student codes of conduct and policies, and determining whether proposed regulations related to speech will withstand First Amendment analysis
- developing clear guidelines for students regarding university-sponsored Internet and e-mail use
- employing risk management plans, clear policies, staff training, and common sense to help reduce liability claims (Barr, 1988)
- developing a "compliance" checklist detailing mandates of federal law related to violence on campus (A checklist to provide guidance in complying with the Campus Security Act is provided in Appendix 15–C.)

These guidelines are not exhaustive, and each college and university must assess its own needs and resources in preventing violence on campus.

Finally, part of preventive law is to anticipate future legal challenges. These include increased protection of individual freedoms under the U.S. Constitution and individual state laws and state constitutions; "hate crimes" against minorities, lesbians, and gays on campus; hate speech and harassment via e-mail on university-sponsored Internet access; pornography via the Internet; increased use and presence of weapons on campus; and tort liability issues dealing with the issue of foreseeability, confidentiality of student records regarding discipline for sex-related crimes, violence against visitors to campus such as children in on- and off-campus university-sponsored programs, violence against students participating in university-sponsored programs in foreign countries, rights of individuals who have committed violent crimes to live in residence halls and seek admission to specific programs, and the "duty to warn" others of the presence of such offenders on campus.

As Ikenberry stated in his presentation to Senate subcommittees regarding the reporting of campus crime, "At one time, the college campus may have been a sanctuary where there were few dangers to physical safety. Sadly, that is no longer true. As with society in general, safety on campus must be an issue of constant concern" (Ikenberry, 1998).

REFERENCES

Arizona Revised Statutes, Title 13. Criminal Code, Ch. 12. Assault and Related Offenses. § 13, 1203–1204.

Arizona Revised Statutes, Title 13. Criminal Code, Ch. 14. Sexual Offenses. § 13, 1401–1404.

Arizona Revised Statutes, Title 13. Criminal Code, Ch. 29. Offenses Against Public Order. § 13, 2923.

Barr, M.J. (1988). *Student services and the law: A handbook for practitioners.* San Francisco: Jossey-Bass.

Bauer v. Kinkaid, 759 F. Supp. 575 (W.D. Mo. 1991).

Bohmer, C., & Parrot, A. (1993). *Sexual assault on campus.* New York: Lexington Books.

Bradshaw v. Rawlings, 612 F.2d 135 (3rd Cir. 1979).

Brzonkala v. Virginia Polytechnic and State University, 935 F.Supp. 772, 779 (W.D. Va. 1996).

Brzonkala v. Virginia Polytechnic and State University, 132 F.3d 949, 968 (4th Cir. 1997).

Calleros, C. (1996–1997). Conflict, apology, reconciliation at Arizona State University: A second case in hateful speech. *Cumberland Law Review, 27,* 91.

Campus Environment Team at Arizona State University. (1995, September). Tempe, AZ: Arizona State University.

The Campus Security Act, 59 *Fed. Reg.* 22314 (1994).

Campus Sexual Assault Victim's Bill of Rights of 1991, Pub. L. No. 102-325 (1992).

City of Oshkosh v. Winkler, 557 N.W. 2d. 464 (Wis. App. 1996).

Cooper v. Delta Chi Housing Corp. of Connecticut, 41 Conn. App. 61, 674 A.2d 858 (1996).

Doe v. Doe, 929 F. Supp 608 (D.Conn. 1996).

Doe v. University of Michigan, 721 F. Supp. 852, 854–856 (E.D. Mich. 1989).

Family Educational Rights and Privacy Act of 1974, 20 U.S.C. § 1232(g) (1976).

Franklin v. Gwinnett County Public Schools, 112 S.Ct. 1028 (1992).

Furek v. University of Delaware, 594 A.2d 506 (Del. 1991).

Griffaton, M.C. Forewarned is forearmed: The Crime Awareness and Campus Security Act of 1990 and the future of institutional liability. (1993). *Case Western Reserve Law Review, 43,* 525.

Hartman v. Bethany College, 778 F. Supp. 286 (N.D.W.Va., 1991).

Heffron v. International Society for Krishna Consciousness, 452 U.S. 640 (1981).

Hegel v. Langsam, 273 N.E.2d 351, 352 (Ohio Ct. of Common Pleas, 1971).

Held v. State University of New York College at Fredonia, 630 N.Y.S.2d 196 [102 Educ. L. Rep. 748] (N.Y. Sup. Ct. 1995).

Hernandez v. Arizona Board of Regents, 866 p. 2d 1330 (1994).

Ikenberry, S. (1998, March 5). Written testimony of president of American Council on Education, before the Senate Appropriations Subcommittee on Labor, Health, and Human Services, and Education.

Illinois Compiled Statutes Annotated. Stalking. Ch. 720. § 5.12–7.3.

Jesik v. Maricopa Community College, 611 P.2d 547 (1980).

Kalesse, D. (1990, September 14). Campus crime: USA's bloody secret. Tranquility is shattered by violence. *USA Today,* p. 1A.

Kaplin, W.A., & Lee, B.A. (1995). *The law of higher education.* San Francisco: Jossey-Bass.

Karibian v. Columbia University, 812 F.Supp. 413 (S.D.N.Y. 1993).

Kelly v. Gwinnell, 476 A.2d 1219 (N.J. 1984).

Krebs v. Rutgers, 797 F.Supp 1246 (D.N.J. 1992).

Lederman, D. (1994, May 4). College must list crimes reported to counselors, U.S. says. *The Chronicle of Higher Education,* p. A32.

Lewis, L., & Farris, E. (1997). *Campus crime and security at postsecondary education institutions,* NCES 97-402. Washington, DC: U.S. Department of Education, National Center for Education Statistics.

Lively, K. (1996, April 26). Education department starts monitoring campus crime reports. *The Chronicle of Higher Education,* p. A49.

Mango, K.A. (1991). Students versus professors: Combating sexual harassment under Title IX of the Education Amendments of 1972. *Commonwealth Law Review, 23,* 355.

Massachusetts General Laws Annotated Part I. Administration of the Government. Ch. 22C, § 33.

Massachusetts General Laws Annotated Part IV. Crimes, Punishment and Proceedings in Criminal Cases. Ch. 265, § 22.

Meritor Savings Bank v. Vinson, 477 U.S. 57 (1986).

Mullins v. Pine Manor College, 449 N.E.2d 331 (Mass. 1983).

Nelson v. Temple University, 920 F. Supp. 633 [108 Educ. L. Rep. 674] (E.D. Pa. 1996).

Nero v. Kansas State University, 253 Kan. 567, 861 P.2d 768 (1993).

Norwood v. Slammons, 788 F. Supp. 1020 (W.D. Ark. 1991).

Office of Civil Rights. (1997, March). Sexual harassment guidance: Harassment of students by school employees, other students, or third parties. Federal Register, *62,* 12034.

Opinions of the Attorney General (Vol. 74). (1989). Opinion No. 89-002. Annapolis, MD: State of Maryland.

Page, R., & Hunnicutt, K.H. (1994). Freedom for the thought that we hate: A policy analysis of student regulation at America's twenty largest universities. *Journal of College and University Law, 21*(1), 1–60.

Paine v. Board of Regents of the University of Texas System, 474 F. 2d 1397 (5th Cir. Tex. 1973).

Pallett v. Palma, 914 F. Supp. 1018 [107 Educ. L. Rep. 165] (S.D.N.Y. 1996).

Pavela, G. (1997). Disciplinary and academic decisions pertaining to students: A review of 1996 judicial decisions. *Journal of College and University Law, 24,* 213.

Peterson v. San Francisco Community College District, 36 Cal.3d 799 (1984).

Pfeiffer v. Marion Center Area School District, 917 F.2d 779 (3rd Cir. 1990).
Policy on student possession and consumption of alcoholic beverages in facilities of the University of North Carolina at Chapel Hill. (1996, January). Chapel Hill, NC: University of North Carolina.
Princeton University campus sexual assault victim's bill of rights. (1996). Princeton, NJ: Princeton University Department of Public Safety, SHARE, and University Health Services.
Purdon, T.S. (1988, April 10). The reality of crime on campus. *New York Times,* p. 49.
R.A.V. v. City of St. Paul, Minnesota, 505 U.S. 377 (1992).
Ray v. Wilmington College, 667 N.E.2d 39 [110 Educ. L. Rep. 1222] (Ohio Ct. App. 1995).
Red & Black Publishing Co. v. Board of Regents, 427 S.E.2d 257 (Ga. 1993).
Restatement (Second) of Torts, § 314–328 (1965).
Robertson v. LeMaster, 171 W.Va. 866, 301 S.E.2d 563 (1983).
Slater v. Marshall, 906 F. Supp. 256 [105 Educ. L. Rep. 501] (E.D. Pa. 1995).
Smith, M.C., & Fossey, R.W. (1995). *Crime on campus: legal issues and campus administration.* Phoenix, AZ: American Council on Education and Oryx Press.
State v. Nye, 943 P.2d 96, 100 (Mont. 1997).
Statement of Rep. Goodling, 136 Cong. Rec. E H 11, 499 (October 22, 1990).
Statement of Sen. Gore, 136 Cong. Rec. S.16615 (October 24, 1990).
Stone v. Arizona Highway Commission, 93 Ariz. 384, 381 P.2d 107 (1963).
Student kills 4, then himself at Iowa campus. (1991, November 2). *Los Angeles Times,* p. A20.
Student Right-To-Know and Campus Security Act of 1990, 20 U.S.C. § 1001, 1990 Amendments, Pub. L. 101–542, Title I, § 101, 1990.
Tarasoff v. Regents of the University of California, 551 P.2d 334 (Cal. 1976).
Title VII of the Civil Rights Act of 1964, 29 CFR § 1604.11 (1993).
Title IX of the Education Amendments of 1972, 20 U.S.C. § 1681(a)(1988).
Title IX of the Education Amendments of 1972, 34 CFR § 106.61 *et seq.* (West, 1998).
University of Michigan Sexual Assault Policy, Policies on Sexual Harassment by Faculty and Staff, Code of Student Conduct (An Urban Campus), The Sexual Assault Prevention and Awareness Center (SAPAC), *Responding to Rape, Helping Survivors Heal Handbook,* Ann Arbor, MI: University of Michigan, July 1996.
University of Washington student conduct code. (1996). Seattle, WA: University of Washington, Office of the Vice President for Student Affairs.
Violence Against Women Act of 1994. 424 U.S.C. § 13981–14040 (1994).
Wellesley College Campus Police. (1997). *Annual Security Report.* Wellesley, MA: Author.
Wisconsin v. Mitchell, 508 U.S. 476 (1993).
Wood, H.A. Liabilities: Actions by and against colleges and universities. (1996, April). *American Jurisprudence, 15A,* section 39. Cumulative Supplement.

APPENDIX 15–A

Campus Environment Team at Arizona State University

I. POLICY STATEMENT SUPPORTING DIVERSITY AND FREE SPEECH
    Arizona State University ("ASU" or "the University") is committed to maintaining hospitable educational, residential, and working environments that permit students and employees to pursue their goals without substantial interference from harassment. Additionally, diversity of views, cultures, and experiences is critical to the academic mission of higher education. Such diversity enriches the intellectual lives of all, and it increases the capacity of a university to serve the educational needs of its community.
    ASU is also strongly committed to academic freedom and free speech. Respect for these rights requires that it tolerate expressions of opinion that differ from its own or that it may find abhorrent.
    These values of free expression justify protection of speech that is critical of diversity and other principles central to the University's academic mission. However, values of free expression are not supported but are undermined by acts of intolerance that suppress alternative views through intimidation or injury. As members of an institution of higher education, we must stand against any assault upon the dignity and value of any individual through harassment that substantially interferes with his or her educational opportunities, peaceful enjoyment of residence, physical security, or terms or conditions of employment (collectively, "protected interests").
    In this spirit, the University adopts an anti-harassment policy, set forth in section III below, which prohibits substantial interference with protected interests, subject to constitutional limitations. In addition, through the work of the Campus Environment Team ("CET," described in section II below), the University will take steps to foster an environment in which discriminatory

---

Courtesy of Arizona State University.

harassment is less likely to occur, an environment that is hospitable to all members of the University community regardless of race, sex, color, national origin, religion, age, sexual orientation, disability, or Vietnam-era veteran status (collectively, "status"). At the same time, the CET will work with others in the University to help safeguard freedom of speech and academic freedom.

Through the efforts of the CET and the many other programs now under way, the University truly hopes to achieve these worthy goals. Ultimately, however, these goals will not be fully met unless every member of the University community takes a personal responsibility for fostering an environment in which diversity can be appreciated and in which all students and employees can reach their fullest potential. No committee or other entity can substitute for the good will, freely given, by the individuals who make up this University.

II. THE CAMPUS ENVIRONMENT TEAM
  A. Creation, Composition, and Support
   1. A Campus Environment Team ("CET") advisory to the President is established on the ASU main campus. The CET is composed of thirteen members who broadly reflect the campus community. The members shall include at least one faculty member, academic professional, student, classified staff employee, and service professional employee. The President will appoint the members of the CET to staggered three-year terms from candidates nominated by any campus organization or any member of the campus community. The Chair of the CET, who normally serves a one-year term, will be appointed by the President. The President will also appoint a Chair-elect, who will become the Chair following the year as Chair-elect. The Chair-elect shall serve in the absence of the Chair. The Director of Equal Opportunity/Affirmative Action, the Dean of Student Life, and the Director of Residence Life/Student Development, a representative from the General Counsel's Office and the Director, DPS will be ex-officio members of the CET. CET members will receive in-service training to enhance the effectiveness of their activities. After three consecutive unexcused absences from CET meetings by a CET member, the Chair of the CET will recommend to the President that the member be removed and a replacement appointed to the CET.
   2. The Office of the President shall set aside appropriate funding for the CET to carry out its duties and fulfill its objectives.
  B. Mission
    The mission of the CET is to: (1) create and maintain a civil and just campus environment that values diversity, (2) promote respect for all individuals regardless of their status, (3) protect free speech and academic

freedom, and (4) promote the pursuit of individual goals without interference from discriminatory harassment.

Beyond its referral activities described below, the CET will not process complaints, nor does it have any authority to impose discipline or to compel attendance at its meetings or cooperation with its efforts. Instead, it will seek to work in tandem with persons who voluntarily approach it for assistance or who voluntarily respond to its invitations for cooperation.

C. Specific Objectives

The specific objectives and activities of the CET are:

1. Education

    The CET should support and participate in efforts to educate the campus community for the purposes of (1) preventing harassment and creating a campus environment that reflects respect for all individuals regardless of status, and (2) preventing infringements of free speech and academic freedom and helping the campus community to understand the University's obligations to protect free speech and academic freedom.

2. Monitoring and Reporting

    The CET should monitor the campus environment by gathering data concerning discriminatory harassment and should report such data annually to the University President. Whenever appropriate, the CET should make recommendations to the President with respect to specific policies and programs that will help carry out the CET's goals.

3. Referrals and Other Responses to Harassment

    a. Referral

    Any member of the campus community who believes that he or she has been subjected to discriminatory harassment, as defined by Arizona Board of Regents or University policy, may obtain assistance from the CET to file a complaint or grievance with the appropriate office or committee, or to secure counseling, mediation, or other relief.

    b. Prevention and Response

    Members of the CET should be "in touch" with the campus environment. CET members may be aware of the potential for discriminatory harassment and may have special knowledge of how to ease tensions when harassment has occurred or is about to occur. The CET should work closely with the University administration to help implement strategies, consistent with free speech and academic freedom, to resolve tensions that may lead to discriminatory harassment and to mitigate such harassment after it has occurred.

III. UNIVERSITY POLICY PROHIBITING HARASSMENT
   A. Harassment Prohibited
   Subject to the limiting provisions of section C below, it is a violation of University policy for any University employee or student to subject any person to harassment on University property or at a University-sponsored activity.
   B. Harassment Defined
   Actions constitute harassment if
   1. they substantially interfere with another's educational or employment opportunities, peaceful enjoyment of residence, or physical security, and
   2. they are taken with a general intent to engage in the actions and with the knowledge that the actions are likely to substantially interfere with a protected interest identified in subsection (1) above. Such intent and knowledge may be inferred from all the circumstances.
   C. Freedom of Speech and Academic Freedom
   Neither this nor any other university policy is violated by actions that amount to expression protected by the state or federal constitutions or by related principles of academic freedom. This limitation is further described in the ASU First Amendment Guidelines, the current version of which supplements this policy and is available in the Office of the General Counsel.
   D. Relationship to the Work of the CET
   If harassment is discriminatory, it falls within the education, monitoring, reporting, and referral functions of the CET. Harassment is discriminatory if taken with the purpose or effect of differentiating on the basis of another person's race, sex, color, national origin, religion, age, sexual orientation, disability, or Vietnam-era veteran status.

FIRST AMENDMENT GUIDELINES—UNIVERSITY POLICY PROHIBITING HARASSMENT

A. INTRODUCTION
   1. General Goals of the CET
   The CET hopes to focus its attention on education, monitoring, and referral as described in section II of the CET policy. It also performs the important function of providing information to victims of alleged viola-

tions of the anti-harassment policy stated in Appendix A of the policy. However, the primary goal of the CET is to encourage a harmonious campus environment in which at least discriminatory harassment is unlikely to occur. If it succeeds in this goal of preventing at least some forms of harassment, disciplinary procedures should play a relatively minor role in the University's efforts to maintain a campus environment that enables all to learn, work, or reside on campus without serious interference.

Whenever appropriate, University officials should respond to harassing behavior through mediation or by counseling and educating the wrongdoer. However, when violations do occur, University officials may seek immediate discipline.

2. The University's Legal Obligations

As a university and employer, ASU has moral, legal, and to a limited extent contractual obligations to maintain reasonable educational, residential, and working environments that permit students and employees to pursue their goals without substantial interference stemming from harassment. Additionally, as a state university, ASU has a constitutional obligation under the Fourteenth Amendment to provide equal educational and employment opportunity and thus to refrain from invidious discrimination. Related legislation reinforces these obligations. Consistent with these obligations, ASU is committed to the goal of achieving diversity within the campus community.

Principles of academic freedom and constitutional guarantees of free speech, however, limit the University's ability to use restrictions on speech as a means of promoting diversity and opposing harassment and discrimination. Thus, as stated in section III of these policies, the anti-harassment policy does not restrict speech protected by state and federal constitutional law or by principles of academic freedom.

The following notes and illustrations provide a general guide to the relevant issues. They are intended to reflect current principles of constitutional law, primarily federal. These guidelines also seek to anticipate how the state constitution will apply to speech in the university context, particularly because it may protect speech to a higher degree than does the federal constitution. State constitutional law in this area, however, is in an early state of development and does not currently provide detailed guidance beyond federal constitutional law. The University will closely monitor developments in both state and federal constitutional law and revise these guidelines accordingly. Finally, these guidelines also reflect long established principles of academic freedom, such as those set forth in the 1940 Statement and in subsequent statements of the American Association of University Professors.

These guidelines certainly do not answer every question that may arise under the anti-harassment policy, but they should remind the University community to be sensitive to the need to avoid a "chilling effect" on academic inquiry and the expression of ideas. The purpose of the guidelines is to provide ample breathing room for protected speech. Accordingly, in any case that presents a serious question regarding freedom of expression, anyone seeking to administer the anti-harassment policy should consult the General Counsel's Office before taking any action that might interfere with protected speech. In appropriate cases, the General Counsel's Office may seek a judicial adjudication before authorizing other action.

3. The University Does Not Necessarily Endorse Ideas Conveyed in Protected Speech

Because the first amendment protects even highly offensive speech in some contexts, readers may find some of the examples in these guidelines to be offensive. By using such examples for illustration, the University does not encourage offensive or insensitive speech; it simply acknowledges the constitutional limitations on its ability to regulate such speech. Indeed, the University is free to express its own views opposing or commenting on offensive speech, even though it cannot restrict the speech.

B. DEFINITION OF DISCRIMINATORY HARASSMENT

As further illustrated in the following subsections, section III applies to conduct or expression if it substantially interferes with another's educational opportunities, peaceful enjoyment of residence, physical security, or terms or conditions of employment and if it is not protected by constitutional guarantees of free speech or principles of academic freedom.

1. Some injurious or intimidating conduct, such as assault or battery, normally has no significant speech content and can be regulated to protect other important interests without infringing upon the right to free speech or academic freedom. For example, unless clearly trivial in scope, and absent some mitigating circumstance such as inadvertence, self-defense, or consent, section III or other University policies normally would apply to such conduct as
    a. touching a person in a manner that a reasonable person would view as hostile, offensive, or intimidating;
    b. taking some action that causes a person to reasonably fear imminent hostile, offensive, or intimidating physical contact;

c. damaging, defacing, or destroying University property or the property of another; or
   d. engaging in extreme and outrageous conduct for the purpose of inflicting severe emotional distress upon another person.
2. Even speech, or conduct combined with speech, can be regulated if it is merely a tool to advance some activity that is unlawful under valid laws independent of this anti-harassment policy. In many cases, the anti-harassment policy has only incidental effects on the communication of ideas, because it is aimed at noncommunicative acts and effects touching upon matters in which the University has a great interest. For example, section III or other University policies normally would apply to the acts of
   a. communicating a threat of physical harm that causes a person to reasonably fear imminent hostile, offensive, or intimidating physical contact;
   b. communicating in a manner that damages, defaces, or destroys University property or the property of another; or
   c. inciting violence or other acts that would be unlawful independent of this policy, if the actor or speaker encourages immediate action and if the conduct or speech is reasonably certain to result in imminent violence or other unlawful action. Such conduct or speech could include:
      (1) directing another person to engage in a battery as defined in subsection 1(a) above, or
      (2) closely confronting a person or persons with threatening or intimidating remarks if in light of all the circumstances the remarks would be reasonably certain to provide a violent breach of the peace.
3. As a further example of harassment described in section B(2) above, the anti-harassment policy also applies to speech or conduct by a University official or other state actor that is merely a vehicle for substantially interfering with a protected interest through discrimination prohibited by the Fourteenth Amendment or related statutes, such as
   a. a professor's stated requirement in a mathematics class that all female students sit in the back of the class on the stereotyped assumption that each of them has a low aptitude for learning mathematics; or
   b. the psychological equivalent of requiring the female students to sit in the back of a mathematics class, such as repeated statements by the mathematics professor that the female students in the class

should not hope or try to match the performance of the male students.
4. Other expression or conduct may be covered by section III, or may be protected speech, depending upon the context of the expression.
   a. The expression even of ideas that are extreme or offensive to many listeners is protected and does not amount to unlawful harassment if offered in a suitable time, place, and manner, such as the expression of ideas for public debate
      (1) in a classroom discussion or a related discussion outside the classroom, if the expression is reasonably germane to the academic subject matter of the course or classroom discussion;
      (2) in academic scholarship or other publication or in a related discussion; or
      (3) in a campus forum, such as an auditorium, a public gathering place outdoors, or a public bulletin board.
   b. Even when expression and related conduct is protected by the First Amendment, the University can impose reasonable regulations on the time, place, and manner of the presentation of the expression. For example, the University could compel students to move or postpone an unscheduled rally that disrupts a meeting or rally held by another group of students who properly reserved the time and location for its own function.
   c. Similarly, even though similar speech might be protected if presented in another forum, threatening or intimidating speech or related conduct may be subject to regulation if it is forced upon specific individuals in a non-public forum who are unwilling targets of the conduct or speech and who cannot reasonably avoid it, such as
      (1) the unwelcome posting of threatening neo-Nazi symbols on the dormitory door of a Jewish student for the purpose of intimidating the Jewish student;
      (2) the act of knocking the books out of the hands of a student each time he tries to enter a classroom; or
      (3) the verbal psychological equivalent of knocking the books out of the student's hands, such as repeated statements at the doorway to a classroom that the student should not enter the classroom.

C. COMMENTARY AND ILLUSTRATIONS
1. Relationships Among Multiple Goals
   The anti-harassment policy, including these interpretive guidelines, reflects an effort to accommodate diverse University goals and obliga-

tions. Members of the University community who have a special allegiance to one goal to the exclusion of others may view the policy as an unacceptable compromise of that goal. The University, however, must take a broader view of its multiple obligations.

In many cases, interests in promoting a hospitable campus environment will be perfectly consistent with interests in free expression and academic freedom. For example, suppose the Director of the School of Art directs his or her faculty to discourage art students from creating even non-obscene art that might be construed as homoerotic. As a result, faculty could suffer a loss of academic freedom, students could suffer loss of freedom of expression, and some students and faculty might suffer serious interference with their educational opportunities or terms or conditions of employment. Administrative measures to eliminate the Director's policy would tend to restore interests in free expression and academic freedom as well as interests in maintaining a campus environment free of harassment. Similarly, suppose that a campus official responsible for preventing and investigating crimes unreasonably detains and searches a minority student on his way to class, causing the student to miss all or part of his class. Suppose further that the detention and search is unreasonable because the official acts largely on the basis of his stereotyped assumptions about the student's propensity to commit crime because of the student's race and ethnic attire. Such conduct by the official might violate the anti-harassment policy by substantially interfering with the student's educational opportunities. It would also place a burden on the student's constitutional interests in being free of unreasonable searches and seizures, in expressing himself through T-shirt slogans or other clothing, and in being free of racial discrimination. A University policy that sought to prevent such conduct could help vindicate all of these concurrent interests.

Even when these interests do not so clearly coincide, the anti-harassment policy primarily seeks to regulate conduct with no significant speech component, raising no First Amendment problems. In some cases, however, efforts by the University to maintain a hospitable campus environment may raise questions about the University's obligations to preserve freedom of speech. These interpretive guidelines are designed to assist an administrator in addressing those questions and in avoiding any violation of state or federal constitutional provisions protecting speech.

The University's constitutional and statutory obligations to provide equal educational and employment opportunities may require it to regulate some conduct and speech. For example, suppose a professor threatens to lower the grades of female students unless they submit to his sexual demands. Although the threats are conveyed through speech in the most general sense, the constitutional protection would not extend to them, because the threats are simply a tool for illegally coercing sexual favors. Moreover, the University may in some circumstances be responsible under the Fourteenth Amendment and related legislation for the professor's harassment, particularly if University officials adopt or implicitly ratify the harassment as University policy by failing to intervene in the harassment after receiving notice of it. Thus, in some circumstances, University regulation of speech and conduct is not only permitted, it is the University's legal obligation, notwithstanding interests in free speech and academic freedom.

This could extend to harassment of students by fellow students: if University officials receive notice that students are harassing another student on the basis of a classification protected under the Fourteenth Amendment or related legislation and fail to take reasonable steps to intervene, they may be guilty of maintaining unequal educational opportunities, in violation of the Fourteenth Amendment and related legislation. On the other hand, if the University restricts protected speech, it will violate the First Amendment.

Thus, the enactment of the University anti-harassment policy should not be viewed as a rejection of interests in free speech; nor should the recognition of First Amendment limitations be viewed as a diminution of the University's commitment to diversity. The University has a wide range of responsibilities that extend to equal opportunity, to freedom of expression, and to maintenance of reasonable educational, working, and residential environments for all members of the campus community. The University will be faithful to all of these obligations if it pursues its goals of diversity, equal opportunity, and non-harassment in a way that fully respects rights to free speech and academic freedom. In some cases, as illustrated by these guidelines, interests in free speech will limit the University's ability to pursue other goals. In those cases, the University is fully committed to honoring those limits.

2. Defacement or Destruction of Property

   Just as a person may burn his own flag but not one stolen from another, a student would be free to display a symbol on his T-shirt but could be disciplined for spray-painting the symbol on a classroom wall or over a poster owned and displayed by another. As discussed in section D(3) below, this presumes that the University would mete out discipline for any defacement or destruction of property, regardless of the presence, absence, or content of any expression associated with it.

3. Free Speech and Academic Freedom in an Academic Context

   Students, faculty, and others are entitled to express any view in an aca-

demic context, even if the content of the speech offends or even shocks some of the speaker's listeners. For example,
a. a student or instructor in a history, sociology, or philosophy class is free to express the shocking view that Hitler's programs and policies during World War II were morally defensible or that slavery and apartheid are just institutions;
b. a staff member could express the view in a campus radio talk show that laws mandating wheelchair access in public buildings should be repealed and that persons who use wheelchairs should be banned from campus;
c. a professor could write an article arguing that women generally have a lower aptitude than men for learning mathematics;
d. a student could write a letter to the editor of a campus newspaper arguing that Native Americans did not belong at the University and should stay on their reservations; or
e. a student could publish his own campus journal in which he argues that homosexual lifestyles are immoral and contrary to religious teachings.

Those who disagree with such speech can, among other things, silently reject the view or respond to it with more speech in such form as class discussion or a letter to the editor of a campus newspaper. However, the University cannot, and should not, seek to regulate the content of intellectual debate.

Nonetheless, like many residential and employment environments on campus, educational facilities are not necessarily full public forums. A university administrator who has assigned a mathematics instructor to teach calculus can demand that the instructor teach that topic rather than use the classroom as a forum for expressing his unrelated political views regarding the Persian Gulf War.

Similarly, for pedagogical reasons, a classroom instructor can exercise a high degree of control over the process of communication in his or her class. The instructor can demand, for example, that students raise their hands and be recognized before speaking, that they speak to the topic raised by the instructor, that they address the instructor rather than speak among themselves, and that they adopt a classroom demeanor that does not disrupt the educational activity of the moment. Although deviations from such rules set down by an administrator or instructor would not necessarily violate University policy, the examples serve to help illustrate the scope of interests in free speech and academic freedom.

4. Time, Place, and Manner Restrictions on Speech
Subject to certain narrow exceptions outlined in section B above, a person enjoys the right to express even offensive ideas in such forums as (1) a written statement posted at appropriate sites after getting approval on a content-neutral basis from the appropriate University office, or (2) a private or public meeting staged at a room or other site properly reserved on campus. Those offended by such expression can, among other things, ignore the speech, avoid it, or respond to it with more speech; however, the University cannot ban the speech simply because it offends others.

On the other hand, the University may adopt content-neutral restrictions on the time, place, and manner of speech to avoid conflicts and disruptions. For example, it could require presentations at West Hall Lawn to be sufficiently limited in scope as to avoid obstructing foot traffic on the bordering sidewalks and to be sufficiently limited in volume as to avoid disrupting work or study in the library or in nearby offices or classrooms. Similarly, if a campus organization has reserved a time and location on campus to celebrate the birthday of Martin Luther King, Jr., the University could prohibit another group that did not reserve the same time and location from disrupting the celebration with a conflicting rally or speech. This interest in freedom from disruption may be enhanced when the event is scheduled inside a room, thus generating expectations of separation from those who do not identify with the goals of the event. It may be stronger still when the event is open only to invited participants, thus generating expectations of privacy.

Time, place, and manner restrictions must be reasonable. For example, if the University prohibits students from posting any notices or affixing any other materials on the hallway walls and exterior doors of dormitory rooms and in the common bathrooms, it must provide other reasonable areas for the posting of public notices. Similarly, if a group of students has reserved West Hall Lawn for a presentation celebrating the birthday of Martin Luther King, Jr., University officials could not prohibit other students in an adjacent area within eyesight of West Hall Lawn from carrying picket signs or handing out leaflets that disparaged King, so long as they do not disrupt the scheduled celebration. Moreover, sometimes those expressing ideas at a rally invite debate and reactions from listeners; in such cases, the expression of competing views in the same time and place would not be expression in an unsuitable time, place, or manner. Nonetheless, such expression could be subject to University regulation if it amounted to exercise of a "heckler's veto," which drowns out the scheduled presentation or otherwise prevents it from proceeding.

5. Slurs and Epithets

Derogatory terms may amount to harassment or may be protected speech, depending upon the context. For example, a drama student writing a play about racism in America would be free to use the ugly, disparaging term "nigger" in her script to drive home her points about racism. Indeed, if he or she were willing to lose credibility and to weather the outpouring of criticism and counterspeech, a person would be free to use such a term in a speech disparaging an ethnic group, gender, or sexual orientation.

On the other hand, a student would violate University policy by referring to another student by the term "nigger," "stupid jerk," or other epithet in such a manner or in such a context as to put the listener in reasonable fear of imminent physical harm. For example, suppose that one or more students stopped an African-American student in an isolated area of campus at night and invoked racially disparaging terms in a threatening manner. In light of the long history of racial violence in our society, the racially disparaging terms in this context could very well put the African-American student in reasonable fear of imminent harm. Under that analysis, the speech and conduct could be regulated without infringing upon interests in free speech.

Because of the high incidence of violence against women and against homosexuals in our society, this analysis might apply with particular force also to disparaging terms directed to women or homosexuals, particularly in a volatile context that presents a risk of physical harm to the target of the speech. Other kinds of confrontations, such as a woman referring to a man as a "male chauvinist pig," or a student calling a professor a "windbag," might not place the listener in reasonable fear of imminent physical harm as frequently, because the incidence of violence historically associated with such disparagement is relatively low. However, the immediate context is more important than the actual language, status of the parties, or historical context. Thus, if the term "male chauvinist pig" or "windbag" were communicated in a threatening manner and in circumstances underscoring the viability and immediacy of the threat, the speech conduct could be regulated.

Other kinds of cases illustrate further that the context may be more important than the term used. For example, in a public forum that unwilling listeners are free to avoid, a speaker has a right to make the highly offensive statement that "women are whores"; the speaker has no duty to make his or her voluntary listeners comfortable or to treat them equally. In another context, however, even the less offensive term "girl" could contribute to harassment. For example, suppose that a professor in a political science class addressed his male students with great respect, but that each time a female student raised her hand the professor paused and said condescendingly: "Oh, no; let's see what the girls have to say." Particularly when frequently repeated, such condescending speech might provide unequal educational opportunity for women, thus violating the Fourteenth Amendment and federal legislation such as Title IX. In these circumstances, the University could constitutionally regulate the speech.

6. Intent and Foreseeability

The general intent requirements in section III(B)(2) of the anti-harassment policy excludes inadvertent harassment, such as purely accidental physical contact. Additionally, the requirement of knowledge of likelihood of serious interference defines a foreseeability test that helps reinforce protection for speech. For example, it helps to ensure that section III distinguishes merely offensive language from threats and intimidation.

D. ROLE OF THE CET

1. Relationship to Anti-harassment Policy

In unusual circumstances, actions might constitute harassment only if discriminatory. For example, suppose that a law professor regularly employed a withering and sometimes humiliating inquisitorial teaching method in class, causing all students to feel uncomfortable and even intellectually threatened. Although others might question the efficacy of her pedagogy and might encourage her to alter it, the instructor's teaching method would generally lie within her academic freedom and would not constitute harassment unless it became physically or emotionally abusive. The same teaching method, on the other hand, would be unquestionably inappropriate if the instructor deliberately used it to discriminate against some students on the basis of an invidious classification. Thus, if the instructor employed the inquisitorial method to humiliate Asian students but engaged in a more popular intellectually collaborative approach with all other students, she would be engaging in state action that deprived some students of equal educational opportunities, in violation of state and federal laws and in violation of the anti-harassment policy. In sum, discriminatory conduct by a state official may be harassing even though the same conduct may be only questionable or offensive if applied equally to all.

In most cases, however, the anti-harassment policy applies to substantial interference with protected interests regardless whether based on invidious discrimination. Nonetheless, it recognizes that discriminatory harassment presents special problems warranting study, educational efforts, and referral services, all of which are provided by the CET. Harassment is "discriminatory" if it is directed toward a person because of that person's "status" as defined in section III(D) of the anti-harassment policy. Discriminatory harassment presents special problems because it implicates the University's obligations under the Fourteenth Amendment and related laws and because it has a high potential for raising tensions on campus generally.

2. Action Because of Another's Status

An assault by several students upon African-American students on campus would be discriminatory harassment if the assault was motivated by the assaulting students' racial hatred of African-Americans or even by their racist assumption that the African-American students, solely by virtue of the color of their skin, must be the same African-Americans who had earlier initiated an altercation with the assaulting students.

On the other hand, a brawl between fraternity members and visitors to campus motivated solely by a dispute over rights to park on fraternity property would not be discriminatory harassment if it did not relate to any protected status. Although the anti-harassment policy would apply to the assaults and batteries, the CET would not be responsible for addressing the problem.

3. University Disciplinary Action

By focusing the efforts of the CET on discriminatory harassment, section III does not suggest that harassment would bring stiffer penalties solely because of the content of speech accompanying the harassment. For example, suppose that student A defaces property leased by another student by spray painting the slogan "Beat U of A" on the door of the other student's dormitory room, that student B defaces the door by spray painting a swastika on it, and that student C defaces the door by spraying black paint on it in a random pattern that does not suggest the expression of any idea other than the intent to accomplish the defacement. If all other factors are equal, students $A, B,$ and C should receive the same punishment for their violations of school policies against interference with physical security or with peaceful enjoyment of residence; otherwise, one might be penalized more harshly because he or she engaged in speech, or engaged in a particular kind of speech, while also engaging in unquestionably objectionable conduct. This principle would not prevent the University from reacting in other ways that differentiated between the three defacements, such as by expressing its disgust for Nazism but not for ASU sports supporters, or by paying University personnel overtime to remove the swastika immediately and letting personnel remove the other defacements in the normal course of work.

Moreover, the University could punish a discriminatory violation of policies more harshly than other violations of policies, if the violation were more egregious for reasons other than the pure speech or thought associated with the violation. For example, if the University is experiencing serious and continuing problems with gangs of students assaulting Arab students, the University could punish such assaults more harshly than it would an isolated assault over a parking space. In this case, elevated punishment would be justified by the University's need to react to a serious and pervasive problem of violent racial or ethnic origin discrimination and not solely by the University's disagreement with the content of the speech uttered by the students as they engaged in their assaults.

4. Activities of the CET

The efforts of the CET to promote diversity and non-harassment are consistent with freedom of speech. In effect, the CET simply enters the marketplace of ideas with its own speech, its efforts to collect information, and its referral activities.

a. Examples of Educational Activities

Following are examples of possible educational activities of the CET:

(1) a public awareness program to inform the campus community of the existence of the CET, its purpose and the University's policy prohibiting discriminatory harassment;

(2) a program to train counselors, resident assistants, employee assistance personnel, student development administrators, staff relations personnel and ombudspersons to deal with harassment;

(3) voluntary or mandatory cultural sensitivity workshops for all administrators and campus police officers;

(4) voluntary or mandatory student orientation sessions on diversity;

(5) a program to design and disseminate brochures, posters, and related materials that encourage members of the University community to appreciate diversity and to report harassment;

(6) collaboration with the administration, curriculum committees, and Faculty Senate to develop and implement courses on diversity;

(7) a program to offer incentives and assistance to faculty to develop new courses and modify old courses to include information on diversity;
(8) a speaker's program on cultural diversity available to all organizations within and outside the campus community.

b. Examples of Monitoring Activities

Following are examples of possible monitoring activities of the CET:
(1) distributing a questionnaire on the campus climate, conducted periodically to assess general feelings and attitudes of faculty, staff, and students;
(2) recommending that exit interviews of University employees who leave their jobs include questions about the campus environment;
(3) conducting studies on specific topics, either periodically or following a significant act of harassment; and
(4) sponsoring public hearings on the campus environment and how it can be improved.

c. Encouragement of Free Speech

In addition to voicing its own opinions in the marketplace of ideas, the CET can help safeguard the freedom of speech and academic freedom of others on campus with its educational, monitoring, and referral activities. The goals of promoting diversity, preventing harassment, and safeguarding free speech are so closely intertwined that they often must be addressed together. Following are examples of ways in which the CET can help protect freedom of speech and academic freedom while also promoting diversity and preventing harassment:
(1) arranging for broad distribution of these policies and First Amendment guidelines;
(2) sponsoring or participating in workshops or other educational activities on issues of free speech and academic freedom;
(3) ensuring that university staff and administrators are aware of issues of free speech and academic freedom when the CET consults with them regarding particular incidents or issues on campus;
(4) when appropriate, directing the attention of potential complainants to issues of free speech and academic freedom when they approach a member of the CET for advice or referral;
(5) informing persons that they have no obligation to meet or cooperate with the CET when a CET member invites them to participate in a CET activity.

E. SEVERABILITY

These guidelines supplement the anti-harassment policy and are designed to give the policy definition and to restrict its scope within constitutional limits. If any portion of the policy, including a portion of these guidelines, is adjudicated to violate state or federal laws, the University intends to abandon the illegal portion and to maintain the severable legal portions.

F. RELATIONSHIP TO OTHER UNIVERSITY OR REGENTS POLICIES

These policies and guidelines supplement provisions in the Student Code of Conduct governing various forms of harassment. In addition, ASU policies against sexual harassment are more detailed on that topic than is this general anti-harassment policy and should be consulted for more specific standards and procedures governing sexual harassment claims. Conversely, under section III(C) of the CET Policies, limitations based on free speech and academic freedom apply to all University policies. Thus, these First Amendment guidelines may help to define such limitations not only to section III of these policies but also to provisions of the Student Code of Conduct, to ASU's specific sexual harassment policies, and to any other policies that raise issues of free speech and academic freedom.

APPENDIX 15–B

General Campus Security Checklist

PERIMETER

___ Is traffic flow through campus minimized?
___ Are fences adequate to discourage entry?
___ Is lighting adequate at entrances, streets?

STUDENTS

___ Are crime statistics furnished to students regularly?
___ Is there an emergency notification procedure?
___ Is there an escort system?
___ Are drug/alcohol rules adequately enforced?
___ Does the institution screen for dangerous applicants?
___ Does it expel dangerous miscreants?
___ Are procedures in place for student complaints about security?

GROUNDS AND BUILDINGS

___ Is shrubbery minimized?
___ Is lighting adequate at buildings, walkways?
___ Are lights monitored for burnouts and failures?
___ Is master key control tight?

Source: Reprinted from *Crime on Campus: Legal Issues and Campus Administration* by M.C. Smith and R. Fossey © 1995 by the American Council on Education and The Oryx Press. Used with permission from the American Council on Education and The Oryx Press, 4041 N. Central Ave., Suite 700, Phoenix, AZ, 85012, 800-279-6799.

___ Are locks changed when needed?
___ Are emergency phones available at remote areas?
___ Do closed-circuit televisions monitor remote places?

POLICING

___ Is the number of patrol offices sufficient?
___ Are officers given adequate original and continuing training?
___ Are incident reports monitored by administration?

HOUSING

___ Are visitors regulated?
___ Do policies punish students or others for door propping and lock stuffing?
___ Are police patrols adequate?
___ Are new employees screened?
___ Does the institution enforce drug/alcohol rules?
___ Are deadbolt locks and peepholes provided?
___ Are keys changed periodically?
___ Are emergency phones accessible?

APPENDIX 15–C

Campus Security Act— Checklist

Date: _____   Year of Annual Report _____

Page #:	Reference	Information to be included	Comments/notes
	668.47(a)(1)	Statement of current campus policies regarding: • Procedures and facilities for students to report criminal actions or other emergencies • Policies re: inst. reponse to those reports, including policies for making timely reports to members of campus community re: the occurrence of crimes from (a)(6) • List of the titles of each person or organization reports should be made to for the purpose of making timely warnings of criminal offenses under (a)(6)	
	668.47(a)(2)	Statement of current policies concerning: • Safety of and access to campus facilities including campus residences • Security considerations used in the maintenance of campus facilities	

Source: Reprinted with permission from *Campus Law Enforcement Journal,* Vol. 27, No. 4, pp. 9–12, © 1997, International Association of Campus Law Enforcement Administrators.

Page #:	Reference	Information to be included	Comments/notes
	668.47(a)(3)	Statement of current policies re: campus law enforcement including: • Enforcement authority of campus personnel • Working relationship of campus personnel with state and local agencies • Arrest authority of security personnel • Policies that encourage accurate and prompt reporting of all crimes to: –campus police –appropriate police agencies	
	668.47(a)(4)	Description of type and frequency of programs designed to encourage students and employees about campus security procedures and practices to encourage responsibility for personal and community safety	
	668.47(a)(5)	Description of programs designed to inform students and employees about crime prevention	
	668.47(a)(6); (f); and Appendix E	Statistics concerning the **occurrence** on campus of criminal offenses reported to local police agencies or to any official of the institution who has significant responsibility for student and campus activities for the 3 calendar years preceding the year of disclosure: **A. Murder** **B. Rape (prior to 8/1/92) or sexual offenses (forcible or nonforcible [on or after 8/1/92])** **C. Robbery** **D. Aggravated assault** **E. Burglary** **F. Motor-vehicle theft**	

Page #:	Reference	Information to be included	Comments/notes
		Statistics concerning the criminal offenses of murder, forcible rape, and aggravated assault that manifest evidence of prejudice based on race, religion, sexual orientation, or ethnicity. Institutions should compile these statistics in accordance with the definitions used in the FBI Uniform Crime Reporting Program.	
	668.47(a)(7)	Statement regarding monitoring and recording, through local police, criminal activity in which students engaged at off-campus locations of student organizations recognized by the institution, including organizations with off-campus housing.	
	668.47(a)(8); (f); and Appendix E	Statistics concerning the number of arrests for the following crimes occurring on campus for the one calendar year preceding the year of disclosure **A. Liquor law violations** **B. Drug abuse violations** **C. Weapons possessions** Institutions should compile these statistics in accordance with the definitions used in the FBI Uniform Crime Reporting Program.	
	668.47(a)(9)	Statement of policy re: the possession, use, and sale of alcoholic beverages and enforcement of federal and state underage drinking laws	

Page #:	Reference	Information to be included	Comments/notes
	668.47(a)(10)	Statement of policy re: the possession, use, and sale of illegal drugs and enforcement of federal and state drug laws	
	668.47(a)(11)	Description of drug or alcohol-abuse education programs as required under 1213 of the HEA (can be a cross-reference to other publications)	
	668.47(a)(12)	Statement of policy regarding campus sexual assault program to prevent sex offenses, and procedures to follow when a sex offense occurs. Statement must include: i. Description of educational programs to promote the awareness of rape ii. Procedures students should follow if a sex offense occurs, including: • Who should be contacted • The importance of preserving evidence for the proof of criminal offense • To whom the alleged offense should be reported iii. Information on the student's option to notify proper law enforcement officials (including campus and local police) and a statement that institutional personnel will assist the student if requested iv. Notification of existing on- and off-campus counseling, mental health, or other student services for victims of sex offenses v. Notification that an institution will change the victim's academic and living situations if changes are requested and are reasonably available	

Page #:	Reference	Information to be included	Comments/notes
		vi. Procedures for campus disciplinary procedures in cases of an alleged sex offense, including a clear statement that: A. The accuser and accused are entitled to the same opportunities to have others present during a disciplinary proceeding. B. Both the accuser and the accused shall be informed of the outcome of any institutional disciplinary proceeding brought alleging a sex offense (the institution's final determination and any sanction against the accused).	

	668.47(b); & (f)(3)	Institution shall distribute the report to:
		1. Current students and employees by appropriate publication and mailings through:
		   i. direct mailing to each individual (campus or U.S. mail) or
		   ii. publications provided directly to each individual
		2. Prospective students and employees upon request provided each is informed of the availability of the security report, given a summary of its contents and given the opportunity to request a copy	

The Legal Response to Violence on Campus 325

Page #:	Reference	Information to be included	Comments/notes
	668.47(c) & (f)	Institution shall comply with requirements for each campus. A branch or other location must issue a separate report if not within a reasonably contiguous geographic area of the main campus. Campus (1) any building or property owned or controlled by an institution within the same reasonable contiguous area and used by the institution's education purpose; (2) any building or property owned or controlled by a student organization recognized by the institution; or (3) any building or property controlled by the institution, but owned by a third party.	
	668.47(d)(1)(i)-(iii)	Reporting (a)(6) data: 1st annual report: Covers 1/1/91–12/31/91 and two preceding calendar years (Must cover at least 8/1/91–12/31/91) 2nd and 3rd annual reports: Due 9/1/93 and 9/1/94, respectively Must cover the most recent calendar year and the two preceding calendar years	
	668.47(e)	Timely reports to community to prevent (a)(6) crimes.	

CHAPTER 16

Strategies for Dealing with Violence

Eugene Deisinger, Charles Cychosz, and Loras A. Jaeger

INTRODUCTION

The growing impact of violence on college campuses is generating new partnerships and new approaches to campus life and community protection. Perhaps most noteworthy among these new partnerships is the growing link between student affairs and campus law enforcement. In recent years, student affairs staff have had to respond to escalating violence in student conduct situations and interpersonal relationships on campus. Roommate disputes that used to be mediated by a student residence hall employee may now be settled with threats, beatings, or even gunfire. Counseling sessions increasingly involve students' attempting to resolve their victimization. Simultaneously, campus law enforcement officials have been faced with more crimes against persons and a demand for crime prevention, victim protection, and community policing. This has resulted in student affairs and campus law enforcement establishing violence prevention as a shared priority. Students and families reinforce the priority given to this issue as they pay closer attention to personal safety in their choice of a college. Indeed, this collaboration seems to extend across administrative units as employees are affected by these issues. As is so often the case, the ubiquitous nature of campus violence increasingly demands an institution-wide approach.

This chapter is designed to identify key roles played by law enforcement, student affairs, and human resources in responding to campus violence. Additionally, the role of the incident response team (IRT) during campus emergencies is discussed. While individual colleges and universities will have to develop approaches that work best for them, depending on the nature of the campus, the IRT is an excellent model to use in providing appropriate resources and strategies when dealing with emergencies. Extensive discussion also is given to dealing with threatening behavior on campus.

KEY ROLES FOR LAW ENFORCEMENT

Campus law enforcement has made rapid changes in the past decade toward professionalization and growth. Universities and colleges in this country employ approximately 11,000 full-time peace officers with arrest powers and another 9,000 security officers without arrest powers (Reaves & Goldberg, 1996). These professionals provide for the physical security and safety of members of the campus community. With three out of four campuses employing personnel with arrest powers, the role of campus law enforcement agencies in dealing with violence cannot be overstated. In their role as emergency responders, law enforcement officers interface frequently with all departments and colleges, but particularly with human resources, the physical plant, residence halls, environmental health and safety, student health services, student counseling services, student affairs, and news services. Campus law enforcement agencies can have significant impact on campus violence by providing proactive services, protecting victims or potential victims from violence, and immediately responding to acts of violence.

Campus law enforcement agencies that aspire to be foremost have developed a variety of proactive approaches to stopping acts of violence before they occur. This section examines eight approaches: a community policing model, law enforcement accreditation, prevention policies and practices, crime prevention programs, motorist-assistance programs, escort services, advisory committees, and access control including environmental design for safety.

Community Policing Model

Community policing, whether referring to a residential community or an institution of higher education, is a philosophy in which law enforcement officials integrate themselves into the area in which they work. Law enforcement officers generally are evaluated on their statistics: number of arrests, citations or warning tickets issued, criminal cases cleared, and calls for service answered.

Under the community policing model, more is asked of the law enforcement officer than number of responses. An officer is assigned a specific geographic area, and he or she begins to develop a relationship with persons in that area. The emphasis is on problem solving and involving those living in that area in the solution.

At the college or university level, the law enforcement agency develops partnerships with departments; colleges; and various staff, student, and faculty groups. Street officers are encouraged to identify problems, especially recurring incidents that are developing into problems. The officer, either alone or with assistance from other law enforcement personnel, begins to bring together problem solvers to jointly address the problem. An example of this problem-solving process is the introduction of graffiti on a university campus. Graffiti can mean the start of gang infiltration into an area, with increased crime activity. The street officer identifies the increase in graffiti incidents and brings together facilities staff and those working or living in the affected area. Facilities staff agree to remove graffiti within 48 hours of the report received at their office. Law enforcement and those in the affected area then immediately report any new acts of graffiti. Through diligence, the graffiti stops and criminal activity that many times follows the graffiti is prevented.

Law Enforcement Accreditation

The Commission on Accreditation for Law Enforcement Agencies (CALEA) was formed in 1979. Today, several hundred law enforcement agencies have voluntarily become nationally accredited. The established accreditation standards help law enforcement agencies: (1) strengthen crime prevention and control capabilities, (2) formalize essential management procedures, (3) establish fair and nondiscriminatory personnel practices, (4) improve service delivery, (5) solidify interagency cooperation and coordination, and (6) boost citizen and staff confidence in the agency (CALEA, 1994a).

Chapter 45 of *Standards for Law Enforcement Agencies* states, "Beyond stating their commitment to crime prevention and community relations, law enforcement agencies should establish policies, goals and objectives by which their commitment can be realized." University law enforcement agencies, even more than city law enforcement agencies, should have several crime prevention programs to offer students, faculty, and staff. Input should be received from the university community as each program is developed, and each of the programs should be evaluated on an annual basis (CALEA, 1994b).

Prevention Policies and Practices

Crime prevention is a relatively new activity in the context of the history of American higher education. Twenty-five years ago, the term was virtually unheard of on campus. Many schools now have a crime prevention specialist working in their law enforcement agency who possesses specific knowledge of crime prevention programs and how to implement them. The International Association of Campus Law Enforcement Administrators has endorsed recommended crime prevention and campus protection practices. Included in these recommendations are the following:

- A rape awareness and education program should be developed and offered to members of the campus community.

- A crime prevention orientation program should be presented to all new faculty and staff as they become members of the institutional community. All new students should be exposed to a campus security and safety-oriented program.
- Specialized crime prevention programs should be developed for presentation to various specialized campus constituencies. Among these groups are the following: commuter students, international students, students with disabilities, student athletes, physical plant or maintenance staff, resident students, fraternity and sorority members, library and other staff who may be subject to confrontational situations.
- The institution should employ both printed and electronic media to convey crime prevention and security information to the campus community.
- One or more effective methods should be developed for distributing warning or notifications to members of the campus community regarding serious crime or security incidents.
- Alcohol and drug programs should be developed that include elements of policy, awareness, education, and enforcement.
- A system of emergency call boxes or telephones should be provided on campus.
- A key management program should be developed.
- Trees and plant material should be trimmed so as not to interfere with lighting or to provide easy access to buildings by climbing by potential perpetrators of crime (International Association of Campus Law Enforcement Administrators, 1996).

Crime Prevention Programs

Many universities now employ full-time crime prevention officers. These crime prevention officers coordinate prevention activities including crime prevention surveys, crime analysis, and crime prevention programming. Most crime prevention programs incorporate personal protection and property management including engraving of valuables, grounds and building inspections, and lighting surveys.

Motorist-Assistance Programs

To reduce random acts of violence, some colleges and universities are now providing motorist-assistance programs to help those stranded on campus. These programs not only provide a valuable service to a stranded motorist, they also provide security to the motorist while needed repairs or assistance are performed.

For example, at Iowa State University the program is called HELPVAN. It operates from 7:30 AM to 3 AM and provides free service if a vehicle needs a jump start, has a flat tire, or has run out of gas. A well-publicized telephone number is used and answered on a 24-hour basis by the department of public safety. From 3 AM to 7:30 AM, the campus public safety department responds to all calls for assistance. When not assisting motorists, the driver of the HELPVAN drives through parking lots on campus with amber top lights activated to help reduce acts of vandalism and theft.

Escort Services

Most schools now have some type of escort service that is available to students, faculty, staff, and visitors to campus. Some are informal escort services managed by student groups, and others are formal services operated out of the campus law enforcement agency.

Campuses that have instituted a formal escort service have established policies governing the operation. These policies include recruitment, background checks, and training of those involved in the escorts. Escort personnel receive training on how to conduct themselves when escorting, basic self-defense, communication with the campus law enforcement agency, and knowledge of service boundaries. Additionally, all escorts are documented, including the time the call was placed, the time the service arrived, the name of the persons escorted, and the time they were delivered to their destination.

Advisory Committees

Advisory committees are an excellent source for input and evaluation of campus law enforcement activities. They help define the responsibilities of law enforcement when dealing with a department on campus, and they can also be called upon during times when a serious, violent crime occurs on campus.

Advisory committees should be established both at the police department and university level. Examples of some advisory committees include a university safety advisory committee, a university traffic advisory committee, and a residence/law enforcement advisory committee. Other ongoing relationships should be established between the law enforcement agency and the library, student affairs staff, and any medical facilities or research facilities located on campus.

Mutual understanding of the role of others on campus is very important to law enforcement officials in establishing good working relationships. For example, how the campus law enforcement agency responds to calls for service in a residence building should be written in a procedural policy. The policy should cover medical emergencies, trespassers, lock-outs, residence hall policy violations, contacting housing staff by law enforcement, alcohol and other drug violations, search of rooms, lost or stolen keys, vandalism, fire alarms, bomb threats, severe weather alert, and commercial soliciting (Keller, 1991).

Access Control and Environmental Design

Access control is the controlling and monitoring of access by persons seeking to enter campus property or buildings. Environmental design for safety means designing areas or structures with crime prevention in mind. The height of the lowest branch of a tree, the maximum shrub height, and the type of plant material used along walkways should be studied. Controlling informal pathways that are not lighted should be a priority.

The National Crime Prevention Institute defines crime prevention through environment design as the proper design and effective use of the built environment that can lead to a reduction in the fear and incidence of crime and an improvement in the quality of life (Crowe, 1991).

University and college law enforcement agencies are now training officers in access control and environmental design for safety. To reduce acts of violence on campus, the school's law enforcement agency should have expertise in security design planning. Law enforcement officials should be included in discussions regarding the campus master plan and capital improvement projects.

Through training, a crime prevention specialist is able to conduct security surveys and understand architectural design, graphic arts, and building design characteristics. The specialist should also understand locking devices, including electronic and key systems. A crime prevention specialist should be able to read and understand blueprints and other architectural drawings.

Victim Assistance

Persons who commit violent acts do so for many reasons. Some have a desire to control and terrify, others may be angry, obsessed with another, under the influence of alcohol or other drug, or mentally ill. Perpetrators of violence often threaten before they act. Law enforcement agencies must make the protection of the victim of a threat a top priority and must be creative in doing so. A variety of tools are available to help potential victims.

Contact with Threat Maker

Campus law enforcement officials can make immediate contact with the threat maker. If an act of violence has not yet occurred, campus law enforcement can place the person on notice. This approach also gives law enforcement the opportunity to develop a relationship with the threat maker and hopefully end the threat or refer the person to the services available on campus.

Escort

The victim of a threat should never travel alone. One way to help ensure the safety of a victim is through escorting. A student can be escorted by law enforcement officers to and from class and other activities. When the threat of violent behavior is imminent, escort or protective custody is the safest solution for the victim.

Cellular Telephones, Noise Makers, Police Radio

The use of cellular telephones, various noise makers, or a police radio have provided yet another level of safety for potential victims. Campus law enforcement officials can loan a cellular telephone to the victim of a threat. The cellular telephone can have the campus law enforcement telephone number programmed to allow quick contact.

Various noise makers are for sale or can be loaned out from the campus law enforcement agency for use by the victim of a threat. A noise maker with high decibel levels will drive most attackers away and also alert persons in the area.

Some campus law enforcement agencies have multiple radio frequencies assigned to them. One of those frequencies can be assigned to emergency services. A radio can be assigned the victim of a threat, and that person can notify the campus law enforcement agency by direct radio communication if a problem should occur. Additionally, some radios are equipped with a security device that sends a distress notice to the law enforcement dispatch center if a button is pushed.

Protective Order/University Trespass Notices

A protective order places the potential or actual perpetrator of violence on notice that any further contact with the victim can result in additional criminal action. Each state has statutes outlining when a person can secure a protective order against another. Most states now have statutes allowing for protective orders in domestic abuse incidents and many are now adding other offenses, such as stalking, to offenses needing this relief. Once the order is issued, it is important that local and campus law enforcement officials are aware of the order. In most states, law enforcement officials can arrest a person for violating a protective order based on probable cause, without a warrant.

Many campuses also have procedures in place that bar persons from the grounds or specific buildings on campus. The notice is generally written by the campus law enforcement head or a high-ranking school administrator. The notice describes what incidents have occurred that cause the person to be barred, lists areas the person cannot enter or occupy, describes what will occur if the notice is violated, and describes a process the person can use to appeal the order.

Victim Advocacy

For several years, victim advocates have been an important resource in assisting a person who has been victimized. Victim advocates are trained to be good listeners. They assist the victim in meeting special needs and empower the victim to make decisions.

In many states, law enforcement agencies must provide victims with information about advocate services that are available to them. When a good working relationship exists between law enforcement and victim advocates, it goes a long way toward empowering victims and helping them regain a sense of control over the situation. The advocate also helps the victim make informed decisions and fully understand the potential consequences of testifying against the perpetrator or failing to testify.

Victim Services

There are many types of victim services available on college campuses and in communities. These include 24-hour crisis help lines that are anonymous and confidential, support groups, peer counseling groups, self-defense classes, and women's centers.

Civil Remedies

The federal Violence Against Women Act of 1994 creates civil liability for crimes of violence that are motivated by gender bias. The act allows the victim of a felonious violent act motivated by gender bias to the recovery of compensatory and punitive damages, injunctive and declaratory relief, and other legal remedies deemed appropriate by the court. Victims today are being encouraged to bring civil action against the perpetrator of a violent act in state court and to recover damage resulting from the act.

Law Enforcement's Response to Violence

Coordination Role

To have an effective response to an act of violence on campus, the law enforcement agency should have a written crisis plan available for immediate reference. The plan should identify the following:

- members of the crisis team
- various alert stages and what occurs at each stage

- appropriate public information (provided to the media)
- on-scene field command along with command at the emergency communication center
- perimeter control of the affected area
- other agencies that can provide assistance
- equipment needs
- detailed maps of the campus
- notification requirements
- deescalation procedures
- debriefing reports

The first half hour after an act of violence has occurred is the most important time. Aiding the victim, ensuring that those in the area are protected, and conducting an immediate investigation should occur in this time frame. All parties involved in the university's response should know who is in charge. For example, if a violent act occurs in an on-campus residence hall, are the police in charge or do they take orders from the director of the hall?

Everyone should know and understand the crisis plan and then follow it. This principle cannot be overstated, because many universities have crisis plans but forget to use them in a crisis. One way of avoiding this problem is to have annual training in which a mock disaster is practiced. For further information about crisis management, see Chapter 13.

The role of the campus law enforcement agency should be to coordinate the immediate notification of an act of violence to key persons on campus. A list should be kept at the dispatch center with names and telephone numbers, including home telephone numbers, of such persons. Depending on the type of incident, both an on-scene command post and an emergency command center at the law enforcement agency may be established. The crisis team generally reports to the law enforcement agency and coordinates with the on-scene command post.

Getting the facts straight from the on-scene command post and writing those facts down has much to do with coordinating activities. One person should be assigned to information gathering. That person should work closely with the campus media staff to enable them to deal with all external media requests.

Information Gathering

Campus law enforcement has a role in gathering information and in sharing that information with key campus personnel. Information should be gathered on the potential target as soon as the potential for a violent act becomes known. Examples of the information that should be gathered are: Where does the person live and work? What is the person's class schedule? What type of vehicle does the person drive? Who are his or her closest friends?

Obviously, information about the potential perpetrator should be gathered. Is there a history of violence or criminal activity? In addition to the above information, detailed information about the description of the person should be gathered. Does the person have access to weapons? Has the person made verbal threats? Does the person have a military background?

THE ROLE OF HUMAN RESOURCE SERVICES

Contemporary human resource functions include an awareness of violence in the workplace (Kinney & Johnson, 1993). Most large employers have followed the lead of the Postal Service and others in working to reduce on-the-job violence. The Occupational Safety and Health Administration acknowledged this issue in 1996 when it established violence prevention guidelines (Occupational Safety and Health Administration, 1996).

Once an employer has identified violence prevention as a goal, a supportive set of policies outline acceptable employee behaviors, corporate responses, and consequences for policy violations. Training can assist employees with difficult customers as well as deescalation and escape strategies. Clear corporate policy also encourages employees to identify off-the-job conflicts that can lead to work site retribution.

Although it is comparatively rare, disgruntled employees have sought retribution from supervisors and coworkers (Feldmann & Johnson, 1996). In a majority of the cases, there has been some advance warning of the concern. In fact, some employers will provide an employee-relations hot line to encourage upset employees to work through conflict without violence. In cases where a threat assessment process identifies a serious possibility of harm, safety becomes a primary concern. Motivated by this commitment to safety, employers can negotiate a separation of the employee from the institution, often in a manner that minimizes anger, hostility, and likelihood of retribution.

Monitoring the work climate, defusing tension with open discussion, and paying attention to morale help diminish these problems. Periods of crisis or high stress may warrant special attention to these matters. Additional information about workplace violence is discussed in Chapter 6.

THE ROLE OF STUDENT AFFAIRS

Student affairs has a significant role in preventing problems as well as resolving individual incidents of violence. While this role traditionally has been vested in "student life," campuses are no longer the sanctuary that they were once perceived to be. Managing "student life" increasingly means establishing and maintaining a close working relationship with professional law enforcement agencies.

Proactive Services

Community-building initiatives are a traditional strength of the student affairs units on campus. Building a sense of belonging, connection, and shared commitment to academic principles is important in defusing violence. Campuses historically have been arenas where civil discourse and consensus decision making are practiced and valued. As campus populations change and links with local communities grow, it is important that these traditional community-building practices be reinvigorated because they provide a strong web of protection in the community.

Campus policies must acknowledge that disputes will occur and provide opportunities to resolve conflict without resorting to violence. Clear behavioral expectations and prompt response to violations is important to reinforcing a community standard. Sample policies are available from a variety of sources including the International Association of Chiefs of Police (1995).

Mediation provides a valuable strategy for engaging people in problem resolution. It is particularly powerful in that mediated resolutions often teach skills that can be transferred to other conflicts. In addition, it is an affirming process in that it invests a sense of control in the participants.

The campus judicial process and the campus code of conduct provide a context for communicating community standards to the campus. Nearly all campuses have a code of conduct with specific prohibitions against interpersonal violence. The campus judicial system then enforces this code, often through the dean of students or other disciplinary agents.

Often this interface between the disciplinary agents and individuals provides early warning of those individuals who are unwilling or unable to conform to the expectations of the community. Lethal violence often is the product of an escalating series of encounters. By paying close attention to repeat offenders, unreasonably distressed individuals and patterns of escalating confrontation and interaction, it may be possible to identify some individuals with a propensity for violence; a persistent, unresolved complaint against an individual or the institution; and the means to do harm. A formalized threat assessment process (discussed later in this chapter) can strengthen the confidence with which these determinations are made.

Mental health services, especially university counseling services, can play a critical role in proactive approaches to violence reduction. University counseling staff tend to be well trained in aspects of interpersonal communication and conflict resolution; indeed, these are fundamental aspects of most therapy experiences for students, faculty, and staff. Counseling staff can use these skills to assist the community in preventing violent incidents, in resolving them more effectively when they occur, and in recovering from them afterwards.

Often conflict arises out of a lack of communication or dialogue between the parties involved. Counseling staff can provide workshops and training materials regarding communication skills, adaptive problem solving, appreciation of diversity, and conflict resolution. These presentations and materials being offered on a regular basis assist in the development of basic skills in achieving nonviolent resolution of differences and grievances. The active and regular presence of these programs also sets the norm that communication (rather than violence) is the path for conflict resolution.

Mental health providers also may be used in actively facilitating communication between individuals or groups who are experiencing significant conflict without apparent resolution. Such facilitation can occur in the safe, private, and neutral domain of the counseling service in a process that emphasizes mutual respect and collaboration as means of resolving conflicts.

If persons appear to be at high risk for acting violently, they may be referred for counseling, which provides a safe forum for the airing of hurts and grievances. In addition, the counselor can help redirect aggressive behavior to more adaptive and effective strategies that help the individual address his or her concerns.

Protection of the Victim

Student affairs contributes at least two key elements to victim protection strategies—victim relocation and protection of identity and confidentiality. Victim relocation may be necessary in response to an immediate threat of personal violence. Residence hall rooms and other student living space controlled by the institution provide excellent settings for short-term housing of individuals. While care must be taken to avoid endangering other residents, rooms often are available with substantial levels of occupant anonymity supported by facility and personal security measures. If such safe housing options are not available on campus or in a community, consideration should be given to developing them.

Student affairs offices increasingly are called upon to cooperate with law enforcement agencies, both on and off campus. It is valuable to establish a working relationship outside the context of a crisis. Given the nature of various student affairs functions, it is common for problems to emerge in a residence hall or a disciplinary hearing before they escalate into a more violent confrontation. Student affairs staff are well positioned to anticipate some of these problems and intervene. In cases where immediate threats of violence occur, it is important to develop a cooperative, timely mechanism for responding. Many campuses are creating collaborative crisis response teams to manage these situations.

THE INCIDENT RESPONSE TEAM

Many campuses are developing incident response teams (IRTs) to improve the administrative response to large, complex incidents such as fire, natural disasters, or terrorism. Quite often, these IRTs also respond to situations involving violence or threats of violence. The purpose of an IRT is to integrate university services—such as public safety, student counseling, university relations, human resources, student affairs, environmental health and safety, legal services, and academic affairs—into a functional unit providing timely, comprehensive, and accessible support services during a wide variety of critical incidents. Critical incidents may include threats or acts of violence, major demonstrations or riots, floods, major fires, tornadoes, airplane crashes or train derailments, major chemical spills, hostage situations, sniper activity, building collapse or explosion, bomb threats, or deadly disease and food contamination occurring within the university community. For many, however, the most common function of the IRT is a response to potential, threatened, or actual violent incidents.

A coordinated response will provide the following:

- anticipation and, where possible, prevention of incidents involving violence or injury
- proactive services for dealing with critical incidents
- coordination of the campus response to critical incidents
- coordination of post-incident recovery services for the university community
- a more systematic and routine approach to critical incidents
- a venue for promptly identifying and supporting university decision makers
- a system for evaluating all critical incidents with the goal of providing improved plans to protect lives and property as well as reduce exposure to vicarious liability
- improved management of public information

The incident response team is a small group that can be assembled quickly to initiate the response to an incident. One particularly important component of IRT is immediate threat assessment. This group is designed to meet frequently and work as a team. The IRT often is supported by a larger set of administrative units or resources services.

The typical IRT team includes representatives from the dean of students, department of public safety, student counseling services, human resources, provost or chief academic officer, and university relations. Because so many incidents involve students, the team may be chaired by a vice president or other senior administrator in student affairs. In any case, the team leader has responsibility to organize and convene the group. In addition, this individual initiates IRT response; is the primary liaison to administrative decision makers; and maintains an updated, centralized database of incidents and resources.

Other members of the team include the dean of students who provides appropriate access to student information, interprets the student conduct code and other policies, and acts as liaison with parents and families. The department of public safety coordinates emergency services, conducts threat assessment, and provides law enforcement liaison. Student counseling services can provide psychological assessment, crisis intervention, and critical incident debriefing. Counseling staff can also facilitate hospitalization when necessary. The chief academic officer interprets academic polices and priorities and provides a liaison with academic units. Human resources staff provide appropriate access to employee information, interpret personnel policies, and provide liaison to union and employee groups. Finally, university relations staff coordinate media relations and release of public information.

The function of an IRT is to provide prompt, coordinated responses to requests for assistance requiring coordination among several university departments. Most noteworthy among these is the responsibility to manage potentially violent or dangerous situations. The team may become involved in the analysis of threats by employees, students, or visitors. In some cases, the team may provide consultation regarding removal or barring of individuals from campus or office strategies for dealing with troubled employees or students. In addition, the IRT should be prepared to advise key administrators and other appropriate personnel during resolution of critical incidents, coordinate dissemination of information regarding critical incidents, and coordinate delivery of post-incident recovery services.

The actual operation of an incident response team is a substantial management task. Crisis management teams must handle several tasks in a logical sequence:

- identifying the need for threat assessment and developing alternative responses
- identifying decision makers
- convening the parties and establishing safety as a priority
- taking action and coordinating responses
- following through
- conducting debriefing and follow-up

Although these steps may seem self-evident, it is not always so clear when dealing with a specific incident. For instance, when a psychologically distressed student utters threats against a faculty adviser in the process of filing a discrimination complaint, questions of who responds to the threat, the complaint, and any other academic or behavioral problems can become quite complicated. Clearly identifying decision-making responsibility and coordinating decisions and actions can be essential for effectively defusing tensions.

THREAT ASSESSMENT

When the crisis involves a specific threat of violence, it is often valuable to engage in a systematic assessment of the seriousness of the threat. Specific protocols for approaching this problem have evolved in the psychological literature. These have been adapted for use in the crisis management environment. Used cautiously, this process can provide a structured decision-making framework for the crisis management team.

The threat assessment process is designed to evaluate the degree of risk posed by a potentially violent individual and provides recommendations for reducing significant risk and increasing the safety of potential victims or targets of violence. In addition to the following summary of the threat assessment process, the reader is encouraged to review the works of Monahan (1995); Flannery (1995); Roth (1994); and Fein, Vossekuil, and Holden (1995). These authors have contributed greatly to the development of more reliable approaches to threat assessment.

The threat assessment process is guided by several fundamental principles and assumptions (Fein et al., 1995). First, acts of violence do not occur in a vacuum or without antecedents. Persons who act out violently show reliable patterns of experiences, conflict, failures, and prior acting out that tend to escalate into acts of physical violence. Gaining awareness and understanding of the history of the perpetrator provides for better prediction of his or her future actions. A second and related principle is that violence is an interaction between three domains: the potential perpetrator, the context or environment in which the individual exists, and triggering events that lead to violence being viewed as the only solution. Violent behavior occurs when risk factors from all three domains coexist. Finally, the most effective assessment of and response to potentially violent subjects occurs through collaboration of well-trained persons from a variety of professions and areas of expertise. In essence, the threat assessment team is best composed of individuals who can understand the potential perpetrator, the perpetrator's environment, and any precipitating factors. Thus, experts in law enforcement, mental health and student affairs, and human resources should be involved in conducting threat assessments on college campuses. The threat assessment process consists of three phases: gathering information, assessing the risk, and developing interventions.

Gathering Information

The team gathers a great deal of information regarding the subject, potential targets of violence, the environment in which the subject and victims exist, and any likely precipitating events. This information comes from a variety of different sources including the potential target(s), the subject, persons knowledgeable about the subject or target(s), about the environment in which the threat has occurred, or about any events that might precipitate acts of violence. Additional information is obtained through a background check of the subject to determine previous problems, concerns, or acts of violence in which the subject has engaged.

Following is a list of several domains in which information should be gathered:

- basis for current concerns regarding the safety of the identified targets (What threats have been made?)
- identification of persons familiar with the subject and/or the situation
- previous history of violent or criminal behavior by the subject
- indications of unstable personality in the subject (history of mental illness, substance abuse, or recent changes in personality)
- the subject's interest in and experience with weapons, including experience in the military
- serious medical conditions or physical illness of the subject
- academic and work history of the subject, especially history of prior conflicts, problems, or violent acts
- social support systems of the subject that may affect the likelihood of acting out
- familiarity of the target to the subject (such as knowledge of the target's daily plans)
- vulnerability of the target (Can the target increase his/her risk?)
- events that may trigger violent behavior (such as loss of job, status, income, or relationships)
- environmental factors that may increase likelihood of violence (such as authoritarian management, excessively stressful or unreasonable expectations placed on the subject, or a permissive attitude regarding violence and inappropriate behavior)

Assessing the Risk

At this stage, the team reviews all information obtained in the information gathering phase and evaluates the interactions of risk factors regarding the subject, the environment, and any precipitating factors. The goal is to determine whether violent behavior is more or less likely to occur within a given time frame. The likelihood, severity, and timing of violent behavior significantly influence the development of plans to redirect the subject away from attempts to resolve conflict through violent means.

Developing Interventions

The primary role of the threat assessment team at this stage is to develop consensus on a course of action and collaborate on an integrated, planned, response to the situation that increases the safety of the persons involved.

Given that a significant predictor for violence is the perception of the perpetrator that violence is the only means of resolving the situation, the primary intervention should be to help the person develop reasonable alternatives to violence. For this to happen, the intervener must have developed a respectful, trusting, relationship with the individual in question. Numerous case studies of violent acts show that the perpetrator became increasingly isolated prior to acting out. Reducing the sense of isolation and the lack of communication and perspective that comes with such isolation is seen as critical to deescalating hostile subjects. Therefore, the primary intervention is nearly always to put someone in contact with the subject and begin laying the groundwork for reducing the sense of isolation and assisting the subject in developing alternatives to violence. It should be noted that it is not an easy or comfortable task to reach out with empathy and regard for someone whose actions may be despised. Careful consideration should be given as to who is best placed and suited to reach out to the hostile individual. The goal of increasing the safety and well-being of those involved must take precedence over punishment of the potential offender. Indeed, a premature focus on punishing the potential offender may result in greater isolation and hostility, precipitating the very acts the team hoped to prevent.

Based on the risk assessment, several interventions may be used to address the subject's concerns. Interventions may include giving referrals for mediation, voluntary or mandatory counseling, or hospitalization for psychiatric care; providing the subject with resources to address his or her concerns; barring the subject from the location of the target; placing the subject on temporary leave; or removing the subject from the environment through incarceration or termination of employment. Again, the interventions used will be those viewed as best protecting the safety of the community, with the accountability of the perpetrator as a secondary goal.

In addition to responding to the subject, the threat assessment team may take steps to decrease the vulnerability of the target through the use of personal safety alarms, escorts, self-defense training, or voluntary or mandatory removal from the threatening subject or environment.

In virtually all cases, it is important to manage the flow of information. A crisis management team must always concern itself with getting timely information to those who need it, restricting access to confidential information, and ensuring that institutional actions are cooperatively directed at health and safety priorities. A vitally important but often overlooked role is facilitating the flow of information from the threatening individual to the team. The best information about an individual's intentions and state of mind often comes directly from that individual. The task of managing a potentially violent individual may begin by simply calling him or her on the telephone.

If there is a public information component, it is vital that this is intentionally managed by a single spokesperson for the institution. While institutions are increasingly aware of this function, it helps to establish relationships and protocol prior to an incident in order to facilitate smoother operations.

REFERENCES

Commission on Accreditation for Law Enforcement Agencies, Inc. (1994a). *Accreditation program overview*. Fairfax, VA: Author.

Commission on Accreditation for Law Enforcement Agencies, Inc. (1994b). *Standards for law enforcement agencies*. Fairfax, VA: Author.

Crowe, T. (1991). *Crime prevention through environmental design: Applications of architectural design and space management concerns*. Stoneham, MA: Butterworth-Heinemann.

Fein, R.A., Vossekuil, B., & Holden, G.A. (1995, September). *Threat assessment: An approach to prevent targeted violence*. Washington, DC: National Institute of Justice.

Feldmann, T.B, & Johnson, P.W. (1996). Workplace violence: A new form of lethal aggression. In H. Hall (Ed.), *Lethal violence 2000* (pp. 311–338). Kamuela, HA: Pacific Institute for the Study of Conflict and Aggression.

Flannery, R.B. (1995). *Violence in the workplace*. New York: Crossroad.

International Association of Campus Law Enforcement Administrators. (1996, April). *Recommended crime prevention and campus protection practices for colleges and universities*. Hartford, CT: Author.

International Association of Chiefs of Police. (1995). *Combating workplace violence*. Alexandria, VA: Author.

Keller, D.P. (1991). Campus crime prevention programs. In D.P. Keller (Ed.), *Crime prevention and security for student residence facilities: A manual for college and university law enforcement officers and housing administrators and staff* (pp. 1–5). Goshen, KY: Campus Crime Prevention Programs.

Kinney, J.A., & Johnson, D.L. (1993). *Breaking point: The workplace violence epidemic and what to do about it*. Chicago, IL: National Safe Workplace Institute.

Monahan, J. (1995). *The clinical prediction of violent behavior*. Northvale, NJ: Aronson.

National Crime Prevention Institute. (1986). *Understanding crime prevention*. Boston: Butterworth.

Occupational Safety and Health Administration. (1996). *Guidelines for preventing workplace violence for health care and social service workers*. Morgantown, WV: Author.

Reaves, B.A., & Goldberg, A.L. (1996). *Campus law enforcement agencies, 1995.* Washington, DC: U.S. Department of Justice, Office of Justice Programs.

Roth, J.A. (1994, February). *Understanding and preventing violence.* Washington, DC: National Institute of Justice.

Violence Against Women Act as Title IV of the Violent Crime Control and Law Enforcement Act of 1994. (1994). Washington, DC: U.S. Government Printing Office.

Chapter 17

Conclusion

John H. Schuh

INTRODUCTION

This book examines issues related to violence in higher education and how college officials can respond. Since the 1960s, crime and violence have been increasing on America's college campuses, be they located in an inner city, suburban setting, or rural area. Violence has touched many campuses and, unfortunately, will touch others in the future.

Educational institutions are no longer viewed as safe havens for students, faculty, or staff. Violence is a community and societal problem that has found its way into institutions of higher education. It is much easier to describe the extent of the problem than to identify solutions to dealing with violence. Nonetheless, some strategies are available for campus leaders to deal with this problem.

In this final chapter, general conclusions from the preceding chapters are presented. Then, approaches to dealing with various aspects of violence are identified, not so much as an attempt to prescribe how to solve the problem but rather as strategies that can be undertaken as an integrated campus approach to the problem.

SOCIETAL CONDITIONS

Conditions related to violence on campus cannot be disconnected from those that cause a societal predisposition toward violent acts. The campus is not an island isolated from societal violence. As incidents of violence occur in society at large, they will also occur on campus.

VIOLENCE IN AMERICA

Treviño, Walker, and Ramírez assert in Chapter 1 that the United States has a history of violence, and that violence has increased over time. They assert that the nature of violent acts has changed in recent years from fist fights to the use of firearms. Perhaps the most revealing assertion is that the age group at greatest risk for being a murder victim is 20 to 24 years of age. Coincidentally, members of this age cohort often are enrolled in postsecondary education. Males aged 20 to 24 are particularly at risk.

To be sure, not all elements of society are at the same risk. Rural areas are safer than large cities. Males are more at risk than females. The Northeast appears to be less prone to violence than the South. Violence is more likely to occur during certain times of the year. But the general conclusion is unmistakable: virtually everyone is at risk.

It can be easy to slip into a debate over whether the United States is more violent than other Western, industrialized countries, or whether stricter gun control legislation is needed. Those debates are beside the point. We live in a violent society, and violence is a national health risk, from which college campuses are not immune.

CAMPUS VULNERABILITY

Physical and psychological factors contribute to the vulnerability of campuses to violence. As pointed out in Chapter 2, there may be no institution in our society that is more open to newcomers and guests than the college campus. Colleges and universities routinely invite outsiders to campus to participate in events, participate in celebrations, or simply to use the resources. Indeed, such is the mission of many colleges.

While these invitations are being issued, no one on campus really knows who is using the college's resources at any time. Nobody keeps track of visitors to the campus. This openness is in stark contrast with business and industry, where outsiders often have to register at a reception desk and wear a badge that identifies them as a visitor.

In a psychological sense, colleges and universities are safe places for people with unorthodox points of view. Where else can members of the community organize a rally to assert that the chief executive officer of the organization should be fired? Where else are outsiders invited to deliver speeches critical of the organization? These are simple examples of what makes institutions of higher education different than other institutions in American society, and potentially more vulnerable to violence.

STUDENTS COMING TO CAMPUS

Fenske and Hood assert in Chapter 3 that school is not a haven for students from crime and violence. They add that the general level of violence has escalated and that this escalation is especially acute for high school–age students. Their research found that an increasing number of students and teachers are direct victims of school violence. Their conclusion is especially poignant—if students do not feel safe in their schools, what can they learn? And, if they bring such an attitude to college, what are the implications for the collegiate environment and their learning in it?

Studying the experiences of students in high school is an excellent technique for college administrators, student affairs practitioners, and faculty as they prepare to meet the needs of the students of the future. What can be learned about the students of the future is that it is becoming increasingly common for them to deal with issues related to violence. If violence becomes more the norm for their experiences in high school, will it become the norm in college?

WEAPON CARRYING

Increasingly, students are carrying weapons. In Chapter 4, Summers and Hoffman assert that 1 student out of every 13 carried a weapon, according to a report from the Centers for Disease Control and Prevention. Clearly, anecdotal evidence exists indicating that weapons are a part of violence on college campuses.

They also cite a report indicating that students who live in urban areas are more likely to carry guns than those who live in rural areas. Students do so to protect themselves but also to impress their friends and for reasons of self-esteem. If those are the reasons students carry weapons in high school, is there any reason to believe that they will cease the practice in college? If firearms contribute to the self-esteem of a 16-year-old, will they not continue to contribute to the self-esteem of an 18-year-old?

One of the reasons that people carry firearms is for self-protection. As people do not feel safe in their communities, they arm themselves. If a fear for their safety accompanies students as they come to college, would it be any wonder that they bring weapons to campus to protect themselves? And if they bring weapons to campus, can violence be far away?

ISSUES OF CIVILITY

As Schuh observes in Chapter 5, hate speech, offensive tee-shirts, and name calling all are a part of contemporary campus life—perhaps not every campus, all the time, but these issues surface with regularity. The consequence, almost inevitably, is for the offended group to demand that the institution respond by taking disciplinary action against the alleged offenders. The offenders use the shield of the First Amendment to protect what they assert as their right of free speech. Administrators, after considerable hand wringing, condemn the offensive act but are unable to do much else.

This scenario has repeated itself many times on college campuses. Sometimes it involves student publications; at other times, outside speakers are responsible for allegedly offensive acts. Never at a loss to join the fun, fraternities are often involved in the activity, through a party, banner, tee-shirt, or other act that results in offending some group of students.

These incidents almost predictably result in people taking sides in the situation. Those who are offended and their supporters take one side of the argument and use the incident as evidence of the chilly campus environment for people aggrieved by the act. Those who allegedly committed the offensive act assert that it was meant in fun or really wasn't all that serious. The clubhouse lawyers assert that whether the act or speech was offensive was beside the point—the First Amendment protects freedom of speech. Compromise is sought by administrators, but the various parties dig in their heels; in the end, everyone is offended.

WORKPLACE VIOLENCE

Data related to violence on the college campus has been collected systematically for only a few years. Indeed, until federal law required this, some colleges never collected these data and did not issue any reports about the extent of violence on campus. Even with the systematic release of such information, it can be difficult to interpret the data with precision due to a variety of factors, including the campus location, the extent to which the institution is a residential or commuter institution, and the definition of certain types of crimes.

The cost of workplace violence is great, both in economic and human terms. In Chapter 6, Hoffman, Summers, and Schoenwald report that each episode of violence costs $250,000. Human costs include fear for a person's well-being, diminished job satisfaction, absenteeism, and employee turnover. Additionally, an institution may have problems attracting students and faculty if it has a reputation as being an unsafe place.

Universities as employers have an obligation to provide a safe workplace. This means that screening employees, providing adequate supervision, and following up on specific incidents are essential activities in providing a safe workplace. Failure to do so may result in considerable legal exposure as well as problems with employee and student morale.

VIOLENCE IN CAMPUS RESIDENCES

Palmer asserts in Chapter 7 that violent acts tend to occur at home, and "at home" in the context of colleges and universities means residence halls, fraternity and sorority houses, and apartment buildings where students live on and around the campus. She concludes that most of the violence that occurs in these settings is perpetrated by residents and their invited guests, so the probability of violence resulting from the random acts of outsiders is fairly low. Violence occurs in campus residences because students live in such proximity to one another, abuse alcohol and other drugs, and accept violence as a conflict-resolution strategy.

Greek letter houses provide special challenges; fraternities have become, rightly or wrongly, the focus of intense scrutiny in recent years. Revelations of substantial alcohol abuse, date rape, and other similar abhorrent activities have resulted from this scrutiny. To date, members of Greek letter organizations have been unable to address these issues with any degree of consistency.

ETHNOVIOLENCE

Prejudice and discrimination are the bases of hate crimes according to Fenske and Gordon in Chapter 8. Hate crimes can range from nastiness and name calling to physical forms of violence including assault, battery, and homicide. Often such crimes are random, and the victim cannot be tied to the perpetrator.

The power of a hate crime lies in the inability to link the crime to a specific event or person. In effect, when a hate crime occurs, people who are members of the same group as the victim feel vulnerable and violated. As Fenske and Gordon point out, an unprovoked hate crime in a public place elicits a basic fear that transcends the time and the place of the event itself.

Although Chapter 8 focuses on students of color, quite obviously its conclusions apply to any group not in power, including women, members of religious groups, gay and lesbian students, and so on. The fear that they are not safe is a likely response from any group that perceives itself as being powerless if one of its members is the victim of a random hate crime.

WOMEN AND CAMPUS VIOLENCE

Women often are the victims of violence. Hunnicutt reports in Chapter 9 that date rape, also known as acquaintance rape, is an all-too-common occurrence on college campuses and is an underreported crime. Women also suffer from sexual harassment by peers and male faculty. Consensual relationships between faculty members and students result in a wide variety of problems for the student involved in the relationship, since almost inevitably the faculty member has more power than the student.

The federal government has undertaken several initiatives related to combating violence against women. Among them are the Student Right-To-Know and

Campus Security Act and the Violence Against Women Act. Hunnicutt concludes, however, that federal laws have not been effective in safeguarding the rights of women in court.

HETEROSEXISM AND CAMPUS VIOLENCE

As is the case with violence against students of color, Evans and Rankin assert in Chapter 10 that hate crimes against lesbian, gay, bisexual, and transgendered (LGBT) individuals are crimes against the entire stigmatized minority community. Campus climate studies, in their view, reveal that institutions of higher education do not provide an empowering atmosphere for LGBT individuals, and violence on campus acts to silence the voices of such individuals.

The consequences of this kind of violence include a heightened sense of vulnerability and fear for personal safety; chronic stress; depression; and feelings of helplessness, anxiety, and anger. In addition, mental health problems can result from experiences of violence or fear of violence.

Evans and Rankin point to several studies that describe the effects of violence against LGBT populations in detail. They present a comprehensive model for dealing with this form of violence.

SEXUAL HARASSMENT OF STUDENTS

Closely related to the discussion of violence against women is sexual harassment. In Chapter 11, Paludi and DeFour characterize sexual harassment as hidden violence, since the behavior often results in the victim's behavior or dress being questioned or the victim being held accountable for an incident or series of incidents.

While typically resulting from a relationship between a male faculty member and a female student, sexual harassment cases span across both genders and across all races. Women of color especially are the targets of sexual harassment, which often has less to do with sex and more to do with the power of one person over another.

Consensual relationships between faculty and students have the potential to result in serious problems. The breakup of the relationship, if that occurs, may have very unpleasant consequences for all parties.

SUBSTANCE ABUSE AND CAMPUS VIOLENCE

The relationship between the use of alcohol and other drugs and campus violence has been demonstrated on numerous occasions. While substance abuse is seen as a rite of passage in the eyes of some, its effects are dramatic, including physical effects on individuals, drunk driving, assault, battery, date rape, and worse.

The manifestation of the relationship between substance abuse and violence is not just the post-game "beer bust" to celebrate an important victory on the gridiron. In Chapter 12, Anderson and Napierkowski point to studies in which students drink heavily (engage in binge drinking) and assert that, as drinking gets heavier, the consequences become more serious.

Anderson and Napierkowski have identified a number of strategies that have been employed to limit substance abuse by college students. They assert that many of the strategies have not been thoroughly evaluated. They recommend that any prevention or intervention strategy should include a rigorous evaluation component.

CRISIS MANAGEMENT

In Chapter 13, Baldridge and Julius report on a study they conducted to examine the key elements in a crisis management plan. Their view is that mistakes often are repeated by colleges and universities. They offer three conclusions about campus violence:

1. More crises will result from violence.
2. Administrators and others will continue to make mistakes.
3. Crisis management skills can be learned.

Baldridge and Julius identify several conditions and factors that, in their opinion, will lead to more crises. Among these are that campuses are becoming larger and more complex; political movements historically have been a part of campus life; computer systems are a source of vulnerability; the media make a private crisis a public crisis; the legal environment accommodates those who seek litigation as an answer; environmental concerns have an impact on campus crises; and natural disasters and civil unrest increasingly threaten campuses. They provide an approach to dealing with crisis on campus.

WORKING WITH THE MEDIA

Because of the very nature of violent incidents, media attention is virtually assured when these problems occur. In Chapter 14, Lauer and Barnes point out that news organizations have different goals than institutions of higher education. News organizations want to capture and hold the attention of the general public and, as problems occur, they want to perpetuate interest in the story.

Media representatives have to be cultivated, although this strategy should not be confused with being able to fool or trick these people. Cultivation really means that relationships of trust need to be established. News reporters and editors need to understand some of the factors that shape the work of colleges and universities. They may or may not understand that legal and policy constraints will not allow the institution to release student records such as grades simply because the information might be useful in writing a story about a high-profile student charged with a crime.

A coordinated approach to working with the media makes the most sense when dealing with difficult problems. The media will find cracks in the institution's position, so the best approach is to make sure that only one person serves as the institution's representative. Without a careful approach to working with the media, the focus of the circumstance may shift from the event itself to how the institution handled it.

LEGAL ISSUES

Clearly, legal issues overlay virtually all aspects of violence that occur on campus. From obvious forms of law breaking (such as violent crimes against persons or property) to more subtle legal issues (including hate speech, ethnoviolence, and the like), legal issues frame the problem and an institution's response.

The federal government has shown an increasing interest in campus violence over the past decade, with the enactment of the Student Right-To-Know and Campus Security Act of 1990, sexual harassment legislation, and various legislative acts dealing with substance use and abuse. Simply raising the drinking age to 21 and the passage of highway legislation, for example, have had a substantial impact on how colleges and universities develop alcohol education programming.

Each chapter in this book is influenced by the legal environment. In Chapter 15, Hunnicutt and Kushibab identify relevant federal legislation related to campus violence and explore state laws, court decisions, and other legal influences on the campus environment. Working closely with legal counsel is absolutely essential in planning to minimize the institution's legal problems and in dealing with legal problems as they arise.

STRATEGIES FOR DEALING WITH VIOLENCE

In Chapter 16, an integrated approach to dealing with campus violence is asserted by Deisinger, Cychosz, and Jaeger. They believe that campus security and safety problems are best addressed by a wide range of offices within a college. More details about their approach are discussed later in this chapter.

DEALING WITH VIOLENCE ON COLLEGE CAMPUSES

A variety of suggestions and approaches have been recommended by the chapter authors in this book. This section integrates some of these ideas into an overall plan designed to deal with campus violence. Clearly, more detail is included in each of the chapters, but these ideas provide a place to start acting on, rather than reacting to, issues related to violence.

Recognize the Problem

Because a campus has not experienced incidents related to violence in the past does not mean that violence will not occur in the future. As the authors of this book have pointed out, we live in a society where violence occurs, where people carry weapons, where students have experienced violence in their schools, and where fear exists. As a consequence, the propensity for violence exists.

The chapter authors have cited violent incident after violent incident throughout the country—large campuses and small, urban and rural, on and off campus. No place is immune from this threat. An institution that does not recognize that violence might occur is engaged in myopic thinking. A constellation of factors make violence an increasing probability on college campuses in the future.

Study Demographics and Population Trends

Harold Hodgkinson (1985) has been an advocate of studying the demographic trends of our country as a means of predicting who will be entering a college at any given time. For example, he writes, "Demographics provides a truly new perception of education systems as people in motion. By knowing the nature of those coming into first grade in the U.S., one can forecast with some precision what the cohort of graduating seniors will be twelve years later, and can reveal with very little error what the entering college class will look like in the 13th year" (p. 1).

If an institution wishes to anticipate incidents related to violence, a study of the demographics of its students, its surrounding neighborhoods, and other clients will begin to paint the picture. As the authors of this book have asserted, students have become increasingly used to violence occurring in their high schools, to carrying weapons as a means of protection, and to living in fear of violence. In some ways, violence begets violence. The more people are used to living in violent circumstances, the more likely it is that they will take violence for granted.

One of the surprising statistics revealed in the 1990 census has to do with prisons. According to Roberts (1994), the fastest-growing category of housing reported in the 1990 census was prisons. The United States has the highest incarceration rate in the world (Roberts, 1994). Whether this means that the United

States has the most punitive judicial systems, the harshest judges, or the most efficient police is beside the point. The fact is that this statistic points to the kind of society in which Americans live. More specific demographic reports can be produced literally by county. Study of these reports will help colleges and universities understand what they may have to contend with in the future.

Broaden the Planning Process

To assume that problems of campus violence should be turned over to the campus police department and that this will solve the problem is to grossly underestimate how to solve the problem. Throughout this book, the authors have recommended a variety of approaches to dealing with issues related to violence involving more than a dozen different offices ranging from the campus police department to the office of academic affairs. They have urged that a wide variety of offices be involved in policy development, contingency planning, and response to violent acts. For example, the institutional resource team that Deisinger, Cychosz, and Jaeger describe is a good illustration of how several departments work together, routinely, to deal with issues related to violence. Their concept is elegant and could be implemented on any campus where staff make the commitment to make the place safer for students, faculty, and staff. Similarly, an integrated approach in dealing with the media has been recommended in several chapters of this book. This advice is well worth heeding.

The process recommended in this book is to look at the root causes of violence, to develop policies and approaches that will minimize the potential for violence, and to develop carefully crafted responses in case incidents of violence occur. That process requires broadly based planning involving many campus departments and agencies. Widespread involvement of various campus constituencies will result in better responses to the potential for violence on campus.

Coordinate with Off-Campus Agencies

In some situations, institutional offices and departments need to coordinate activities and plans with off-campus agencies. Among these are police and sheriff's departments, fire departments, the office of the mayor, the department of public health, the district attorney, and state or federal agencies. This sort of activity is best accomplished in advance of any crisis. People need to get to know one another, understand the parameters of their work, and learn the extent to which they can provide resources to deal with the problem.

Similarly, having a good working relationship with various media will be enormously helpful. Knowing reporters or news editors, understanding the constraints under which they work, and realizing that they have a job to accomplish is very useful for campus officials. Similarly, reporters and editors need to realize that the campus has certain protocols for providing public information; when information is not immediately forthcoming, it may not be a conspiracy to cover up facts.

Establishing these relationships is an essential element in planning for campus violence. Campus officials are urged to begin this work immediately if it has not occurred to date.

Know Your Students and Community

Institutions are encouraged to study demographics and population trends so as to better anticipate future challenges related to violence. Institutions should also know their students and the surrounding community. Is the campus residential or commuter? Do lots of students live in surrounding neighborhoods, or do they commute some distance to campus? Do they live at home or own their own homes? Is there a large international student presence on campus? From which countries? Are many communitywide events held on campus? Is the campus generally busy or quiet at night and on the weekends? Answers to these questions and others will help in contingency planning. Two examples illustrate the value of knowing the student body.

1. After a group of U.S. citizens were taken hostage in Iran in 1979, contingency planing had to be developed at campuses that enrolled international students from Iran. Fear was expressed that these students, who had nothing to do with the hostage-taking, could be at risk. The campuses that were able to identify these students and contact them were far better able to develop a contingency plan to protect them than the campuses that did not know who these students were or how to contact them.
2. As the Gulf War began in the early 1990s, students who looked as though they were from the Middle East were subject to some degree of threat as pro-war demonstrations were held on campuses. As was the case with the hostage taking more than a decade earlier, institutions that were able to identify these students and provide protection were far better able to deal with the problem than those that did not know who the students were or how to contact them.

These examples demonstrate that knowing the student body is very helpful in anticipating problems and responding to them. This kind of activity—which would involve collaboration on the part of campus police services, student affairs, the international students office, and the registrar—also illustrates the collaborative nature of problem solving. None of these offices would be solely responsible for the institution's response to the situation. Working together, they can diffuse the problem before it gets out of hand.

Work with Special Populations

Some members of the campus community are at greater risk than others in terms of their vulnerability to violence. Fenske and Gordon present an excellent approach to dealing with random acts of violence perpetrated toward people of color. Other chapters identify some of the unique problems encountered by members of other nondominant populations.

In each of these cases, a specific theme emerges. Members of nondominant populations need to be empowered to be full partners in the university community. Their vulnerability stems from their position of limited power and influence on the campus. When random acts of violence occur toward some members of their population, all members are affected. The very randomness of such an act can lead all members of the group to come to the same conclusion: "Except for luck, I could have been a victim of the violent act."

Random acts of violence cannot be stopped. But institutions can have affirmative policies in place, continue to offer a wide variety of high-quality educational programs, and continually assure members of special populations that they are welcome members of the academic community and are full partners in the educational enterprise.

Know Legal Responsibilities

Issues of violence are inextricably intertwined with the law, court cases, and the legal system. Hunnicutt and Kushibab have outlined basic legal principles related to violence on the college campus. Still, this book is not intended as a complete primer on the vast array of legal issues that may be adjudicated as a result of violent acts. Rather, the intention is to point out that institutions have certain general obligations to all members of the campus community, visitors, and others who use the institution's resources. Among these is the obligation to comply with the laws related to campus crime and violence. Incidents must be tabulated and reported in compliance with federal law. Members of the community need to be warned of the potential for violence in accordance with the law. Steps must be taken to ensure that the workplace is safe.

Legal responsibilities permeate the institution. Any person who is a supervisor, works with students, plans programs, or supervises facilities, for example, has some legal obligation. These staff members need to be aware of their legal requirements and then need to act accordingly to comply with the law. Institutions need to make sure that these obligations are well understood. Not understanding one's legal obligations is not a compelling defense for failing to do so. Moreover, understanding the variety of institutional policies that are related to the issues described in this book is essential. A variety of policies have been described in this book that deal with violence-related issues, from sexual harassment policies to working with special populations. Institutions would do well to remind faculty and staff of these policies and make sure that they are applied throughout the institution.

Take Special Precautions

A variety of suggestions are presented in this book to minimize the potential for campus violence. These range from adopting specific policies to taking general steps to reduce the potential for violence. Institutions are urged to adopt the approaches recommended throughout this book.

Among them is carefully screening new employees to make sure that they do not have a history that would make them a risk. Another recommendation is making sure that the campus police department is a contemporary law enforcement agency that works closely with a variety of other institutional offices and agencies to address potential problems of violence or to react to situations as they occur.

In their chapter on crisis management, Baldridge and Julius lay out a comprehensive approach to managing disasters. Their thesis is that too many institutions do not have a crisis management plan in place and then are surprised when a disaster occurs and they repeat the mistakes that have been made at other institutions.

Make Education a Priority

Some institutions look at federal laws and various reporting requirements and grudgingly comply, if for no reason other than to avoid penalties. However, a different approach should be adopted. Issues related to violence should be discussed frankly and openly. Members of the campus community ought to understand the risks related to the campus and the campus's response to minimizing those risks. Educational activities ought to be undertaken to make sure that people understand how to protect themselves from the risks on the campus.

A typical approach to dealing with issues of violence is to avoid discussing them. When incidents occur, they are "put behind" the campus as quickly as possible. A certain defensiveness permeates the campus's response to the incident. This is common and predictable. But if dealing with issues of violence can be turned into an educational opportunity, the campus will be strengthened.

We do not suggest for a moment that turning violent incidents into learning opportunities can be accomplished without pain. Violent incidents extract a toll on a campus. Students feel uneasy. Members of the staff wonder if the college is a safe place to work. Admissions officers are faced with tough questions from prospective students. Members of the governing board may wonder about the campus's leadership. The public relations side of the situation is very difficult. On the other hand, a "hunker down" approach leaves lingering doubts in the minds of many. Taking an educational approach and applying the principles described by Baldridge and Julius will have a much more positive, long-term effect on the institution.

Have Policies in Place

A number of chapters discuss model policies to deal with issues related to sexual harassment, consensual relationships, and so on. The policies are included in the book to encourage colleges and universities to hold conversations to discuss their positions regarding these issues.

The wrong time to discuss and debate these issues is in the middle of a crisis. Rather, institutions should contemplate such policies when there is no crisis at hand. Elements of policies can be debated more theoretically during times of calm, and they can then be adopted or rejected for reasons other than the need to smooth troubled waters.

If an institution does not have these kinds of policies in place, discussion about them should be initiated. Such action could be forthcoming from the office of student affairs, the faculty senate, or even the chief executive's office. While our bias clearly is toward adopting a package of polices as described in this book, at least if the institution chooses not to put them in place, a rationale for making such a decision would be available rather than "we never thought about this before."

Practice Responding To Disasters

People who earn their living at managing disasters practice how to respond. Responses to mock disasters at airports, for example, are practiced from time to time. Similarly, those on the campus ought to practice how they would respond to a violent incident.

One of the ways to practice a disaster is to use an actual incident that has occurred at another campus and to use the facts of the case to discuss how to respond to the incident. Every person involved in planning for a response should be part of the conference. If not all people are available, having deputies fill in is fine—incidents do not wait for all members of the campus management team to be in place before they occur. Campuses are urged to practice at least once each year to identify where problems in the campus plan exist and to remind people that these kinds of situations occur.

Do Not Be Surprised When Incidents Occur

Finally, those who are responsible for providing leadership on a college campus ought to realize that violent incidents do occur on college campuses, they occur on a random basis, and they often have little or no relationship to the place of the event. Instead of asking the question, "How could this have happened here?" a better question is "Did we take every precaution possible to prevent the incident?"

Engaging in hand wringing or assigning blame is not helpful. By definition, random acts of violence cannot be predicted or prevented. They can be minimized. People can be educated. But, in the end, some tragedies simply cannot be stopped. The most important ingredient is that campuses should do the best job possible of preparing and planning for violent events, and responding as affirmatively as possible.

A FINAL WORD

This book concludes with several observations about the general topic of violence on campus. First, it is a sad commentary on our times that there is a need for a book on campus violence. Clearly, the idyllic days of the campus as a refuge from the trials of the world are over, if they ever existed. It is an absolute shame that students and faculty cannot retreat into a world of ideas where they can learn and sharpen their minds before experiencing the problems of contemporary society. Clearly, the evidence presented in this book suggests that the college campus is not a tranquil sanctuary for reflection and learning.

Second, there is much that college campuses can do to prepare for violence. Thorough organization, careful planning, and a recognition that virtually no campus is immune from violence will serve institutions well as they prepare for violence. Much can be done, as the book's authors have asserted, to minimize the potential for and the effects of violence.

Finally, in the end, there is no way to stop violence from occurring. That, unfortunately, is a fact of life in today's world. No one can predict with absolute assurance where violence will rear its ugly head next. Such may be the price we pay for living in contemporary society.

REFERENCES

Hodgkinson, H.L. (1985). *All one system*. Washington, DC: Institute for Educational Leadership.
Roberts, S. (1994). *Who we are: A portrait of America*. New York: Random House.

Index

A

Access to campus, 18–20
  access control, 332
  to buildings, 19
  and environmental design, 332
  to special events, 19–20
Accreditation, law enforcement agencies, 329
Accuracy in Campus Crime Reporting Act of 1997, 119–120, 162
  status of, 277
Addiction, and substance abuse, 211
Administrators, vulnerability of, 21–22
Age, and violent behavior, 35–36
Aggravated assault
  definition of, 60
  UCR statistics on, 5–6
Aggression
  primary aggression, 2
  reactive aggression, 2
  and violence, 2
Alcohol abuse
  binge drinking, 116–117
  and culture of institution, 216–217
  disciplinary actions, 292–293
  versus life principles, 218
Alcohol abuse and violence
  perpetrator/victim risks, 112–113
  secondhand effects, 113
  sexual violence, 117, 214–215
  societal view, 208
  theories of, 215–216
  See also Substance abuse and violence
American Teacher: Violence in America's Public Schools, The, 37
Assault
  aggravated assault, 60
  criminal penalties, 287
  dating aggression, 154–155
Assault alert, 260–262, 270
  fact sheet, 260–261
  format/information in, 262

B

Battering, women on campuses, 153–154
Bias crime. *See* Hate crimes
Binge drinking, 116–117
Black Student Union (BSU), 140, 142

Bombs
  bomb-detection improvements, 103–104
  example incidents, 103
Buckley Amendment, 129, 250–251, 283
Buildings on campus, easy access to, 19

C

Campus Environment Team, 141
Campus Security Act. *See* Student Right-to-Know and Campus Security Act of 1990
Campus Sexual Assault Victim's Bill of Rights, 163, 277–278
  provisions, 277–278
Campus violence
  and campus patrol, 59
  costs of, 94–97
  examples of, 17–18, 59
  increase in crime, 88–89, 91–92
  recent statistics on, 91–92
  and societal violence, 90
  statistical misinterpretations, 93–94
  statistical misrepresentations, 92–94
  weapon carrying by students, 53–65
  as workplace violence, 89–90
Campus vulnerability
  access to buildings, 19
  access to campus, 18–19
  and campus events, 20–21
  controversy and campus life, 24
  faculty/administration exposure, 21–22
  gun carrying by students, 61
  library/laboratory, 20
  police security needs, 26–27
  stresses of students, 25–26
  substance abuse, 24–25
  surrounding areas of campus, 22–23
  trust/respect of students, 23
Cannabis, 211
Change, theoretical model for, 182
Civility. *See* Free speech
Civil remedies, 286–287, 334
Civil Rights Act of 1871, 283–284
Civil Rights Act of 1964
  Title IX, 279–282
  Title VII, 190, 279–282
Clery Bill, 128
Codes of conduct, prevention aspect, 292–295
Communication
  and crisis management personnel, 251–252
  devices for potential victims, 333
  about federal legislation, 250–251
  news media, 248–250
  See also Crisis communications plan; News media
Communications committee, 256–257
Communications Decency Act, 77
Communications officer, role of, 253–254, 263, 269, 271
Community-based policing model, 63
Community policing model, 328–329
Consensual relationships
  discouragement policies, 196
  student/faculty, 195–196
Conservatives, view of crime, 58
Constitutional rights, and handling of violent students, 39–40
Continuing traumatic stress syndrome, 1
Contrapower sexual harassment, 197
Controversy, on campuses, 24

Costs of violence, 3–4, 60, 94–97
  of gun injuries, 60
  human costs, 94–95
  legal costs, 95–97
Crime
  fear of, 60–63
  nonviolent crimes, 61
  violent crimes, 61
Crime Awareness and Campus Security Act of 1990, 162, 276–277, 293
  provisions, 276–277
Crime Clock, 14
Crimes against persons, UCR statistics on, 7–11
Crime statistics, 4–15
  aggravated assault, 6
  campus violence, 91–94
  crimes against persons, 7–11
  forcible rape, 7
  gun ownership statistics, 59–60
  location of crimes, 13–14
  murder, 7
  nonnegligent manslaughter, 7
  property crime, 4–5
  robbery, 6–7
  school violence, 36–37
  time of crimes, 13–14
  Uniform Crime Report (UCR), 1995 statistics, 4–15
  victim-offender relationships, 11–12
  violent crime, 4–11
Criminal penalties, 287–290
  for assault, 287
  for hate crimes, 288–290
  for rape, 287
  for stalking, 287
Criminals, as college students, 32–33
Crisis
  critical situations, types of, 106
  stages of shock, 107
Crisis communication plan, 253–259
  assault alert, 262, 270
  basic procedures, 256
  communications committee, 256–257
  communications officer role, 253–254, 263, 269, 271
  example of, 259–271
  fact sheet, 256, 260, 261
  facts to media, 254
  follow-up briefings, 259
  handling reporters, 255
  importance of, 255–256
  integration of elements, 247–248
  of law enforcement agencies, 334–335
  objectives of, 260
  and police, 259, 263
  policies/procedures for, 253
  press conference, 258–259
  release of information, 254
  spokesperson in, 254–255, 257–258
  statements to media, 257
Crisis management
  and areas of vulnerability, 232–234
  contingency plan, 106–107
  crisis audit, phases of, 241–245
  crisis communication plan, 253–259
  crisis leadership, 107–108
  examples of emergencies, 230, 234–235
  first response, 252–253
  importance of, 231
  incident response team, 339–341
  meaning of, 231
  recovery phase, 245
  and routine management, 245–246

stages in response to crime,
  240–241
Crisis management problems
  action logs, lack of, 237–238
  cover-up, 236
  inadequate insurance, 238–239
  inhumane behavior, 240
  interagency relations failures,
    237
  lack of crisis management plan,
    239
  media management guidelines,
    238
  media mismanagement, 236–237
  muddled command, 236
  response time, 235
Critical situation, definition of, 106
Cross-cultural view
  crime rates, 56–57
  violent death among youths, 35–36

D

Date rape, 152–153
  and age, 152
  and alcohol abuse, 152–153
  nonreporting of, 153
  prevention of, 293
  and roofies, 215
Dating aggression, 154–155
  forms of, 154, 187
  male/female characteristics, 154
Depressant drugs, 210
Disciplinary actions
  alcohol use/possession, 292
  employee discipline process,
    99–100
  for off-campus misconduct,
    291–292
  presenting to media, 251
  for social misconduct, 291

Discrimination. *See* Hate crimes
Diversity of population
  campus policy statement, 294–295,
    300–317
  and student housing violence, 115
  See also Hate crimes; Hate crimes
    prevention
Double-jeopardy, 291
Drug-Free Schools and Communities
  Act, 40
Drugs
  cannabis, 211
  depressant drugs, 210
  factors in effects on body,
    208–209
  hallucinogens, 210
  inhalants, 211
  narcotics, 210
  stimulant drugs, 210
  See also Substance Abuse and
    violence

E

Education Amendments of 1972,
  156
  Title IX, 189
Education about violence, 359–360
  as prevention, 293–294
Eiseman v. New York, 33
Elderly, risk as crime victim, 57–58,
  62
E-mail, free speech issue, 76–77
Equal Employment Opportunity
  Commission (EEOC), on sexual
  harassment, 190
Escort services, 331, 333
Ethnicity, meaning of, 134
Ethnoviolence. *See* Hate crimes
Events on campus, easy access to,
  19–20

F

Fact sheet, for crisis communication, 256, 261
Faculty
   female sexual harassment of males, 196–197
   heterosexism toward, 172, 175
   policy on amorous relationships, 164
   sexual harassment by, 156–160, 193–195
   student-professor consensual relationship, 160–162, 195–196
   vulnerability of, 21–22
Family Educational Rights and Privacy Act, 129, 250, 282
   provisions of, 282
Family members
   homicide among, 57
   murder from household gun, 60
Fear
   of crime, 61–63
   and gun carrying, 61–62
   and self-protective measures, 62
Federal actions
   anticrime actions, 58
   campus security actions, 30
   Civil Rights Act of 1964, Title VII/Title IX, 279–282
   Crime Awareness and Security Act of 1990, 276–277
   crime reporting act, 119–120
   Family Educational Rights and Privacy Act, 282
   hate crime actions, 127
   Ramstad Amendment, 277–278
   school violence actions, 40–42
   Student Right-to-Know and Campus Security Act of 1990, 30, 87, 96, 128, 162, 274–276, 283–284
   Violence Against Women Act, 278–279
   See also individual laws
Federal Bureau of Investigation (FBI)
   Campus Crime Statistics, 120
   Uniform Crime Report, 4–15
First Amendment, 70–71
   See also Free speech
Fraternities, 80–81
   hazing, 79–80
   housing and violence, 116–117
   and substance abuse, 116–117
   violence by, 80–81
Free speech
   campus attempts to promote civility, 70
   campus attitudes toward, 23
   campus policy statement, 294–295, 300–317
   First Amendment, 70–71
   hate crimes, 288–289
   hate speech, 73–76
   Internet/e-mail, 76–77
   offensive, examples of, 69–70, 81–82
   sexual harassment cases, 158–159
   speakers on campus, 72–73
   and student publications, 70–72

G

Gangs, 43
Gender, and school violence, 44
Gun control, pro-gun/anti-gun arguments, 65
Gun-Free School Zones Act of 1990, 40
Guns
   economic cost of gun injuries, 60
   gun carrying motivations, 60–64
   and homicide rate in U.S., 56–57

number carried by juveniles, 56
number of households with, 56
number in U.S., 59–60
and residential students, 117
and rise in homicides, 60

H

Hallucinogens, 210
Hate crimes, 123–145
  bases of, 124–125
  on campus, 130, 136–139
  and campus diversity movement, 137–138
  causal chain effect of, 125–126
  characteristics of, 124–127
  Clinton on, 129
  definition of, 123, 169
  disturbing elements of, 123
  examples of, 125–126, 137, 145
  federal actions, 127, 128–129, 288–289
  and homosexuals. *See* Heterosexism
  incidence of, 129–133
  legislative challenges, 288–289
  majority group victims, 124
  message sent by, 169
  by multiple offenders, 126–127
  multiplicative effect, 133
  perceptions by race, 132
  rate and ethnic background, 132–133
  state laws, 288, 289–290
  theoretical bases of, 133–134
  types of crimes, 130
  underreporting of, 130–131
Hate crimes prevention
  campus environment team approach, 139–140
  diversity policy, 140–142
  diversity programming, 142–143
  examples of, 140–141
  intergroup dialogues, 143–145
Hate Crimes Statistics Act, 123, 128
Hate speech
  examples of issues, 73–74
  free speech issue, 73–76
  meaning of, 73
  speech codes, 74
  suggested institutional response, 75–76
Hazing, 79–80
  violent acts in, 79–80
Heterosexism, 115
  anti-LGBT violence, 171
  campus climate review, 171–175
  compulsory form, 170
  emotional effects on victims, 177–179
  examples of, 169–170
  faculty members, 172, 175
  and fears of ostracism, 175
  interventions for, 170–171, 179–183
  and isolation of LGBT persons, 175–176, 177
  perpetration of, 170
  prevention of assault, 294
  types of violent acts, 173–174
  university derived forms, 176–177
Hidden violence
  sexual harassment, 188–189
  student housing, 118
High school. *See* School violence
Hill, Anita, 155
Hiring practices
  negligent hiring, 95
  safe hiring guidelines, 97–99
  warning signs, 98–99
Homicide
  among relatives, 57

cross-cultural view, 56–57
and rise in gun ownership, 60
and teenage boys, 57
UCR statistics on, 7
Hostile environment, sexual harassment, 189–190

I
Incident response team (IRT), 339–341
actions of, 339, 340
composition of, 339–340
functions of, 340
Information gathering
in threat assessment, 342–343
Information release, legal aspects, 282
Inhalants, 211
In loco parentis, 285
Insurance, problems related to, 238–239
Integrated communication, elements of, 247–248
Internet, free speech issue, 76–77

J
Jewish students, victimization of, 115

K
King, Rodney, 63, 234

L
Laboratory, and vulnerability to crime, 20
Leadership, and institution of change, 182–183

Legislation. *See* Federal actions; State actions
Lesbian, gay, bisexual, transgendered (LGBT) persons. *See* Heterosexism
Liability of employer, 95–97
negligence issues, 95–97
state-based actions, 286–287
tort action, 284–286
See also Tort action
Liberals, view of crime, 58
Library, and vulnerability to crime, 20
Life sentence, three-strikes policy, 58
Location of campus, and security risks, 22–23
Location of crimes
time of crimes, 13–14
UCR statistics on, 11–13

M
Male faculty
sexual harassment by, 156–160, 193–196
student-professor consensual relationship, 160–162, 195–196
Management, campus violence prevention, 99
Management of crisis. *See* Crisis management
Manslaughter, UCR statistics on, 7
Media violence
depictions of crime, 64
news programs and crime reporting, 64
scope of, 35
and violent behavior, 64–65
Monitoring the Future, 38
Motorist-assistance programs, 330–331
Multiculturalism, meaning of, 134

Multidimensional framework initiative, drug/alcohol abuse prevention, 222–224
Murder. *See* Homicide

N

Narcotics, 210
National Crime Victimization Survey, 42
National Educational Goals Panel (NEGP), school violence data, 40–42
National Household Education Survey, 42
National Longitudinal Survey of 1972, school violence data, 42–49
Negligence of employer, 95–97
　negligent entrustment, 96
　negligent hiring, 95
　negligent retention, 95
　tort action, 284–286
Neighborhoods
　community-based policing model, 63
　costs of violence, 3–4
　economic improvement and crime reduction, 65
　fear/violence correlation, 63–64
　neighborhood crime watch, 59
　response to campus crime, 263
Nero v. Kansas State University, 34
News media
　aggressive reporters, 255
　attracting audiences, 248–249
　business aspects of, 248
　declining communication with, 255
　disciplinary process, presenting to, 250–251
　goals of, 248
　media management guidelines, 238
　media relationship building, 249–250
　meetings with university officials, 251–252
　mismanagement of crisis, 236–237
　presenting legal facts to, 250–251
　reports, examples of, 264–269
　at town hall meeting, 263
　See also Crisis communication plan
Noise problems, student housing, 113–114
Nonreporting of crime, 118–120
　date rape, 153
　encouraging victim reporting, 119–120
　federal proposal, 119–120
　hate crimes, 130–132
　rape, 92–93
　reasons for, 118–119
　sexual harassment, 157
Nontraditional students, 32–34
　with criminal background, 32–33
　profile of, 32
Nonviolent crimes, forms of, 61

O

Occupational Safety and Health Administration, 336
Off-campus misconduct, disciplinary actions, 291–292

P

Peer sexual harassment, 155–156, 197
　examples of, 156
　harasser characteristics, 155
　legal aspect, 282
Penalties. *See* Criminal penalties

Police
  on campus, 26–27
  communication with, 259, 263
  in neighborhoods, 63
  See also Security
Policies, of institution, 290
Politics, crime as issue, 58–59
Prejudice. *See* Hate crimes
Press conference, 258–259
  See also News media
Prevention of violence, 97–108, 291–297
  access control, 332
  advisory committees, 331–332
  bomb awareness, 103
  campus security checklist, 318–325
  community policing, 328–329
  crime prevention officers, 330
  crime prevention specialist, role of, 329–330
  crisis contingency plans, 106–107
  crisis leadership, 107
  discipline/penalties, 291–292
  discipline vs. punishment, 99–100
  escort services, 331
  and future legal challenges, 296–297
  guidelines for, 295–296
  hate crimes prevention, 139–143
  hiring practices, 97–99
  and human resources services, 336
  injury/illness prevention, example program, 102–103
  and law enforcement accreditation, 329
  from legal standpoint, 295–297
  motorist-assistance programs, 330–331
  policies/procedures, 329–330
  quality management practices, 99
  safety and employee termination, 100–101
  security measures, 104
  strike policy, 104
  student affairs, 336–339
  student codes of conduct, 292–295
  student education, 293–294
  and substance abuse, 218–225
  support systems for victims, 105
  training as, 105–106
  and treats, 332–334, 341–344
  violence action plan, 101–102
Primary aggression, 2
Primary victimization, 118–119
Prisons, cost of, 58
Property crime, statistics on, 4–5
Protective order, 333–334
Psychiatric reports, and hiring, 98–99
Publications of students, free speech issue, 70–72

Q

Quid pro quo sexual harassment, 189–190

R

Race, meaning of, 134
Racial/ethnic minorities
  crime associated with, 59
  fights in school, 43–44
  hate crimes, 123–145
  sexual harassment of, 191, 294
  victimized in student housing, 115
Ramstad Amendment, 163
  provisions, 277–278, 283–284
Rape, 151–153
  adjustment phases for victim, 105
  aftereffects, 152
  campus issues, 78–79

as community crime vs. campus crime, 93
criminal penalties, 287
date rape, 92–93, 152–153
dimensions of, 151
federal actions, 162–163
frequency of, 89, 151–152
inadequate handling by universities, 77–79
overlap of offenses, 93
post-rape support, 105, 293
reporting vs. internal handling of, 93
reporting to police, 92
student education about, 293–294
UCR statistics on, 7
Reactive aggression, 2
Reporting crimes
crime reporting act, 119–120
versus internal handling, 93
sexual harassment, 193
See also Nonreporing of crime
Residences. *See* Student housing violence
Robbery, UCR statistics on, 6–7

S

Safe and Drug-Free Schools and Communities Act of 1994, 40
Safe technology, 65
School Crime Supplement of the National Crime Victimization Survey, 42
Schools and Staffing Survey, 42
School violence
and college-bound high schoolers, 30–31
data sources on, 37–38, 41–49
examples of, 39–40
fear and gun carrying, 62–63
high school to college violence, 30–31
and individual rights, 38–40
rise in, 36–37
Secondary victimization, 118–119
Security
accreditation of, 329
campus security checklist, 318–319
and campus vulnerability, 26–27
community policing model, 328–329
coordination role, 334–335
crime prevention specialist, 329–330
crisis communication, 334–335
federal actions, 30
information gathering role, 335–336
security legislation checklist, 320–325
violence prevention measures, 104
Sexual harassment, 154–162
campus characteristics, 192
contrapower harassment, 197
emotional reactions to, 192–193, 197–198
famous cases, 155
female sexual harrassment of men, 196–197
free speech issues, 158–159
frequency of, 187, 191–192
as hidden issue, 188–189
hostile environment, 189–190
investigation of complaints, 200–201
legal factors, 156, 190, 279–282
and male aggression, 195
male attitudes towards other men, 195
male attitudes about women, 194
by male faculty, 156–160, 193–195

male-to-male sexual harassment, 197
of men by female faculty, 196–197
nonreporting of, 157
peer sexual harassment, 155–156, 197
policy statement on, 200, 294
pro-active campus actions, 200–201
public/private harassers, 194–195
quid pro quo harassment, 189–190
reporting, student characteristics, 193
research issues, 198–200
and student-professor consensual relationship, 160–162, 195–196
students at risk, 191–192
training programs as protection, 201
Sexual violence
and alcohol abuse, 117, 214–215
criminal penalties, 287
emotional impact on women, 188
See also Rape
SHARE program, 293
Shock, stages of, 107
Social misconduct, discipline for, 291
Sororities, housing and violence, 116–117
Sovereign immunity, 284–285
Speakers on campus
controversial speakers, campus actions, 72–73
free speech issue, 72–73
Special events, security risks, 20–21
Spokesperson, communication with media, 254–255, 257–258
Stalking, criminal penalties, 287
State actions, 284–290
civil liability, 286–287
criminal penalties, 287–290

hate crimes, 288, 289–290
institutional liability, 284–287
Statistics. *See* Crime statistics
Stimulant drugs, 210
Stress, causes for students, 25–26
Strike policy, 104
Student affairs, and crime prevention, 336–339
Student housing violence
crime prevention measures, 120–121
and diversity of population, 115
fraternity/sorority houses, 116–117
hidden violence, 118
and nonreporting of crimes, 118–120
proximity factors, 113–114
and substance abuse, 112–113
and weapons, 117
Student Right-to-Know and Campus Security Act of 1990, 30, 87, 96, 128, 162, 274–276, 283–284
controversial aspects, 275–276
presenting provisions to media, 250
provisions of, 274–276, 283
sanctions, 276
security checklist, 320–325
Students Against Racism (SAR), 140–141, 142
Substance abuse
and addiction, 211
college students use of drugs, 212–214
developmental aspects, 216–217
vs. life principles, 218
meaning of, 211
Substance abuse and violence, 24–25, 91
data sources on, 212, 219
and fraternities, 116–117
and gun ownership, 56

high school students, 208
student housing violence, 112–113
theories of, 215–216
types of violence, 212
Substance abuse and violence prevention, 218–225
best practices Sourcebook, 221
increased efforts, 220–221
multidimensional framework initiative, 222–224
recommendations for campus prevention program, 224–225
scope of programs for, 219–220
treatment/referral services, 220
and university functions, 219
Support systems
for campus victims, 105
for drug/alcohol abusers, 220
sexual assault, 105, 293
student affairs as, 338
for threatened victim, 332–334
victim advocacy, 334
victim services, 334

T

Tailhook incident, 155
Teenage boys, risk as crime victims, 57–58
Television violence. *See* Media violence
Termination of employees, safety factors, 100–101
Terrorism, on campus, 91
Threats, 332–334, 341–344
assessment of, 341–343
communication devices, use of, 333
escort of victim, 333
information gathering, 342
interventions for threats, 343–344
legal contact with threat maker, 332–333
principles related to, 341
protective order, 333–334
risk assessment, 342–343
Three-strikes policy, 58
Tort action, 284–286
in loco parentis, 285
sovereign immunity, 284–285
special relationship situation, 285–286
student as invitee, 286

U

Uniform Crime Report (UCR)
crime statistics (1995), 4–15
handgun statistics, 59–70
Unions
strike policy, 104
and workplace violence, 96–97
United States
crime statistics, 4–15
gun ownership statistics, 59–60
violence in, 1–2, 56–57, 231–232, 347–348
violent death, cross-cultural view, 35–36
violent death among youths, 35–36

V

Victims
campus support system, 105
nonreporting of crimes, 118–120
offender relationship, 11–12, 57, 60
primary and secondary, 118–119
reactions of, 119
Violence
and aggression, 2

costs of, 3–4, 60, 94–97
crime statistics, 4–15
definition of, 2, 90–91, 207
financial costs of, 3
learning of, 1
measurement of, 3
media influences, 64–65
violent acts, types of, 90–91
Violence Against Women Act, 162–163, 278–279
civil liability for crimes, 334
provisions of, 278–279
Violent crimes, forms of, 61
Voices of Discovery program, 143–145

W

Weapon carrying by students, 53–65
examples of, 54
and incidence of crime, 54–55
survey data, 53–55
See also Guns
Weathermen, 91

White House Conference on Hate Crimes, 129
Women
risk as crime victim, 58
sexual victimization, effects of, 187–188
Women and campus crime
battery/date aggression, 153–154
examples of, 149–150
federal actions, 162–163
rape, 151–153
sexual harassment, 154–162
suggested university policies, 164–165
types of crimes, 150
Workplace violence
campus violence as, 89–90
human resources services, 336
liabilities for, 95–97
prevention of, 336

Y

Youth Risk Behavior Survey, 42

Z

Zero-tolerance, 30

About the Authors

ABOUT THE EDITORS

Allan M. Hoffman, EdD, CHES, is dean of the College of Health Sciences and a professor at the University of Osteopathic Medicine and Health Sciences in Des Moines, Iowa. He is an experienced educator and human services administrator. He has held executive-level positions in higher education, having served as dean, associate dean, executive officer, and faculty member. He has been a visiting scholar, adjunct professor, clinical professor, and professor at several institutions including the University of Southern California School of Medicine, the University of LaVerne, State University of New York (SUNY)–Buffalo, and California State University. Dr. Hoffman has published extensively and authored several books including *Schools, Violence and Society* (Praeger Publishers, 1996), and he coauthored a book for the College and University Personnel Association focusing on effective management strategies for higher education institutions.

Dr. Hoffman has served on the editorial board of the *Journal of Allied Heath* and as a manuscript reviewer for the *Journal of Public Health*. He has provided consulting services to schools, colleges, and business organizations in the United States, Puerto Rico, Canada, Scandinavia, Mexico, the West Indies, England, and elsewhere. His consulting efforts focus on violence in schools, colleges, and the workplace. He served as a resource person to the Presidential Task Force on Health Care Reform. He earned his BS magna cum laude from the University of Hartford, and received two MA degrees and an EdD from Teachers College, Columbia University, where he was named a Kellogg Fellow and held an appointment at the Institute for the Study of Higher Education. Dr. Hoffman is the recipient of numerous awards and honors associated with his teaching and his efforts to prevent violence and resolve conflict creatively. Dr. Hoffman holds national board certification as a certified health education specialist (CHES) from the National Commission on Health Education Credentialing.

John H. Schuh, PhD, is professor of educational leadership and policy studies at Iowa State University. He received his PhD from Arizona State University and previously held administrative and faculty positions at Arizona State University, Indiana University, and Wichita State University.

The recipient of a Fulbright scholarship to study higher education in Germany, Dr. Schuh has been recognized for contributions to the literature by the American College Personnel Association and the National Association of Student Personnel Administrators. He has served on several editorial boards and currently is editor of the New Directions for Student Services Sourcebook Series and associate editor of the *Journal of College Student Development*.

Robert H. Fenske, PhD, is professor of higher education in the Division of Educational Leadership and Policy Studies at Arizona State University. He earned his PhD from the University of Wisconsin and has served on the faculty of the University of Minnesota, as director of research for the Illinois Board of Higher Education, and as senior research psychologist at the American College Testing Program.

In 1987–1988 he was awarded a research fellowship at the National Center for Education Statistics, and in 1993–1994 he was visiting senior research scholar at the Arizona Board of Regents. He has been involved in institutional research and policy analysis as well as academic pursuits related to his research interests. He has published widely in such areas as student financial aid and access to higher education.

ABOUT THE CONTRIBUTORS

David S. Anderson, MA, PhD, serves as an associate research professor with the Center for the Advancement of Public Health, the Institute of Public Policy, at George Mason University. In this capacity, he serves as a project director and researcher for numerous national, state, and local projects. He teaches graduate and undergraduate courses on drug and alcohol issues and conducts needs assessments, evaluation, analysis, and writing on numerous projects. His work has emphasized college students, school and community leaders, youth, program planners, and policy makers. His specialty areas include prevention, strategic planning and mobilization, communication and education, and evaluation. He has produced, moderated, and served as a guest on several television programs.

Dr. Anderson coauthors two national surveys on college drug/alcohol prevention efforts (The College Alcohol Survey, conducted every three years since 1979, and The Drug and Alcohol Survey of Community, Junior and Technical Colleges). He is codirector of the Promising Practices: Campus Alcohol Strategies project, which is identifying alcohol abuse prevention strategies. This project has resulted in the publication of a 300-page sourcebook. He has developed numerous risk assessment guides and resource materials, and has conducted extensive training throughout the United States and Micronesia. Early in his career, he served as a student affairs administrator at Ohio University and Ohio State University. He received his BA from Duke University, with a major in psychology and a minor in business administration. His MA, from Ohio State University, is in student personnel administration. His PhD in public policy/public affairs is from Virginia Polytechnic Institute and State University.

J. Victor Baldridge, PhD, is a senior partner at Pacific Management, a management consulting firm. Dr. Baldridge holds three graduate degrees from Yale University in sociology, organizational theory, and administrative science. He has taught management for more than 20 years at Stanford University, Fresno State University, University of California–Los Angeles (UCLA), and the University of San Francisco; and he has served as vice president at Fresno State University. Dr. Baldridge has been a principal investigator on millions of dollars of research grants at Stanford, UCLA, and the University of San Francisco. During the first half of his career, he focused almost exclusively on applying management theories to college and university administration.

Dr. Baldridge has published 19 books and 50 articles. He is working on two new books, one entitled *Workplace Violence: A Handbook for Action,* and the other entitled *Fire on the Mountain: Disaster Management in Action.*

Carolyn N. Barnes has been director of marketing at Texas Woman's University since 1997. Previously, she was director of publications at the university, where she has worked since 1992. Ms. Barnes has been a communications professional for more than 20 years. She began her career in journalism, spending 4 years as a newspaper feature writer and reporter covering breaking news. She entered higher education communications in 1980; since that time she has worked in almost every facet of the profession: media relations, public relations, periodicals, special events, and marketing. Ms. Barnes holds a BA in journalism from the University of North Texas. She has received 14 national and regional awards for writing, public relations projects, magazines, periodicals, and special events.

Charles Cychosz, PhD, is manager of safety and health development in the Department of Public Safety at Iowa State University, where he has developed crime and violence prevention activities. He also serves on the board of directors for the Partnership for a Drug-Free Iowa and chairs the executive committee of the

Network of Colleges and Universities Committed to the Elimination of Drug and Alcohol Abuse.

Darlene C. DeFour, MA, PhD, is associate professor of psychology at Hunter College, City University of New York. Dr. DeFour's expertise is in racism and sexual harassment in academia. Along with Michele Paludi, she investigates sexual harassment complaints in academia and provides training programs on sexual harassment awareness for faculty, administrators, and college students.

Eugene Deisinger, PhD, is captain of the Behavioral Sciences Unit in the Department of Public Safety at Iowa State University. He is a police psychologist with extensive experience in threat assessment, crisis intervention, and critical incident debriefing. He serves on the university critical incident response team, providing consultation regarding issues of violence and assessment of potentially violent subjects. He also provides training on issues related to workplace violence.

Nancy J. Evans, MFA, MSEd, PhD, is an associate professor in the Department of Educational Leadership and Policy Studies at Iowa State University. She has edited, coedited, or coauthored several books including *Encouraging the Development of Women; Beyond Tolerance: Gays, Lesbians and Bisexuals on Campus; The State of the Art of Professional Preparation and Practice: Another Look;* and *The Development of College Students: Theory, Research and Practice.* Her most recent research examines the experiences of gay, lesbian, and bisexual students on college campuses. She has received the Annuit Coeptis and Contribution to Knowledge awards from the American College Personnel Association. She earned her PhD degree from the University of Missouri-Columbia.

Leonard Gordon, MA, PhD, is associate dean of academic programs and professor of sociology in the College of Liberal Arts and Sciences at Arizona State University. Dr. Gordon earned his BA and PhD from Wayne State University and his MA from the University of Michigan with training in history and, at the doctoral level, in sociology.

For further background and expansion of issues explored in Chapter 8, which Dr. Gordon coauthored with Dr. Fenske, see "Race Relations and Attitudes at Arizona State University," in *The Racial Crisis in American Higher Education,* edited by P.G. Altbach and K. Lomotey, and "Student Movements" in *The Encyclopedia of Sociology,* edited by Borgatta and Borgatta.

Stafford L. Hood, MS, PhD, is associate professor of counseling/counseling psychology in the Division of Psychology in Education at Arizona State University. Dr. Hood earned his PhD at the University of Illinois at Champaign-Urbana.

He has served on the faculty of Northern Illinois University in educational psychology and as an assessment specialist and program evaluator for the Illinois State Board of Education.

Dr. Hood's publications in scholarly journals and books have addressed bias in testing, program evaluation, and faculty at major research universities. He has also served as a program evaluation and testing consultant to state departments of education, school districts, universities, and regional educational laboratories.

Kay Hartwell Hunnicutt, PhD, JD, is an associate professor in the Division of Educational Leadership and Policy Studies, College of Education, at Arizona State University. She earned her PhD in educational administration from Southern Illinois University in Carbondale, a JD degree from Arizona State University, and her MA and BA from Murray State University in Kentucky. In addition to teaching five courses in education law including graduate courses in higher education law and student affairs administration and the law, she is a practicing attorney specializing in employment law, civil rights, and education law.

Dr. Hunnicutt has published articles and chapters in textbooks related to hate speech on campus, sexual harassment, and employment and personnel legal issues. She has lectured in Scotland, Australia, and Germany and has provided legal consultation regarding policy development to private and public colleges and universities throughout the United States.

Loras A. Jaeger, MA, is director of public safety at Iowa State University. He serves on numerous committees and task forces at Iowa State University, as well as on the Iowa Lieutenant Governor's Task Force on Stranger and Non-Stranger Sexual Assault.

Daniel J. Julius, EdD, is associate vice president for academic affairs at the University of San Francisco and senior fellow for the Higher Education Research Institute at Stanford University. Dr. Julius earned his BA from the Ohio State University and holds graduate degrees in higher education administration and in organizational behavior and industrial relations from Columbia University. He has been assistant vice chancellor (systemwide) at the California State University System and director of personnel services (systemwide) for the Vermont State College System.

Dr. Julius has published 7 books and approximately 75 articles. He has been a visiting scholar at Stanford University and at the International Labor Organization (Geneva) and a visiting professor at Shanghai University, the University of Hawaii, San Diego State University, the University of New Hampshire, and the University of Manitoba, Canada. He has lectured in England, Scotland, Canada, China, France, and Ireland.

Peter Kushibab, JD, is an assistant general counsel with the Maricopa County Community College District in Tempe, Arizona. He has taught courses on legal research and writing, as well as jurisprudence and ethics, in the Justice and Legal Studies Department at Phoenix College. As chief counsel of the education unit of the Arizona Attorney General, he represented that state's board of education and superintendent of public instruction.

Larry D. Lauer is associate vice chancellor for communications and public affairs at Texas Christian University. He has worked with executives to solve communication problems for over 25 years. His experience includes large corporations, small businesses, colleges and universities, human service and arts agencies, media organizations, and city and state governments. He is the author of more than 20 articles on the communication dimensions of management and is the author of the book *Communication Power: Energizing Your Nonprofit Organization.* Mr. Lauer received an MA in communications from American University in Washington, DC. His work in communications and organizations has taken him to England, Wales, Scotland, France, Italy, Germany, Canada, and Mexico.

Carol Napierkowski, MA, PhD, is associate professor in the Department of Counseling and Educational Psychology at West Chester University in Pennsylvania. She teaches counselor education courses in the school of counseling and higher education counseling programs. Dr. Napierkowski has experience as a school counselor at both the elementary and secondary levels. She has research interests in violence prevention in both secondary and higher education and has presented on this topic at regional and national conferences.

Carolyn J. Palmer, PhD, is associate professor in the Department of Higher Education and Student Affairs at Bowling Green State University. She began her research on campus violence during her 15 years as a student affairs (housing) practitioner at the University of Illinois, Champaign-Urbana, where she completed her doctoral degree in quantitative and evaluative research. Her studies on violence, particularly in student housing, have been summarized in her book *Violent Crimes and Other Forms of Victimization in Residence Halls,* as well as in journal articles and book chapters. She has made 47 presentations at meetings of professional organizations of student leaders, housing administrators, student affairs administrators, campus police, college and university attorneys, and criminal court prosecutors and judges.

Michele A. Paludi, MA, PhD, is principal of Paludi & Associates, consultants in sexual harassment, located in Schenectady, New York. This firm offers education and training in academic and workplace harassment, the development of policies and procedures, and expert witness testimony. Dr. Paludi is chair of the U.S. Department of Education's subpanel on the prevention of violence, sexual harassment, and alcohol and drug abuse in higher education.

Gilbert Ramírez, DrPH, is associate professor and vice chair of the Department of Public Health and Preventive Medicine and associate director of the Graduate Programs in Public Health at the University of North Texas Health Science Center at Fort Worth. He served as the codirector of the San Antonio Cochrane Center and as a member of the editorial board of the international Cochrane Collaboration Diabetes Collaborative Review Group.

Dr. Ramírez was codirector of the Center for Cross-Cultural Research and assistant professor of health research at Southwest Texas State University. He serves as a member of the National Advisory Board for Alternative Medicine (National Institutes of Health) and as a statistical consultant to the Centers for Disease Control and Prevention. He has published and lectured extensively throughout the United States and abroad on methods of research synthesis and meta-analysis. Dr. Ramírez obtained his doctorate in public health from the University of Texas Health Science Center at Houston, his MPH from the University of Hawaii, his MA in health care facility management from Webster University, and his BS in environmental health from East Tennessee State University.

Sue Rankin, MS, PhD, is a senior diversity planning analyst and coordinator of lesbian, gay, bisexual, and transgender equity at The Pennsylvania State University. She has presented and written several papers on the impact of heterosexism on the academy and on intercollegiate athletics.

Ira Schoenwald, PhD, is professor of public administration and associate vice president for faculty affairs at California State University, Dominguez Hills. He earned his PhD in government from Claremont Graduate School. In addition to experience in human resources in various management positions, he has had more than 20 years' consulting experience in both private and public sector organizations in the areas of human resources and management. He has worked with Rockwell International, Shell Oil, the U.S. Army and Air Force, as well as municipal governments in southern California, including Torrance, Santa Monica, and Los Angeles. Dr. Schoenwald has published papers in psychology and management practices and has implemented training programs in a variety of work environments such as the health care industry and manufacturing companies.

Randal W. Summers, PhD, is an adjunct professor in the Business Administration Program at the University of Phoenix. He is also a general partner in the consulting firm of Summers and Associates. He has been a manager in government, health care, and industry specializing in human resource development, organizational development, and curriculum development. Summers received his doctorate in educational psychology from the University of Alberta and his MS in counseling and psychology from North Dakota State University. He has been a contributing author in a number of books on higher education and violence. Summers is a member of the American Society for Training and Development in Orange County, California, and a licensed psychologist.

Fernando M. Treviño, PhD, MPH, is professor and chair of the Department of Public Health and Preventive Medicine and executive director of the Graduate Program in Public Health at the University of North Texas Health Science Center at Fort Worth. Dr. Treviño served as executive director of the American Public Health Association in Washington, DC, and also has served as the executive editor of the *American Journal of Public Health*. He was dean of the School of Health Professions and professor of health administration at Southwest Texas State University.

Dr. Treviño has served on numerous national committees and panels including the U.S. Preventive Services Task Force, the National Committee on Vital and Health Statistics, and the Institute of Medicine's Access to Health Care Monitoring Panel. He holds a PhD in preventive medicine and community health from the University of Texas Medical Branch at Galveston, an MPH from the University of Texas Health Science Center at Houston, School of Public Health, and a bachelor's degree in psychology from the University of Houston. Dr. Treviño has published and lectured widely on national statistical data policy and Mexican-American and minority health issues.

Sharon L. Walker, RN, MS, MPH, is a doctoral student in sociology at the University of North Texas at Denton and works at the Survey Research Center in the School of Community Service. Ms. Walker has extensive experience in the field of occupational health and safety and physical and vocational rehabilitation with workers' compensation clients. Her graduate work in public health and in sociology has focused on the nature of violence in society, specifically with children and young adults. Her current research examines the application of a violence prevention program with students. Ms. Walker received her MPH from the University of North Texas Health Science Center at Fort Worth.